ADVANCED DYNAMICS
Modeling and Analysis

ADVANCED DYNAMICS
Modeling and Analysis

A. FRANK D'SOUZA

Department of Mechanical Engineering
Illinois Institute of Technology
Chicago, Illinois

VIJAY K. GARG

Research and Test Department
Association of American Railroads
Chicago, Illinois

PRENTICE-HALL, INC., *Englewood Cliffs, New Jersey 07632*

Library of Congress Cataloging in Publication Data

D'Souza, A. Frank.
  Advanced dynamics.

  Includes bibliographical references and index.
  1. Dynamics   I. Garg, Vijay Kumar
II. Title.
QA805.D79 1984   531'.11'0151   83-4549
ISBN 0-13-011312-3

Editorial/production supervision and
  interior design: Ellen Denning
Manufacturing buyer: Anthony Caruso

Printed in the United States of America

10 9 8 7 6 5 4 3

ISBN 0-13-011312-3

PRENTICE-HALL INTERNATIONAL INC., *London*
PRENTICE-HALL OF AUSTRALIA PTY. LIMITED, *Sydney*
EDITORA PRENTICE-HALL DO BRASIL, LTDA., *Rio de Janeiro*
PRENTICE-HALL CANADA INC., *Toronto*
PRENTICE-HALL OF INDIA PRIVATE LIMITED, *New Delhi*
PRENTICE-HALL OF JAPAN, INC., *Tokyo*
PRENTICE-HALL OF SOUTHEAST ASIA PTE. LTD., *Singapore*
WHITEHALL BOOKS LIMITED, *Wellington, New Zealand*

CONTENTS

3.   DYNAMICS OF PARTICLES: NEWTON'S LAW, ENERGY, AND MOMENTUM METHODS   41

4.   DYNAMICS OF RIGID BODIES: NEWTON'S LAW, ENERGY, AND MOMENTUM METHODS   78

PREFACE

A large number of books on the subject of dynamics have been written during the past decades, and therefore, those who venture to add to their numbers should offer some explanation. Many of these books with engineering orientation are intended for a first course in dynamics at the undergraduate level. Some of the books on advanced dynamics have been authored by physicists who are primarily interested in classical dynamics as a preliminary to studying particle physics and quantum mechanics.

This book is intended to serve as a text for engineering students at a first graduate-level course in dynamics. Most of the material can also be used for a senior-level undergraduate course. Engineers are interested in classical dynamics primarily for the purpose of obtaining mathematical models of dynamic systems which are then employed in the analysis of the dynamic behavior and in design synthesis. This book attempts to combine classical dynamics and methods of analysis.

For this reason, the book is divided into two parts. Part I is devoted to the description and illustration of the principles of dynamics which are employed in the derivation of the equations of motion (i.e., mathematical models of dynamic systems). Part II covers some of the methods of analysis that can be employed in the investigation of the dynamic behavior of engineering systems. The emphasis is on dynamic response and stability of motion.

A brief survey of the contents is as follows. Part I of the book contains five chapters. Chapter 1 is the introduction, in which the classical dynamic concepts of particles and rigid bodies, inertial coordinates, and Newtonian and Lagrangian dynamics are discussed. Chapter 2 is aimed at kinematics and includes discussions on various coordinate systems and their transformations. Chapter 3 presents particle dynamics, including Newton's laws, and the energy and momentum methods. Two-body central force motion, and the orbits of planets and satellites, are also discussed. Chapter 4 deals with the dynamics of rigid bodies from the Newtonian viewpoint. Chapter 5 presents Lagrangian dynamics, and includes the principle of virtual work, Hamilton's principle, the Lagrangian equations of motion and Euler's angles for rigid bodies.

Part II of the book consists of four chapters. In Chapter 6, a discussion of the response of dynamic systems is presented. Chapter 7 deals with the numerical solution of the equations of motion for a dynamic system. Both implicit and explicit solution methods are included. In Chapter 8, the theory of linear vibrations is presented, and both single- and multiple-degree-of-freedom systems are discussed. Chapter 9 deals with the stability of motion, and stability considerations for both autonomous and nonautonomous systems are presented.

Chapter 8 on linear vibrations is included because it discusses some of the techniques of linear analysis and also because some students may not have the opportunity to take a separate course in mechanical vibrations. This chapter may be omitted depending on the interest and aims of the students. At the graduate level, all the chapters of Part I and most of the chapters of Part II could be covered in a one-semester course. At the senior undergraduate level, all the chapters of Part I and about two chapters of Part II, such as Chapters 7 and 8, would be adequate for a one-semester course, depending on the instructor's goal.

To gain maximum benefit from the book, the reader should have some knowledge of elementary dynamics. A working knowledge of calculus, ordinary differential equations, vector and matrix algebra, and Laplace transformation is an adequate mathematical background. Vector and/or matrix notation is employed throughout most of the presentation. The two appendices present elements of vector and matrix analysis, respectively, for the benefit of those who need to review this material.

The development of this book has been influenced by several existing books mentioned in the references. We extend our thanks to our students and colleagues who have offered constructive criticism and many valuable suggestions. Thanks are also due to Mehran Farahmandpour for his help in preparing the illustrations. We acknowledge the assistance of the staff of Prentice-Hall, Inc., especially that of Charles Iossi, Engineering Editor, and Ellen Denning, Production Editor. Finally, we are indebted to our wives, Cecilia and Pushpa, for their support and encouragement.

A. Frank D'Souza
Vijay K. Garg

INTRODUCTION

1.1 CLASSICAL DYNAMICS OF PARTICLES AND RIGID BODIES

The discipline of dynamics is concerned with the study of motion. In general, two viewpoints may be adopted: microscopic and macroscopic. In our study of dynamics, we neglect quantum mechanics effects and employ macroscopic models. It should be noted that macroscopic models invoke the continuum hypothesis according to which matter is continuously distributed in space occupied by a body. The branch of dynamics that employs macroscopic models is referred to as classical dynamics, in contrast to quantum mechanics, which employs microscopic models. Our central aim is to study the classical dynamics of solid bodies; we do not attempt any investigation of fluid mechanics.

In the case of solid bodies, two important approximations can often be made in practical applications and a body is then referred to either as a particle or a rigid body. When a body does not rotate but only translates, we may approximate its motion by describing the motion of a single representative point of the body. The body is then called a particle and mathematically represented by a point mass. Even when a body rotates, especially when the dimensions of the body are small compared to the distance it travels, and we are interested only in its translation, the rotation may be ignored and the body represented by a particle. For example, for the study of the orbit of the earth around the sun, the earth's rotation may be ignored and it may be approximated as a particle.

When a body rotates and this rotation is to be studied, its finite size is to be considered, but we may still be able to ignore the small deformations associated with its flexibility. The body is then called a rigid body. Hence, for a rigid body, the distance between any two of its particles remains constant for all time and for all configurations. As a consequence of approximating a body as either a particle or a rigid body, its motion is described by ordinary differential equations with time as an independent variable. On the other hand, the motion of a flexible body is described by partial differential equations with time and space coordinates as independent variables.

Whether a solid body may be accurately represented as a particle, a rigid body, or a flexible body depends on the purpose of the study. For example, to determine an optimal nominal trajectory, a rocket may be considered as a particle. For the purpose of guidance and control, this approximation is too simplistic since the rocket's attitude and orientation are important, so the rocket may be approximated as a rigid body. However, the forces acting on a rocket can produce bending moments. In order to study the bending of a rocket and investigate the stresses, the rocket is considered as a flexible body. The bending-mode shapes may then be superimposed on the rigid-body motion. This book is concerned with the classical dynamics of particles and rigid bodies.

1.2 RELATIVISTIC AND NONRELATIVISTIC DYNAMICS

According to Einstein's general theory of relativity, also referred to as Einstein's gravitational theory, the mass, m, of a body is related to its velocity, $\vec{v}$, by the equation

$$m = \frac{m_0}{(1 - |\vec{v}|^2/c^2)^{1/2}} \tag{1.1}$$

where c is the speed of light and m_0 is the mass of the body at rest when $|\vec{v}| = 0$. Our assumption is that the speed, $|\vec{v}|$, of the body is much less than the speed of light (i.e., $|\vec{v}| \ll c$), and we are concerned only with nonrelativistic dynamics. Consequently, mass is an inherent constant property of a body and is independent of its motion or passage of time. However, in nonrelativistic dynamics, the mass of an open system may change when it gains or loses mass. For example, the mass of a rocket will change as its fuel is depleted, but this change in mass is quite distinct from and unrelated to the relativity effect.

In nonrelativistic dynamics, Newton's viewpoint of completely independent and absolute time and Euclidean geometry is adopted. Hence, space and time are independent. Euclidean space is a normed, linear vector space that is homogeneous and isentropic. The metric which is a measure of distance is given by the norm. The norm or distance between any two points of the space with coordinates (x_1, y_1, z_1) and (x_2, y_2, z_2) is defined by

$$d = [(x_1 - x_2)^2 + (y_1 - y_2)^2 + (z_1 - z_2)^2]^{1/2} \tag{1.2}$$

Since the space is homogeneous, this distance is invariant of the origin of the coordinate system and since it is isentropic, the distance is also invariant of the orientation of the coordinate system. Forces of kinematical nature, such as Coriolis and centrifugal forces, can be eliminated by employing an inertial frame of reference, which is discussed later. However, gravitational forces cannot be eliminated by kinematical transformation in Euclidean space.

In relativistic dynamics, the concept of independent time and space is discarded in favor of a four-dimensional space–time continuum. A four-dimensional Riemannian space is employed in which the gravitional forces disappear. This space is not linear but curved, and its metric is related to the gravitational mass at every point. This book, then, is restricted to the non-relativistic, classical dynamics of particles and rigid bodies.

1.3 INERTIAL COORDINATES, KINEMATICS, AND KINETICS

As mentioned earlier, in nonrelativistic dynamics it is assumed that there is an absolute space, which is Euclidean, and an absolute time, which is independent of space. A coordinate system must be employed to measure and describe a motion. It was Galileo who showed that there exist preferred reference systems for which the acceleration has its simplest possible form. Such a reference frame is called an inertial coordinate system or Galilean reference frame. The acceleration when measured with respect to inertial coordinates is called absolute acceleration.

An inertial frame of reference may be defined as a coordinate system that does not rotate and whose origin is either fixed in space or if it translates, then is moves in a straight line at a constant velocity. Suppose that there exist two coordinate systems that do not rotate but translate at a constant velocity with respect to each other, and one of them is inertial; then the other system is also inertial.

Suppose that the origin of a coordinate system is chosen as a point on the surface of the earth and the coordinate system does not rotate with respect to the earth. But as the earth rotates, this coordinate system would rotate with it with respect to fixed space. It would also translate with the earth and not in a straight line. This coordinate system is then obviously not inertial. But in case the additional acceleration terms due to the rotation and translation of the earth, which are obtained in Chapter 2, are negligibly small compared to the relative acceleration of a body with respect to this coordinate system, no noticeable error is introduced by assuming that this coordinate system is inertial. In case the acceleration term due to the rotation of the earth is not negligible, it may be possible to fix the origin of the coordinate system at the center of the earth and to fix its orientation in inertial space. Now, if the acceleration term due to the translation of the earth is negligibly small compared to the relative acceleration, we can assume that this coordinate system is inertial. In some cases it may become necessary to choose the center of the sun for the origin of the coordinate system and its orientation fixed in space in order for it to be considered as inertial. In other cases, it may become necessary to choose a distant "fixed" star, such as Canopus, for the origin of the coordinate system. Hence, it becomes obvious that the concept of an inertial coordinate system is purely hypothetical, as even the distant "fixed" stars are not really fixed in space. We treat relative acceleration as absolute when the additional terms due to the rotation and translation of the coordinate system are negligibly small.

The study of dynamics may be conveniently divided into two parts, kinematics and kinetics. Kinematics is concerned with the geometry of motion and deals with relationships among displacement, velocity, acceleration, and time without any reference to the cause of motion. On the other hand, kinetics deals with relationships among forces, mass, and motion of the body. It is concerned with the cause of motion. Chapter 2 is devoted exclusively to the study of kinematics, and some discussion of kinematics of rigid bodies is also given in Chapter 4.

1.4 NEWTONIAN DYNAMICS

The fundamental ideas of classical dynamics have been developed over a number of years. These developments fall into two classes, Newtonian dynamics and Lagrangian dynamics. The development of Newtonian dynamics began with Galileo, who introduced the concept of acceleration and stated his law of inertia. The inertia of a body is its resistance to a change in its uniform motion. The mass of the body is used as the quantitative measure of inertia. The concept of an inertial frame of reference was also recognized by Galileo. Later, in 1687, Newton formulated his three laws for single particles and his law of gravitation. Euler extended these concepts to the study of dynamics of rigid bodies. The branch of classical dynamics based on direct application of Newton's laws is called Newtonian dynamics. Newtonian dynamics is studied in Chapter 3 for a system of particles and in Chapter 4 for rigid bodies. The first two laws of motion have been stated by Newton for a single particle. Newton's laws are discussed in the following.

Newton's first law.   If there are no forces acting on a particle, the particle will remain at rest, if originally at rest, or will move in a straight line at constant velocity, if originally in motion.

Let $\vec{F}$ be the resultant force acting on a particle and $\vec{v}$ be its velocity vector measured with respect to an inertial coordinate system. Newton's first law can be stated mathematically as:

$$\text{If } \vec{F} = \vec{0}, \text{ then } \vec{v} = \text{constant} \qquad (1.3)$$

where a special value of the constant may be zero.

Newton's second law.   If the resultant force acting on a particle is not zero, the particle will move so that the resultant force vector is equal to the time rate of change of the linear momentum vector.

Letting $\vec{F}$ be the resultant force acting on a particle, $\vec{v}$ its velocity, and m its mass, Newton's second law can be stated as

$$\vec{F} = \frac{d}{dt}(m\vec{v}) \tag{1.4}$$

where $\vec{v}$ is measured with respect to an inertial coordinate system. Neglecting relativistic effects and not including those particles of an open system that may gain or lose mass, the value of mass does not depend on time and (1.4) can be written as

$$\vec{F} = m\frac{d\vec{v}}{dt}$$
$$= m\vec{a} \tag{1.5}$$

where a is the absolute acceleration which is measured with respect to an inertial coordinate system. Hence, it is seen from (1.5) that Newton's second law may be expressed alternatively as follows: If the resultant force acting on a particle is not zero, the particle will have an acceleration proportional to the magnitude of the resultant force and in the direction of this force.

Newton's third law.   When two particles exert forces on one another, the forces lie along the line joining the particles and the force vectors are the negative of each other.

Hence, when a force exerted on a particle is the result of an interaction with another particle, the forces of action and reaction are equal in magnitude, opposite in direction, and collinear.

Let $\vec{F}_{ij}$ be the force exerted on the ith particle by the jth particle and $\vec{F}_{ji}$ be vice versa. Then, according to Newton's third law,

$$\vec{F}_{ij} = -\vec{F}_{ji} \tag{1.6}$$

Newton's law of gravitation.   This law states that two particles of mass, m_1 and m_2, mutually attract each other with equal and opposite forces, $\vec{F}$ and $-\vec{F}$, whose magnitude

$$F = \frac{Gm_1m_2}{r^2} \tag{1.7}$$

where r is the distance between the two particles, and G is a universal constant called the constant of gravitation. The direction of the force is along the line joining the two particles, as shown in Fig. 1.1.

In formulating his first two laws of motion, Newton was undoubtedly influenced by Galileo and in formulating his law of gravitation by Kepler. In his turn, Kepler formulated his laws of planetary motion from observations of

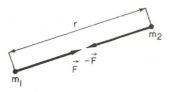

Figure 1.1 Newton's law of gravitation.

the orbits of planets around the sun. Newton's three laws of motion, together with his law of gravitation, form the basis of Newtonian dynamics. They are valid for most engineering applications where the speeds encountered are much smaller than the speed of light. However, there are some exceptions where relativistic dynamics must be employed. For example, electromagnetic forces between moving particles do not follow Newton's third law. Also, the motion of certain planets, such as the anomalous motion of the perihelion of Mercury, can be explained by the general theory of relativity.

1.5 LAGRANGIAN DYNAMICS

The second approach to the formulation of the equations of motion is known as Lagrangian dynamics and is also referred to as analytical mechanics. It was developed about a hundred years after Newton formulated his laws. Lagrangian dynamics requires the concept of virtual displacement and it is formulated by Lagrange's equations of motion by employing kinetic energy and work. The introduction of generalized coordinates instead of the physical coordinates makes the method very versatile. The equations of motion are derived from Hamilton's principle, which is a variational principle and leads to the extremization of a functional.

These techniques have their roots in the development of the calculus of variations by Bernoulli, Euler, and others. Hamilton's concept of regarding the generalized coordinates and the generalized momenta as independent canonical variables led him to transform Lagrange's equations of motion, which are second order in the generalized coordinates to a set of first-order equations in the canonical variables. The study of Lagrangian dynamics is covered in Chapter 5.

1.6 SUMMARY

The major aim of this chapter has been to outline the scope of the book and to state certain classifications and definitions. This study is restricted to non-relativistic classical dynamics of particles and rigid bodies. First, classical dynamics is defined as that branch of dynamics that employs macroscopic models, in contrast to quantum mechanics, where microscopic models are employed. In many practical applications, a solid body may be approximated as a particle or a rigid body. Since we deal only with nonrelativistic dynamics, space and time are assumed to be independent and Euclidean space is employed. Inertial coordinates are defined and Newtonian and Lagrangian dynamics is discussed.

REFERENCES

1. Goldstein, H., *Classical Mechanics*, Addison-Wesley Publishing Company, Inc., Reading, Mass., 1950.
2. Meirovitch, L., *Methods of Analytical Dynamics*, McGraw-Hill Book Company, New York, 1970.
3. Halfman, R. L., *Dynamics*, Vol. 1, Addison-Wesley Publishing Company, Inc., Reading, Mass., 1962.

2

KINEMATICS

2.1 INTRODUCTION

This chapter is concerned with kinematics and deals with the geometry of motion, including the use of various coordinate systems and the study of relationships among displacement, velocity, acceleration, and time. The cause of motion is not considered in this chapter but will be studied in later chapters dealing with the kinetics of motion. The choice of an appropriate coordinate system is very important from the point of view of obtaining the equations of motion in as simple a form as possible.

The acceleration has a simple form when it is expressed in terms of an inertial coordinate system. However, in many applications, it is much more convenient to employ a noninertial coordinate system. For example, in the study of the orbit of one particle around another in the two-body central force problem, the motion takes place in a plane and the equations of motion take on a simple form when a polar coordinate system is employed. In the study of motion of a rigid body, it will be observed that it is more convenient to employ a noninertial coordinate system fixed to the rotating and translating body because the mass moments of inertia remain time invariant with respect to such a coordinate system.

In this chapter we first discuss the use of various coordinate systems, including Cartesian, tangential and normal, and polar coordinates. Then the transformation is studied between two sets of coordinate systems, where one set is rotated with respect to the other. The motion is then expressed in terms of translating and rotating system of coordinates.

2.2 INERTIAL CARTESIAN COORDINATE SYSTEM

We consider a coordinate system xyz whose origin O is fixed and the coordinate axes do not rotate. It was discussed in Chapter 1 that such an inertial coordinate system is only hypothetical. As shown in Fig. 2.1, the position vector of a particle P at time t is denoted by the vector $\vec{r}$ joining the origin O and point P.

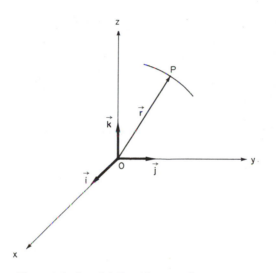

Figure 2.1   Inertial Cartesian coordinate system.

Resolving the position vector $\vec{r}$ of the particle into rectangular components, we get

$$\vec{r}(t) = x(t)\vec{i} + y(t)\vec{j} + z(t)\vec{k} \qquad (2.1)$$

where $\vec{i}, \vec{j}$, and $\vec{k}$ are unit vectors as shown in Fig. 2.1 and the coordinates x, y, and z are functions of time. Differentiating (2.1) once the velocity vector is given by

$$\vec{v}(t) = \frac{d\vec{r}}{dt} = \dot{x}\vec{i} + \dot{y}\vec{j} + \dot{z}\vec{k} \qquad (2.2)$$

since the vectors $\vec{i}, \vec{j}$, and $\vec{k}$ are fixed in magnitude and direction and hence

$$\frac{d\vec{i}}{dt} = \frac{d\vec{j}}{dt} = \frac{d\vec{k}}{dt} = 0$$

In (2.2), a dot over a symbol denotes the first derivative with respect to time t. Differentiating (2.2) once more, the acceleration is obtained as

$$\vec{a} = \frac{d\vec{v}}{dt} = \ddot{x}\vec{i} + \ddot{y}\vec{j} + \ddot{z}\vec{k} \tag{2.3}$$

where $\ddot{x}, \ddot{y}$, and $\ddot{z}$ denote the second derivatives with respect to t. Hence, we note that the acceleration assumes a simple form when expressed in terms of an inertial coordinate system. In many applications, the motion of the particle cannot be considered separately along the x, y, and z directions, respectively, on account of the coupling caused by the forces.

2.3 MOTION RELATIVE TO A FRAME IN TRANSLATION

We now employ a coordinate system that is in translation without rotation with respect to an inertial coordinate system. In Fig. 2.2, let $Oxyz$ be an inertial coordinate system whose origin O is fixed. Let $O_1 x_1 y_1 z_1$ be a coordinate system whose origin O_1 has a motion whose velocity and acceleration are denoted by $\vec{v}_1$ and $\vec{a}_1$, respectively. The axes $O_1 x_1$, $O_1 y_1$, and $O_1 z_1$ always remain parallel to the axes Ox, Oy, and Oz, respectively; that is, the axes $O_1 x_1 y_1 z_1$ do not change their orientation.

Noting that the position vector $\vec{r}_p$ of particle P with respect to xyz is the sum of the position vector $\vec{r}_{p/1}$ of P with respect to $x_1 y_1 z_1$ and the position

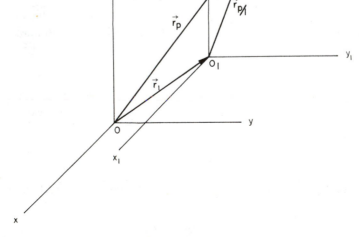

Figure 2.2   Coordinate system $O_1 x_1 y_1 z_1$ in translation.

vector $\vec{r}_1$ of O_1, we get

$$\vec{r}_p = \vec{r}_{p/1} + \vec{r}_1 \tag{2.4}$$

Differentiating (2.4) with respect to time, the velocity $\vec{v}_p$ and acceleration $\vec{a}_p$ are given, respectively, by

$$\begin{aligned}
\vec{v}_p = \dot{\vec{r}}_p = \dot{\vec{r}}_{p/1} + \dot{\vec{r}}_1 \\
= \vec{v}_{p/1} + \vec{v}_1
\end{aligned} \tag{2.5}$$

and

$$\begin{aligned}
\vec{a}_p = \dot{\vec{v}}_p = \dot{\vec{v}}_{p/1} + \dot{\vec{v}}_1 \\
= \vec{a}_{p/1} + \vec{a}_1
\end{aligned} \tag{2.6}$$

The motion with respect to an inertial coordinate system is called the absolute motion. Equation (2.6) expresses that the absolute acceleration of P may be obtained by adding vectorially the acceleration $\vec{a}_{p/1}$ of P relative to frame $x_1 y_1 z_1$ and the acceleration $\vec{a}_1$ of the origin O_1 of frame $x_1 y_1 z_1$. Equation (2.5) may also be given similar interpretation concerning velocities.

Example 2.1

A motor boat has a speed of 3 m/s with respect to a river that is flowing east at a constant speed of 2 m/s. The boat desires to follow a straight path from point C to point D, where CD is 20° east of north (Fig. 2.3). Determine the absolute velocity of the boat and the direction of its relative velocity with respect to the river.

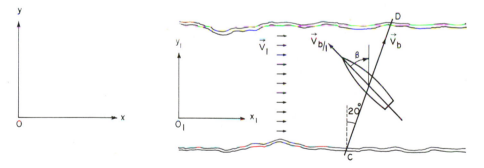

Figure 2.3   Boat crossing a river.

In this case, the boat has plane motion. Let xyz be an inertial coordinate system whose origin O is fixed. The origin of the coordinate system $x_1 y_1 z_1$ has a velocity $\vec{v}_1 = 2\vec{i}$ m/s (i.e., the velocity of the river), and its orientation remains fixed. The velocity $\vec{v}_b$ of the boat with respect to xyz must be directed along CD: that is,

$$\vec{v}_b = v_b(\sin 20°\vec{i} + \cos 20°\vec{j})$$

Let the velocity $\vec{v}_{b/1}$ of the boat with respect to $x_1 y_1 z_1$ (i.e., the relative velocity of boat with respect to the river) make an angle β to the y direction as shown in Fig. 2.3. Then, from (2.5), we get

$$\vec{v}_b = \vec{v}_{b/1} + \vec{v}_1$$

That is,

$$v_b (\sin 20° \vec{i} + \cos 20° \vec{j}) = 3 (-\sin \beta \vec{i} + \cos \beta \vec{j}) + 2\vec{i}$$

Equating the coefficients of $\vec{i}$ and $\vec{j}$ in the foregoing equation, we obtain

$$v_b \sin 20° = -3 \sin \beta + 2 \quad \text{and} \quad v_b \cos 20° = 3 \cos \beta$$

Eliminating angle β from the foregoing two equations, we get

$$v_b^2 - 1.368v_b - 5 = 0$$

that is, $v_b = 3.022$ m/s or -1.6543 m/s.

The admissible value of the absolute velocity of the boat is $v_b = 3.022$ m/s and it is directed along CD. With this value of v_b, we find that the angle which the relative velocity makes with the y direction as shown in Fig. 2.3 is $\beta = 18.79°$.

2.4 TANGENTIAL AND NORMAL COORDINATES

The velocity of a particle is a vector tangent to its path. Sometimes, it is convenient to express the acceleration in terms of its components directed along the tangent and normal to the path, respectively. Figure 2.4(a) shows the path of a particle in space. At time t when the particle is at A, let $\vec{i}_t(t)$ be a unit vector tangent to the path at A and pointing in the direction of motion. The velocity of the particle at that instant of time may be expressed as

$$\vec{v} = v\vec{i}_t \tag{2.7}$$

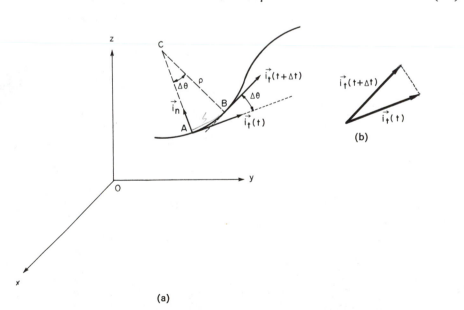

(a)

Figure 2.4   Tangential and normal coordinate system.

Differentiating (2.7) once with respect to t, the acceleration is given by

$$\vec{a} = \frac{d\vec{v}}{dt} = \frac{dv}{dt}\vec{i}_t + v\frac{d\vec{i}_t}{dt} \tag{2.8}$$

It is noted that the magnitude of $\vec{i}_t$ is unity and it remains constant but its direction changes with time and in the following we obtain an expression for $d\vec{i}_t/dt$ that appears in (2.8). At time $t + \Delta t$, let the particle be at position B and $\vec{i}_t(t + \Delta t)$ be the unit vector tangent to the path corresponding to that position, as shown in Fig. 2.4(a). It may be observed from Fig. 2.4(b) that

$$\vec{i}_t(t + \Delta t) = \vec{i}_t(t) + (\Delta\theta)(1)\vec{i}_n(t) \tag{2.9}$$

where $\Delta\theta$ is the angle between the two unit tangent vectors, and $\vec{i}_n$ is a unit vector along the principal normal to the path. Hence, we get

$$\frac{d\vec{i}_t}{dt} = \lim_{\Delta t \to 0} \frac{\vec{i}_t(t + \Delta t) - \vec{i}_t(t)}{\Delta t} = \lim_{\Delta t \to 0} \frac{\Delta\theta}{\Delta t}\vec{i}_n \tag{2.10}$$

Letting Δs denote the change in the path length, we have

$$\frac{d\vec{i}_t}{dt} = \frac{d\theta}{ds}\frac{ds}{dt}\vec{i}_n \tag{2.11}$$

But $ds/dt = v$ and $d\theta/ds$ is equal to $1/\rho$, where ρ is the radius of curvature of the path at A as shown in Fig. 2.4, with C being the instantaneous center of curvature. Hence, we obtain

$$\frac{d\vec{i}_t}{dt} = \frac{v}{\rho}\vec{i}_n \tag{2.12}$$

and (2.8) may be written as

$$\vec{a} = \frac{dv}{dt}\vec{i}_t + \frac{v^2}{\rho}\vec{i}_n \tag{2.13}$$

The tangential component of the acceleration may be positive or negative; that is, it may point in the direction of motion or against the direction of motion, depending on the sign of dv/dt. The normal component of the acceleration is directed toward the instantaneous center of curvature of the path. In plane motion there is only one straight line perpendicular to the tangent at a given point of the path, whereas in three-dimensional motion there is an infinite number of such straight lines. In the latter case, the unit vector $\vec{i}_n$ is in the direction of the principal normal at a given point of the path. Consider a plane at point A in Fig. 2.4 containing the tangent to the curve at A and parallel to the tangent to the curve at B. As point B approaches A, this plane is called the osculating plane at A. The principal normal lies in the osculating plane and is perpendicular to the tangent. The binormal at A is perpendicular to the osculating plane at A and a unit vector $\vec{i}_b$ in that direction completes the right-hand triad $\vec{i}_t$, $\vec{i}_n$, and $\vec{i}_b$. However, the acceleration has no component along the binormal.

Example 2.2

A grinding wheel of outside radius r_o is attached to the shaft of an electric motor (Fig. 2.5). At time t, the angular velocity of the motor is $\omega_m(t)$ and its angular acceleration is $\alpha_m(t) = \dot{\omega}_m$. Determine the acceleration of a point on the circumference of the wheel at time t.

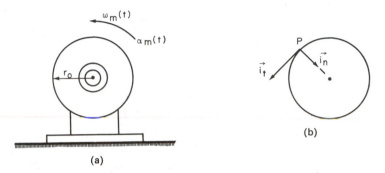

(a)

(b)

Figure 2.5   Grinding wheel.

It should be noted that a point P on the circumference of the wheel has plane motion and the instantaneous center of curvature remains fixed at the center of the shaft. At time t, the velocity of a point P on the circumference of the wheel is given by

$$v_p = \omega_m r_o$$

$$\frac{dv_p}{dt} = \dot{\omega}_m r_o = \alpha_m r_o$$

Substituting this result in (2.13), the acceleration of P is given by

$$\vec{a}_p = \alpha_m r_o \vec{i}_t + \omega_m^2 r_o \vec{i}_n$$

Example 2.3

A train is traveling along a curve of radius r_o. At time t, its speed is $v(t)$. Determine the maximum rate at which the speed may be decreased at that time if the total acceleration of the train is not to exceed $0.2g$.

We note that for this problem it is convenient to employ tangential and normal coordinate system. Employing (2.13), we obtain

$$a^2 = \left(\frac{dv}{dt}\right)^2 + \left(\frac{v^2}{\rho}\right)^2$$

Here, we have $a_{\text{max}} = 0.2g$ and $\rho = r_o$. Hence, it follows that the maximum rate at which the speed may be decreased is given by

$$\max \frac{dv}{dt} = \left[(0.2g)^2 - \left(\frac{v^2}{r_0}\right)^2\right]^{1/2}$$

Example 2.4

A block of mass m_1 is constrained to move on a straight bar AB. A mass m_2 is suspended from mass m_1 and is free to move about the pivot O_1 as shown in Fig. 2.6. Determine the acceleration of mass m_2.

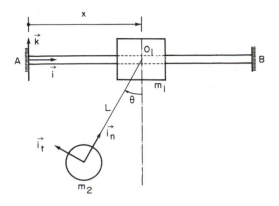

Figure 2.6   Motion of masses m_1 and m_2.

By employing (2.6), the absolute acceleration $\vec{a}_2$ of mass m_2 may be written as

$$\vec{a}_2 = \vec{a}_{2/1} + \vec{a}_1$$

where $\vec{a}_1$ is the acceleration of m_1 and $\vec{a}_{2/1}$ is the acceleration of mass m_2 relative to mass m_1. Now, $\vec{a}_1 = \ddot{x}\vec{i}$ and $\vec{a}_{2/1}$ can be expressed in terms of its components in the tangential and normal directions. Hence, employing (2.6) and (2.13), we obtain

$$\vec{a}_2 = \frac{dv}{dt}\,\vec{i}_t + \frac{v^2}{\rho}\,\vec{i}_n + \ddot{x}\vec{i}$$

Since, $v = \dot{\theta}L$, $dv/dt = \ddot{\theta}L$, and $\rho = L$, the foregoing equation can be written as

$$\vec{a}_2 = \ddot{\theta}L\vec{i}_t + \dot{\theta}^2 L\vec{i}_n + \ddot{x}\vec{i}$$

This problem is considered again in Example 2.9 by employing a translating and rotating coordinate system.

2.5 POLAR AND CYLINDRICAL COORDINATES

2.5.1 Plane Motion

We first consider plane motion where it is convenient to represent the position of a particle by means of its polar coordinates, r and θ, as shown in Fig. 2.7(a). Let $\vec{i}_r$ and $\vec{i}_\theta$ be two unit vectors at A in the radial and transverse directions, respectively. As the particle moves from A to B, the magnitudes of the unit vector $\vec{i}_r$ and $\vec{i}_\theta$ remain constant at unity but their directions change to $\vec{i}_r(t + \Delta t)$ and $\vec{i}_\theta(t + \Delta t)$ with time Δt as shown in Fig.2.7(a). First, we obtain expressions for the time rate of change of these unit vectors. From Fig. 2.7(b) it can be seen that

$$\vec{i}_r(t + \Delta t) = \vec{i}_r(t) + (\Delta\theta)(1)\vec{i}_\theta(t) \tag{2.14}$$

$$\vec{i}_\theta(t + \Delta t) = \vec{i}_\theta(t) - (\Delta\theta)(1)\vec{i}_r(t) \tag{2.15}$$

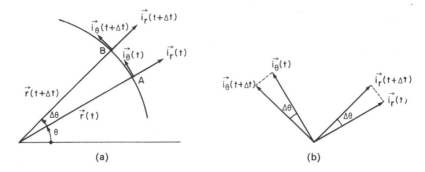

Figure 2.7 Polar coordinates.

Using these results, we obtain

$$\frac{d\vec{i}_r}{dt} = \lim_{\Delta t \to 0} \frac{\vec{i}_r(t + \Delta t) - \vec{i}_r(t)}{\Delta t} = \lim_{\Delta t \to 0} \frac{\Delta \theta}{\Delta t} \vec{i}_\theta$$

$$= \dot{\theta} \vec{i}_\theta \tag{2.16}$$

$$\frac{d\vec{i}_\theta}{dt} = \lim_{\Delta t \to 0} \frac{\vec{i}_\theta(t + \Delta t) - \vec{i}_\theta(t)}{\Delta t} = \lim_{\Delta t \to 0} - \frac{\Delta \theta}{\Delta t} \vec{i}_r$$

$$= -\dot{\theta} \vec{i}_r \tag{2.17}$$

Expressing the position vector $\vec{r}$ of a particle as the product of scalar r and the unit vector $\vec{i}_r$, we obtain

$$\vec{r} = r\vec{i}_r \tag{2.18}$$

Differentiating (2.18) with respect to t and using (2.16), the velocity is given by

$$\vec{v} = \dot{\vec{r}} = \dot{r}\vec{i}_r + r\frac{d\vec{i}_r}{dt}$$

$$= \dot{r}\vec{i}_r + r\dot{\theta}\vec{i}_\theta \tag{2.19}$$

Differentiating (2.19) with respect to t and employing (2.16) and (2.17), the acceleration is obtained as

$$\vec{a} = \ddot{r}\vec{i}_r + \dot{r}\frac{d\vec{i}_r}{dt} + \dot{r}\dot{\theta}\vec{i}_\theta + r\ddot{\theta}\vec{i}_\theta + r\dot{\theta}\frac{d\vec{i}_\theta}{dt}$$

$$= (\ddot{r} - r\dot{\theta}^2)\vec{i}_r + (r\ddot{\theta} + 2\dot{r}\dot{\theta})\vec{i}_\theta \tag{2.20}$$

A circular motion is a special case where $\dot{r} = 0$ and in this case it follows that

$$\vec{v} = r\dot{\theta}\vec{i}_\theta \tag{2.21}$$

$$\vec{a} = -r\dot{\theta}^2\vec{i}_r + r\ddot{\theta}\vec{i}_\theta \tag{2.22}$$

2.5.2 Three-Dimensional Motion

In some applications it is advantageous to employ cylindrical coordinates to represent the motion of a particle. Let R, θ, and z be the cylindrical coordinates and $\vec{i}_R, \vec{i}_\theta$, and $\vec{k}$ be the unit vectors in their respective directions as shown

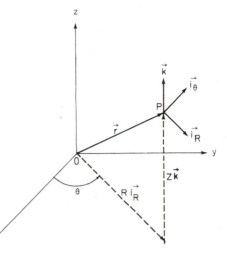

Figure 2.8 Cylindrical coordinate system.

in Fig. 2.8. The position of the particle is expressed as

$$\vec{r} = R\vec{i}_R + z\vec{k} \tag{2.23}$$

and employing the results obtained in the foregoing, the velocity and acceleration are given by

$$\vec{v} = \dot{\vec{r}} = \dot{R}\vec{i}_R + R\dot{\theta}\vec{i}_\theta + \dot{z}\vec{k} \tag{2.24}$$

$$\vec{a} = \dot{\vec{v}} = (\ddot{R} - R\dot{\theta}^2)\vec{i}_R + (R\ddot{\theta} + 2\dot{R}\dot{\theta})\vec{i}_\theta + \ddot{z}\vec{k} \tag{2.25}$$

Example 2.5

A mechanism is shown in Fig. 2.9, where a slotted rod OA rotates about O with displacement $\theta = c \sin \omega t$. A slider S is constrained to move in the slot and along a curve BCD whose equation is given by $r = a/(1 + \theta)$. Determine the velocity and acceleration of slider S at any instant of time t.

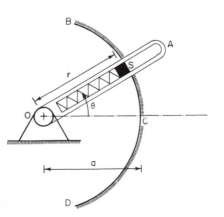

Figure 2.9 Slider mechanism.

We note that the slider has plane motion and that it is advantageous to employ polar coordinates for this problem. Expressing r as a function of time, we have

$$r = \frac{a}{1 + c \sin \omega t}$$

Hence, we obtain

$$\dot{r} = \frac{-ca\omega \cos \omega t}{(1 + c \sin \omega t)^2}$$

$$\ddot{r} = \frac{ca\omega^2}{(1 + c \sin \omega t)^3} (\sin \omega t + c \sin^2 \omega t + 2c \cos^2 \omega t)$$

Also, we have $\dot{\theta} = c\omega \cos \omega t$ and $\ddot{\theta} = -c\omega^2 \sin \omega t$. Substituting these results in (2.19), the velocity of the slider is given by

$$\vec{v}(t) = \frac{-ca\omega \cos \omega t}{(1 + c \sin \omega t)^2} \, \vec{i}_r + \frac{ac\omega \cos \omega t}{1 + c \sin \omega t} \, \vec{i}_\theta$$

Substitution of the foregoing expressions in (2.20) yields the acceleration of the slider as

$$\vec{a} = \left[\frac{c^2 a \omega^2 [(1/c) \sin \omega t + \sin^2 \omega t + 2 \cos^2 \omega t]}{(1 + c \sin \omega t)^3} - \frac{c^2 a \omega^2 \cos^2 \omega t}{1 + c \sin \omega t} \right] \vec{i}_r$$

$$+ \left[\frac{-ca\omega^2 \sin \omega t}{1 + c \sin \omega t} + \frac{-2c^2 a \omega^2 \cos^2 \omega t}{(1 + c \sin \omega t)^2} \right] \vec{i}_\theta$$

2.6 ROTATIONAL TRANSFORMATION OF COORDINATES

We consider two Cartesian sets of axis $Oxyz$ and $Ox_1y_1z_1$ which are rotated with respect to each other as shown in Fig. 2.10. A vector $\vec{r}$ may be decomposed using each of the coordinate systems as

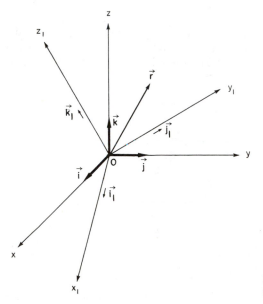

Figure 2.10 Rotating coordinate system $O_1x_1y_1z_1$.

$$\vec{r} = x\vec{i} + y\vec{j} + z\vec{k} \tag{2.26}$$

$$\vec{r} = x_1\vec{i}_1 + y_1\vec{j}_1 + z_1\vec{k}_1 \tag{2.27}$$

Taking the scalar product of both (2.26) and (2.27) with $\vec{i}_1$, we obtain

$$
\begin{aligned}
x_1 &= x(\vec{i} \cdot \vec{i}_1) + y(\vec{j} \cdot \vec{i}_1) + z(\vec{k} \cdot \vec{i}_1) \\
&= x \cos(x, x_1) + y \cos(y, x_1) + z \cos(z, x_1) \\
&= C_{i_1 i} x + C_{i_1 j} y + C_{i_1 k} z
\end{aligned}
\tag{2.28}
$$

where $C_{i_1 i}$, $C_{i_1 j}$, and $C_{i_1 k}$ are the direction cosines between axes x_1 and x, x_1 and y, and x_1 and z, respectively. Similarly, it follows that

$$y_1 = C_{j_1 i} x + C_{j_1 j} y + C_{j_1 k} z \tag{2.29}$$

$$z_1 = C_{k_1 i} x + C_{k_1 j} y + C_{k_1 k} z \tag{2.30}$$

Equations (2.28), (2.29), and (2.30) may be written in the matrix notation as

$$
\begin{Bmatrix} x_1 \\ y_1 \\ z_1 \end{Bmatrix} = [C] \begin{Bmatrix} x \\ y \\ z \end{Bmatrix}
\tag{2.31}
$$

Expressing x, y, and z in terms of the components along the x_1, y_1, and z_1 axis, we obtain

$$
\begin{Bmatrix} x \\ y \\ z \end{Bmatrix} = [C]^T \begin{Bmatrix} x_1 \\ y_1 \\ z_1 \end{Bmatrix}
\tag{2.32}
$$

where the superscript T denotes the matrix transpose. Hence, it is noted that

$$[C]^{-1} = [C]^T \tag{2.33}$$

that is, the inverse of matrix $[C]$ is its transpose and such a matrix is called an orthogonal matrix. Furthermore, from matrix algebra, the determinant of a product of two matrices is equal to the product of the determinants of the two matrices and we obtain

$$|[C][C]^T| = |[C]||[C]^T| = |[I]| \tag{2.34}$$

But the determinant of an identity matrix is unity and the determinant of a transposed matrix is equal to the determinant of the matrix. Hence, it follows from (2.34) that $|[C]|^2 = 1$; that is, the determinant of $[C]$ may assume the value $+1$ or -1. The value of $+1$ is chosen in order to transform a right-hand triad into another right-hand triad, and matrix $[C]$ is an orthonormal matrix. It should be noted, of course, that the transformation considered here is only a rotational transformation between two sets of rectangular axes and is a special case of coordinate transformations.

2.7 ROTATING COORDINATE SYSTEMS

In many applications, a noninertial rotating coordinate system is employed to express the equations of motion. In this section we develop explicit expressions for the direction cosines between an inertial coordinate system xyz and a rotating one denoted by $x_1 y_1 z_1$, as shown in Fig. 2.11. The coordinate system $x_1 y_1 z_1$ is obtained by first rotating about axis x through angle θ_1 to obtain $\xi_1 \xi_2 \xi_3$, then rotating about axis ξ_2 through angle θ_2 to obtain $\eta_1 \eta_2 \eta_3$, and finally rotating about axis η_3 through angle θ_3 to obtain $x_1 y_1 z_1$. First consider the rotation about axis x through angle θ_1. The relationship between xyz and $\xi_1 \xi_2 \xi_3$ coordinate systems is given by

$$\xi_1 = x$$
$$\xi_2 = y \cos \theta_1 + z \sin \theta_1 \qquad\qquad (2.35)$$
$$\xi_3 = -y \sin \theta_1 + z \cos \theta_1$$

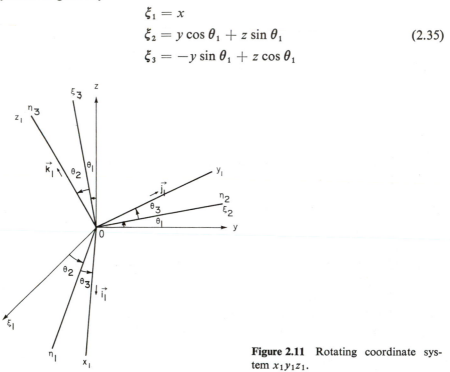

Figure 2.11   Rotating coordinate system $x_1 y_1 z_1$.

These equations may be written in the matrix notation as

$$\begin{Bmatrix} \xi_1 \\ \xi_2 \\ \xi_3 \end{Bmatrix} = \begin{bmatrix} 1 & 0 & 0 \\ 0 & \cos \theta_1 & \sin \theta_1 \\ 0 & -\sin \theta_1 & \cos \theta_1 \end{bmatrix} \begin{Bmatrix} x \\ y \\ z \end{Bmatrix} \qquad\qquad (2.36)$$

$$= [C_1(\theta_1)] \begin{Bmatrix} x \\ y \\ z \end{Bmatrix} \qquad\qquad (2.37)$$

where $[C_1(\theta_1)]$ denotes the transformation matrix of (2.36). We now consider the rotation of $\xi_1\xi_2\xi_3$ axes through angle θ_2 about ξ_2 to obtain $\eta_1\eta_2\eta_3$ axes. It follows that

$$
\begin{Bmatrix} \eta_1 \\ \eta_2 \\ \eta_3 \end{Bmatrix} = \begin{bmatrix} \cos\theta_2 & 0 & -\sin\theta_2 \\ 0 & 1 & 0 \\ \sin\theta_2 & 0 & \cos\theta_2 \end{bmatrix} \begin{Bmatrix} \xi_1 \\ \xi_2 \\ \xi_3 \end{Bmatrix}
$$

$$
= [C_2(\theta_2)] \begin{Bmatrix} \xi_1 \\ \xi_2 \\ \xi_3 \end{Bmatrix} \tag{2.38}
$$

Finally, consider the rotation of $\eta_1\eta_2\eta_3$ axes through angle θ_3 about η_3 to obtain $x_1 y_1 z_1$ axes. This relationship is described by

$$
\begin{Bmatrix} x_1 \\ y_1 \\ z_1 \end{Bmatrix} = \begin{bmatrix} \cos\theta_3 & \sin\theta_3 & 0 \\ -\sin\theta_3 & \cos\theta_3 & 0 \\ 0 & 0 & 1 \end{bmatrix} \begin{Bmatrix} \eta_1 \\ \eta_2 \\ \eta_3 \end{Bmatrix}
$$

$$
= [C_3(\theta_3)] \begin{Bmatrix} \eta_1 \\ \eta_2 \\ \eta_3 \end{Bmatrix} \tag{2.39}
$$

Now, combining (2.37), (2.38), and (2.39), we obtain

$$
\begin{Bmatrix} x_1 \\ y_1 \\ z_1 \end{Bmatrix} = [C_3(\theta_3)][C_2(\theta_2)][C_1(\theta_1)] \begin{Bmatrix} x \\ y \\ z \end{Bmatrix}
$$

$$
= [C] \begin{Bmatrix} x \\ y \\ z \end{Bmatrix} \tag{2.40}
$$

A question arises whether the rotations can be represented by vectors $\vec{\theta}_i$ ($i = 1, 2, 3$) directed along the axis, x, ξ_2, and η_3, respectively. The answer to this question is negative because as seen from (2.40), angular displacements are compounded by the law of matrix multiplication which is not commutative; that is, $[C_2][C_1]$ is in general not equal to $[C_1][C_2]$. Hence, a finite angular displacement θ is a directed line segment but it is not a vector since it does not satisfy the commutative law of vector algebra; that is, $\vec{\theta}_1 + \vec{\theta}_2$ is in general not equal to $\vec{\theta}_2 + \vec{\theta}_1$.

Example 2.6

We consider the rotation of a body about an inertial axis xyz whose origin O is fixed as shown in Fig. 2.12. In Fig. 2.12(a), the body is first rotated by 90° about the z axis and this is followed by 90° rotation about the y axis. In Fig. 2.12(b), the order of rotation is reversed and the body is first rotated by 90° about the y axis and then by 90° about

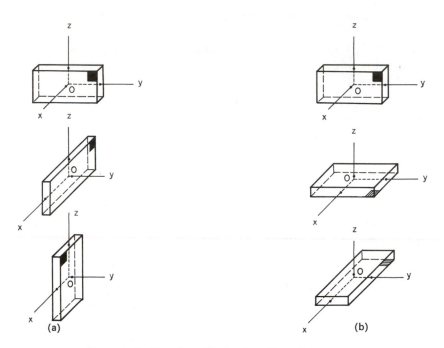

Figure 2.12 Rotation of body about inertial axes.

the z axis. It can be seen that the final orientations of the rotated body are not the same in cases (a) and (b). Here, the order of rotation is very important.

Example 2.7

This example considers the motion of an airplane. The inertial axes system xyz is fixed in space. Axes $x_1 y_1 z_1$ constitute a body coordinate system whose origin O_1 is the center of mass of the plane and which yaws, pitches, and rolls with the plane (Fig. 2.13).

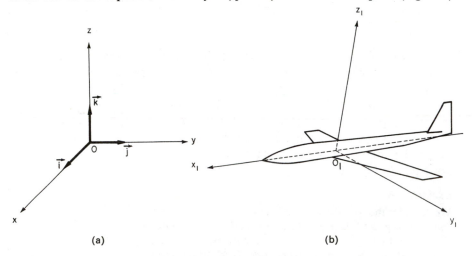

Figure 2.13 Airplane body coordinate system $x_1 y_1 z_1$.

The velocity of the plane at time t, as observed with the xyz coordinate system, is given by $\vec{v}(t) = v_x\vec{i} + v_y\vec{j} + v_z\vec{k}$. If this velocity is also resolved in the body coordinate system as $\vec{v}(t) = v_{x_1}\vec{i}_1 + v_{y_1}\vec{j}_1 + v_{z_1}\vec{k}_1$, determine the components v_{x_1}, v_{y_1}, and v_{z_1} in terms of the components v_x, v_y, and v_z.

The yaw of the plane takes place as a rotation of the plane about the z axis as shown in Fig. 2.14. The system $\xi_1\xi_2\xi_3$ yaws with the plane through the yaw angle ψ about the z axis. The transformation of vectors between the xyz and $\xi_1\zeta_2\xi_3$ coordinate systems is given by

$$\begin{Bmatrix} \xi_1 \\ \xi_2 \\ \xi_3 \end{Bmatrix} = \begin{bmatrix} \cos\psi & \sin\psi & 0 \\ -\sin\psi & \cos\psi & 0 \\ 0 & 0 & 1 \end{bmatrix} \begin{Bmatrix} x \\ y \\ z \end{Bmatrix} \qquad (2.41)$$

$$\begin{Bmatrix} x \\ y \\ z \end{Bmatrix} = \begin{bmatrix} \cos\psi & -\sin\psi & 0 \\ \sin\psi & \cos\psi & 0 \\ 0 & 0 & 1 \end{bmatrix} \begin{Bmatrix} \xi_1 \\ \xi_2 \\ \xi_3 \end{Bmatrix} \qquad (2.42)$$

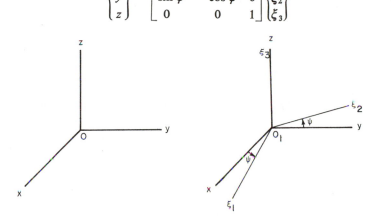

Figure 2.14 Yaw rotation.

The pitch of the plane is shown in Fig. 2.15 as a rotation θ about the ξ_2 axis. The reference frame $\eta_1\eta_2\eta_3$ translates, yaws, and pitches with the plane. The transformation equations between $\xi_1\xi_2\xi_3$ and $\eta_1\eta_2\eta_3$ coordinate systems are

$$\begin{Bmatrix} \eta_1 \\ \eta_2 \\ \eta_3 \end{Bmatrix} = \begin{bmatrix} \cos\theta & 0 & -\sin\theta \\ 0 & 1 & 0 \\ \sin\theta & 0 & \cos\theta \end{bmatrix} \begin{Bmatrix} \xi_1 \\ \xi_2 \\ \xi_3 \end{Bmatrix} \qquad (2.43)$$

The roll of the plane is defined in Fig. 2.16 as a rotation about the η_1 axis through angle ϕ. The $x_1y_1z_1$ frame is then fixed to the plane and it translates, yaws, pitches, and rolls with the plane. The transformation equations between $\eta_1\eta_2\eta_3$ and $x_1y_1z_1$ coordinates are given by

$$\begin{Bmatrix} x_1 \\ y_1 \\ z_1 \end{Bmatrix} = \begin{bmatrix} 1 & 0 & 0 \\ 0 & \cos\phi & \sin\phi \\ 0 & -\sin\phi & \cos\phi \end{bmatrix} \begin{Bmatrix} \eta_1 \\ \eta_2 \\ \eta_3 \end{Bmatrix} \qquad (2.44)$$

Hence, the transformation equation between xyz and $x_1y_1z_1$ coordinate system becomes

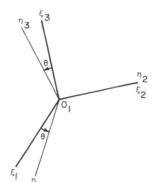

Figure 2.15 Pitch rotation.

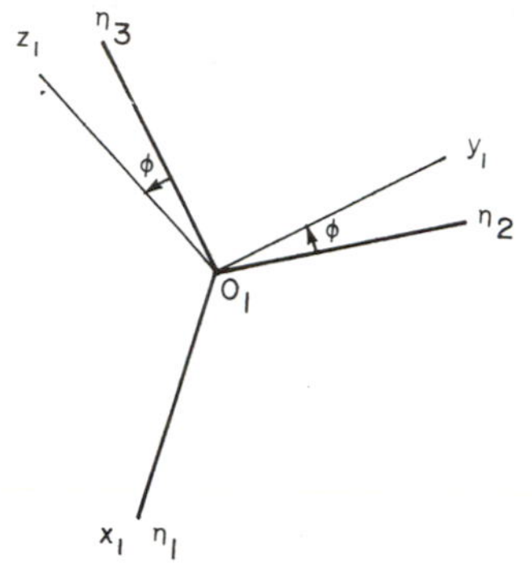

Figure 2.16 Roll rotation.

$$\begin{Bmatrix} x_1 \\ y_1 \\ z_1 \end{Bmatrix} = [C] \begin{Bmatrix} x \\ y \\ z \end{Bmatrix}$$

where the $[C]$ matrix is obtained from (2.41), (2.43), and (2.44) as

$$[C] = \begin{bmatrix} 1 & 0 & 0 \\ 0 & \cos\phi & \sin\phi \\ 0 & -\sin\phi & \cos\phi \end{bmatrix} \begin{bmatrix} \cos\theta & 0 & -\sin\theta \\ 0 & 1 & 0 \\ \sin\theta & 0 & \cos\theta \end{bmatrix} \begin{bmatrix} \cos\psi & \sin\psi & 0 \\ -\sin\psi & \cos\psi & 0 \\ 0 & 0 & 1 \end{bmatrix} \quad (2.45)$$

The desired velocity components can now be obtained from the equation

$$\begin{Bmatrix} v_{x_1} \\ v_{y_1} \\ v_{z_1} \end{Bmatrix} = [C] \begin{Bmatrix} v_x \\ v_y \\ v_z \end{Bmatrix} \quad (2.46)$$

Again, it should be emphasized that the order in which the rotations are defined is very important in combining the transformation equations.

2.7.1 Infinitesimal Rotations and Angular Velocity Vector

While finite angles of rotation cannot be represented by vectors, we now show that infinitesimal rotations can be represented in that manner. For infinitesimal rotation through angle $\Delta\theta$, we let $\cos\Delta\theta = 1$ and $\sin\Delta\theta = \Delta\theta$ as $\Delta\theta \longrightarrow$

0. Thus for infinitesimal rotations, the transformation matrices $[C_1]$, $[C_2]$, and $[C_3]$ defined by (2.37), (2.38), and (2.39) respectively, can be represented by

$$[C_1(\Delta\theta_1)] = \begin{bmatrix} 1 & 0 & 0 \\ 0 & 1 & \Delta\theta_1 \\ 0 & -\Delta\theta_1 & 1 \end{bmatrix} \tag{2.47}$$

$$[C_2(\Delta\theta_2)] = \begin{bmatrix} 1 & 0 & -\Delta\theta_2 \\ 0 & 1 & 0 \\ \Delta\theta_2 & 0 & 1 \end{bmatrix} \tag{2.48}$$

$$[C_3(\Delta\theta_3)] = \begin{bmatrix} 1 & \Delta\theta_3 & 0 \\ -\Delta\theta_3 & 1 & 0 \\ 0 & 0 & 1 \end{bmatrix} \tag{2.49}$$

In (2.47), (2.48), and (2.49) only the first-order terms in $\Delta\theta_i$ have been retained. It can be easily shown that

$$[C] = [C_3(\Delta\theta_3)][C_2(\Delta\theta_2)][C_1(\Delta\theta_1)]$$
$$= \begin{bmatrix} 1 & \Delta\theta_3 & -\Delta\theta_2 \\ -\Delta\theta_3 & 1 & \Delta\theta_1 \\ \Delta\theta_2 & -\Delta\theta_1 & 1 \end{bmatrix} + O[(\Delta\theta)^2] \tag{2.50}$$

where $O[(\Delta\theta)^2]$ denotes terms of second or higher order. If these terms are neglected as $\Delta\theta \longrightarrow 0$, then in this special case of infinitesimal rotations, the matrix multiplication commutes and the order of multiplication becomes immaterial. In this case, the rotations can be represented by a vector $\Delta\vec{\theta}$ as shown in Fig. 2.17. Our main interest is in representing angular velocities by vectors. The angular velocity vector $\vec{\omega}$ of the rotating frame $x_1 y_1 z_1$ with respect to the fixed frame xyz is given by

$$\vec{\omega} = \lim_{\Delta t \to 0} \frac{\Delta\vec{\theta}}{\Delta t} \tag{2.51}$$

The direction of the vector $\vec{\omega}$ is along the instantaneous axis of rotation of the frame $x_1 y_1 z_1$ with respect to the fixed frame xyz. This angular velocity vector can be decomposed into components along the axes of $x_1 y_1 z_1$ in the form

$$\vec{\omega} = \omega_{x1}\vec{i}_1 + \omega_{y1}\vec{j}_1 + \omega_{z1}\vec{k}_1 \tag{2.52}$$

where the components are given by

$$\omega_{x1} = \lim_{\Delta t \to 0} \frac{\Delta\theta_{x1}}{\Delta t}$$

$$\omega_{y1} = \lim_{\Delta t \to 0} \frac{\Delta\theta_{y1}}{\Delta t} \tag{2.53}$$

$$\omega_{z1} = \lim_{\Delta t \to 0} \frac{\Delta\theta_{z1}}{\Delta t}$$

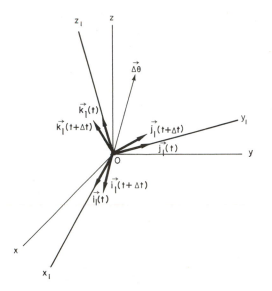

Figure 2.17   Infinitesimal rotation.

As the frame $x_1 y_1 z_1$ rotates, the unit vectors $\vec{i}_1, \vec{j}_1$, and $\vec{k}_1$ change their directions with respect to the fixed frame xyz. The angular displacement $\Delta\vec{\theta}$ which carries $\vec{i}_1(t), \vec{j}_1(t)$, and $\vec{k}_1(t)$ into $\vec{i}_1(t + \Delta t), \vec{j}_1(t + \Delta t)$, and $\vec{k}_1(t + \Delta t)$, respectively, can be represented by using the rotation matrix $[C]$ of (2.50) as

$$\begin{Bmatrix} i_1(t + \Delta t) \\ j_1(t + \Delta t) \\ k_1(t + \Delta t) \end{Bmatrix} = [C(\Delta\theta)] \begin{Bmatrix} i_1(t) \\ j_1(t) \\ k_1(t) \end{Bmatrix} \tag{2.54}$$

Hence, we obtain

$$\begin{Bmatrix} \dfrac{\Delta i_1}{\Delta t} \\[2mm] \dfrac{\Delta j_1}{\Delta t} \\[2mm] \dfrac{\Delta k_1}{\Delta t} \end{Bmatrix} = \begin{Bmatrix} \dfrac{i_1(t + \Delta t) - i_1(t)}{\Delta t} \\[2mm] \dfrac{j_1(t + \Delta t) - j_1(t)}{\Delta t} \\[2mm] \dfrac{k_1(t + \Delta t) - k_1(t)}{\Delta t} \end{Bmatrix}$$

$$= \frac{[C(\Delta\theta)] - [I]}{\Delta t} \begin{Bmatrix} i_1(t) \\ j_1(t) \\ k_1(t) \end{Bmatrix} + \frac{O[(\Delta\theta)^2]}{\Delta t} \tag{2.55}$$

In the limit as $\Delta\theta$ and Δt both tend to zero, the remainder vanishes and we get

$$\begin{Bmatrix} \dfrac{di_1}{dt} \\[2mm] \dfrac{dj_1}{dt} \\[2mm] \dfrac{dk_1}{dt} \end{Bmatrix} = \begin{bmatrix} 0 & \omega_{z1} & -\omega_{y1} \\ -\omega_{z1} & 0 & \omega_{x1} \\ \omega_{y1} & -\omega_{x1} & 0 \end{bmatrix} \begin{Bmatrix} i_1 \\ j_1 \\ k_1 \end{Bmatrix} \tag{2.56}$$

Using this skew-symmetric matrix $[\omega]$, (2.56) can be represented as

$$\begin{Bmatrix} \dfrac{di_1}{dt} \\[2mm] \dfrac{dj_1}{dt} \\[2mm] \dfrac{dk_1}{dt} \end{Bmatrix} = [\omega] \begin{Bmatrix} i_1 \\ j_1 \\ k_1 \end{Bmatrix} \tag{2.57}$$

Alternatively, using the vector notation, (2.57) may be represented as

$$\frac{d\vec{i}_1}{dt} = \vec{\omega} \times \vec{i}_1$$

$$\frac{d\vec{j}_1}{dt} = \vec{\omega} \times \vec{j}_1 \tag{2.58}$$

$$\frac{d\vec{k}_1}{dt} = \vec{\omega} \times \vec{k}_1$$

We now consider a vector $\vec{r}(t)$, which is expressed in the $x_1 y_1 z_1$ coordinate system as

$$\vec{r} = x_1 \vec{i}_1 + y_1 \vec{j}_1 + z_1 \vec{k}_1 \tag{2.59}$$

Differentiating this vector with respect to t, we obtain

$$\dot{\vec{r}} = [\dot{x}_1 \vec{i}_1 + \dot{y}_1 \vec{j}_1 + \dot{z}_1 \vec{k}_1] + \left[x_1 \frac{d\vec{i}_1}{dt} + y_1 \frac{d\vec{j}_1}{dt} + z_1 \frac{d\vec{k}_1}{dt} \right] \tag{2.60}$$

Employing (2.58) in (2.60), the latter equation can be expressed as

$$\dot{\vec{r}} = (\dot{\vec{r}})_{x_1 y_1 z_1} + \vec{\omega} \times (\vec{r})_{x_1 y_1 z_1} \tag{2.61}$$

In the foregoing equation, the subscript denotes that the vector has been expressed in terms of the $x_1 y_1 z_1$ coordinate system. The first term on the right-hand side of (2.61) denotes the rate of change of $\vec{r}$ relative to the system $x_1 y_1 z_1$ and the second term is the rate of change of $\vec{r}$ caused by the rotational motion of $x_1 y_1 z_1$. Hence, when a vector is expressed in terms of a rotating coordinate system, (2.61) provides the rule of obtaining its derivative with respect to time.

2.8 MOTION IN TERMS OF TRANSLATING AND ROTATING FRAME

The result developed in the preceding section is now employed for the determination of expressions for the velocity and acceleration of a particle whose position vector is expressed in terms of a coordinate system that is translating and rotating with time.

In Fig. 2.18, xyz is an inertial coordinate system whose origin O is fixed. The system $x_1 y_1 z_1$ rotates at an angular velocity vector $\vec{\omega}$ and its origin O_1 has velocity $\vec{v}_1$ and acceleration $\vec{a}_1$ with respect to the inertial coordinate system. Let vector $\vec{r}$ denote the position of a particle P relative to the $x_1 y_1 z_1$ coordinate system; that is,

$$\vec{r} = x_1 \vec{i}_1 + y_1 \vec{j}_1 + z_1 \vec{k}_1 \tag{2.62}$$

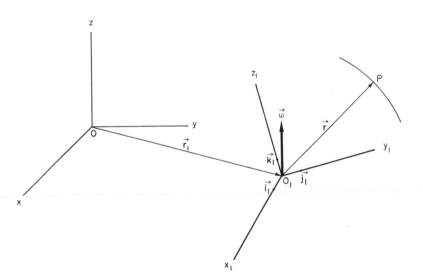

Figure 2.18   Translating and rotating coordinates $O_1 x_1 y_1 z_1$.

However, the position of P with respect to xyz coordinate system is given by

$$(\vec{r})_{xyz} = \vec{r}_1 + \vec{r} \tag{2.63}$$

where $\vec{r}_1$ denotes the position of the origin O_1 of the $x_1 y_1 z_1$ frame with respect to xyz. The absolute velocity of P with respect to the inertial coordinate system is obtained by differentiating (2.63) with respect to t as

$$\vec{v} = \vec{v}_1 + \dot{\vec{r}} + \vec{\omega} \times \vec{r} \tag{2.64}$$

where it should be noted that $\vec{r}$ has been expressed in terms of the $x_1 y_1 z_1$ coordinate system and is given by (2.62). In (2.64), the first term on the right-hand side is the velocity of the origin O_1 of $x_1 y_1 z_1$, the second term is the velocity relative to $x_1 y_1 z_1$, and the third term is the velocity due to rotational motion of $x_1 y_1 z_1$ and is the velocity of a point coinciding with P instantaneously. The last two terms have been obtained by employing the rule given by (2.61). Employing this same rule, the absolute acceleration of P with respect to the inertial coordinate system xyz is obtained as

$$\vec{a} = \vec{a}_1 + [\ddot{\vec{r}} + \dot{\vec{\omega}} \times \vec{r}] + [\dot{\vec{\omega}} \times \vec{r} + (\vec{\omega} \times \vec{\omega}) \times \vec{r}$$
$$+ \vec{\omega} \times \dot{\vec{r}} + \vec{\omega} \times (\vec{\omega} \times \vec{r})]$$
$$= \vec{a}_1 + \ddot{\vec{r}} + 2\vec{\omega} \times \dot{\vec{r}} + \dot{\vec{\omega}} \times \vec{r} + \vec{\omega} \times (\vec{\omega} \times \vec{r}) \qquad (2.65)$$

where again it should be noted that vector $\vec{r}$ has been expressed with respect to the $x_1 y_1 z_1$ coordinate system. In (2.65), $\vec{a}_1$ is the acceleration of the origin O_1 of $x_1 y_1 z_1$, $\ddot{\vec{r}}$ is the acceleration of P relative to $x_1 y_1 z_1$, $2\vec{\omega} \times \dot{\vec{r}}$ is called the Coriolis acceleration, and $\dot{\vec{\omega}} \times \vec{r} + \vec{\omega} \times (\vec{\omega} \times \vec{r})$ is the acceleration of the point that at that instant of time coincides with P. The last term $\vec{\omega} \times (\vec{\omega} \times \vec{r})$ is called the centripetal acceleration and is directed toward the instantaneous axis of rotation.

Hence, (2.64) and (2.65) give the absolute velocity and absolute acceleration, respectively, with respect to an inertial frame of a particle whose motion is observed with respect to a translating and rotating coordinate system. In case the coordinate system $x_1 y_1 z_1$ has only rotational motion without translation (i.e., its origin O_1 is fixed with respect to an inertial frame), then we set $\vec{v}_1 = 0$ and $\vec{a}_1 = 0$ in (2.64) and (2.65), respectively.

Example 2.8

In some applications, the dynamic loads acting on mechanisms and structures due to inertia forces are much greater than the statically applied loads. We consider a mechanism shown in Fig. 2.19. The two arms, each carrying a load W at its end, rotate in the xy plane about the z axis, which is vertical. The weight of each arm is w per unit length. Determine the maximum shear force and maximum bending moment in the arms.

Coordinate system xyz is inertial with origin at O. We employ a body coordinate

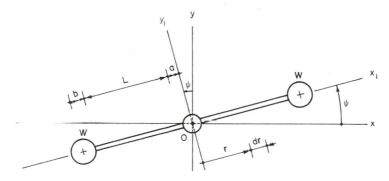

Figure 2.19   Rotating mechanism.

system $x_1 y_1 z_1$ which rotates with the body about the z or z_1 axis with angular velocity $\omega = \dot{\psi}\vec{k}_1$ and has the same fixed origin O. Hence, in (2.65) we have

$$\vec{r} = r\vec{i}_1, \quad \vec{a}_1 = 0, \quad \dot{\vec{r}} = 0, \quad \ddot{\vec{r}} = 0$$

and that equation is simplified to

$$\begin{aligned} \vec{a} &= \dot{\vec{\omega}} \times \vec{r} + \vec{\omega} \times (\vec{\omega} \times \vec{r}) \\ &= \ddot{\psi}\vec{k}_1 \times r\vec{i}_1 + \dot{\psi}\vec{k}_1 \times (\dot{\psi}\vec{k}_1 \times r\vec{i}_1) \\ &= \ddot{\psi}r\vec{j}_1 - \dot{\psi}^2 r\vec{i}_1 \end{aligned} \qquad (2.66)$$

Now, an inertia force is the product of mass and acceleration and is in a direction opposite to that of the acceleration. Hence, the inertia forces acting on weight W, located along the positive x_1 axis, due to the acceleration of (2.66) are shown in Fig. 2.20(a). We employ the usual sign convention employed in strength of materials courses for the shear force and bending moment, as shown in Fig. 2.20(b). We first consider the $x_1 y_1$ plane and shear force and bending moment caused by the inertia forces. Only the component of the inertia force in the y_1 direction causes the shear force and bending moment in the arms. The component in the x_1 direction is the normal force. The shear force due to inertia force on W is given by

$$\vec{V}_{w,y_1} = \vec{j}_1 \frac{W}{g} \ddot{\psi}(a + L + b) \qquad (2.67)$$

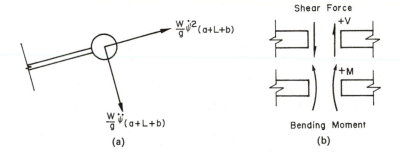

Figure 2.20 (a) Inertia forces; (b) shear force and bending moment.

The shear force due to inertia force on the arm is

$$\begin{aligned} \vec{V}_{a,y_1} &= \vec{j}_1 \int_r^{a+L} \left(\frac{w}{g}\, dr\right) \ddot{\psi} r \\ &= \vec{j}_1 \frac{w}{g} \ddot{\psi} \frac{r^2}{2}\bigg|_{r=r}^{a+L} \end{aligned}$$

This shear force is maximum when $r = a$ and is given by

$$\max \vec{V}_{a,y_1} = \vec{j}_1 \frac{w}{2g} \ddot{\psi}(2aL + L^2) \qquad (2.68)$$

Combining (2.67) and (2.68), the maximum shear force in the arm due to inertia forces becomes

$$\max \vec{V}_{y_1} = \vec{j}_1 \left[\frac{W}{g} \ddot{\psi}(a + L + b) + \frac{w}{2g} \ddot{\psi}(2aL + L^2) \right] \tag{2.69}$$

The maximum bending moments due to inertia forces on W and the arm are given, respectively, by

$$\vec{M}_{w,z_1} = -\vec{k}_1 \frac{W}{g} \ddot{\psi}(a + L + b)(L + b) \tag{2.70}$$

and

$$\vec{M}_{a,z_1} = -\vec{k}_1 \int_a^{a+L} \left(\frac{w}{g} \ddot{\psi} \, dr \right) r(r - a)$$

$$= -\vec{k}_1 \frac{\ddot{\psi} w}{g} \left[\frac{(a + L)^3 - a^3}{3} - \frac{a}{2}(2aL + L^2) \right] \tag{2.71}$$

Hence, the maximum bending moment in the arm due to the inertia forces is obtained by adding (2.70) and (2.71) as

$$\max \vec{M}_{z_1} = -\vec{k}_1 \left[\frac{W}{g} \ddot{\psi}(a + L + b)(L + b) + \frac{\ddot{\psi} w}{g} \left\{ \frac{(a + L)^3 - a^3}{3} - \frac{a}{2}(2aL + L^2) \right\} \right] \tag{2.72}$$

Now we consider the $x_1 z_1$ plane and shear force and bending moment caused by statically applied loads due to the own weights as shown in Fig. 2.21.

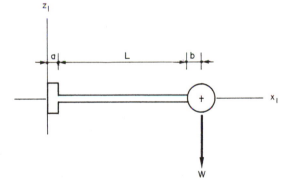

Figure 2.21   View of mechanism in the $x_1 z_1$ plane.

The maximum shear force and bending moment due to the own weight are given, respectively, by

$$\max \vec{V}_{z_1} = \vec{k}_1(W + wL) \tag{2.73a}$$

and

$$\max \vec{M}_{y_1} = -\vec{j}_1 \left[W(L + b) + \frac{wL^2}{2} \right] \tag{2.73b}$$

The total maximum shear force and bending moment in the arms are obtained by combining (2.69) and (2.73a), and (2.72) and (2.73b), respectively, as

$$\max \vec{V} = \vec{j}_1 \left[\frac{W}{g} \ddot{\psi}(a + L + b) + \frac{w}{2g} \ddot{\psi}(2aL + L^2) \right] + \vec{k}_1(W + wL) \tag{2.74a}$$

$$\max \vec{M} = -\vec{j}_1 \left[W(L + b) + \frac{wL^2}{2} \right] - \vec{k}_1 \left[\frac{W}{g} \ddot{\psi}(a + L + b)(L + b) \right.$$

$$\left. + \ddot{\psi} \frac{w}{g} \left\{ \frac{(a + L)^3 - a^3}{3} - \frac{a}{2}(2aL + L^2) \right\} \right] \tag{2.74b}$$

It can be seen that if the weights are small and the value of acceleration is high, the dynamic loads become much bigger than the static loads. In the theory of linear elasticity, the normal and shear stresses can be obtained by considering the loads separately, including the axial load, and then employing superposition.

Example 2.9

A block of mass m_1 is constrained to move on a straight bar AB. A mass m_2 is suspended from mass m_1 and is free to move about the pivot O_1 as shown in Fig. 2.22(a). Determine the acceleration of mass m_2. The problem was considered in Example 2.4 by employing tangential and normal coordinates. In this example, we employ a translating and rotating coordinate system.

In Fig. 2.22(b), xyz is an inertial coordinate system whose origin is fixed at O. The origin of $x_1 y_1 z_1$ is at the moving point O_1 and the frame rotates about the y_1 or y axis along with the mass m_2. The angular velocity of frame $x_1 y_1 z_1$ is given by $\vec{\omega} = \dot{\theta}\vec{j}_1$. The acceleration of m_2 is obtained by employing (2.65). We note that

$$\vec{a}_1 = \ddot{x}\vec{i}, \qquad \vec{r} = -L\vec{k}_1, \qquad \dot{\vec{r}} = 0, \qquad \ddot{\vec{r}} = 0$$

Hence, letting $\vec{a}_2$ be the acceleration of mass m_2, (2.65) yields

$$\vec{a}_2 = \ddot{x}\vec{i} + \dot{\vec{\omega}} \times \vec{r} + \vec{\omega} \times (\vec{\omega} \times \vec{r})$$

$$= \ddot{x}\vec{i} + \ddot{\theta}\vec{j}_1 \times (-L\vec{k}_1) + \dot{\theta}\vec{j}_1 \times (\dot{\theta}\vec{j}_1 x - L\vec{k}_1)$$

$$= \ddot{x}\vec{i} - \ddot{\theta}L\vec{i}_1 + \dot{\theta}^2 L\vec{k}_1 \tag{2.75}$$

In the foregoing equation, the first term is expressed in terms of the xyz coordinate system, whereas the second and third terms are expressed in the $x_1 y_1 z_1$ system. The acceleration can be expressed completely either in the xyz or the $x_1 y_1 z_1$ coordinates by employing the rotational transformation matrices discussed in Section 2.7. For example, choosing the $x_1 y_1 z_1$ coordinate system and the transformation matrix $[C_2]$ of (2.38), we get

$$\{a_2\} = \begin{bmatrix} \cos\theta & 0 & -\sin\theta \\ 0 & 1 & 0 \\ \sin\theta & 0 & \cos\theta \end{bmatrix} \begin{Bmatrix} \ddot{x} \\ 0 \\ 0 \end{Bmatrix} + \begin{Bmatrix} -\ddot{\theta}L \\ 0 \\ \dot{\theta}^2 L \end{Bmatrix}$$

or

$$\vec{a}_2 = (\ddot{x}\cos\theta - \ddot{\theta}L)\vec{i}_1 + (\ddot{x}\sin\theta + \dot{\theta}^2 L)\vec{k}_1 \tag{2.76}$$

Alternatively, if the xyz coordinate system is employed, then employing the inverse of this transformation matrix, we obtain

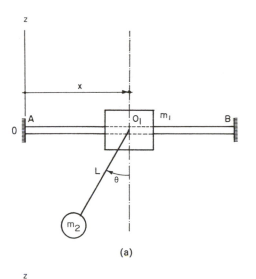

(a)

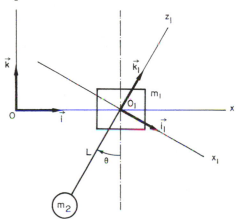

(b)

Figure 2.22 System of Example 2.9.

$$\{a_2\} = \begin{Bmatrix} \ddot{x} \\ 0 \\ 0 \end{Bmatrix} + \begin{bmatrix} \cos\theta & 0 & \sin\theta \\ 0 & 1 & 0 \\ -\sin\theta & 0 & \cos\theta \end{bmatrix} \begin{Bmatrix} -\ddot{\theta}L \\ 0 \\ \dot{\theta}^2 L \end{Bmatrix}$$

or

$$\vec{a}_2 = (\ddot{x} - \ddot{\theta}L\cos\theta + \dot{\theta}^2 L\sin\theta)\vec{i} + (\ddot{\theta}L\sin\theta + \dot{\theta}^2 L\cos\theta)\vec{k} \qquad (2.77)$$

It is noted that (2.75) yields the same result for the acceleration that was obtained in Example 2.4 by employing normal and tangential coordinate systems.

Example 2.10

A mechanism shown in Fig. 2.23 rotates about the vertical axis with angular velocity $\dot{\phi}$ and angular acceleration $\ddot{\phi}$. A mass m is pivoted at point C on the arm OC. Determine the velocity and acceleration of mass m.

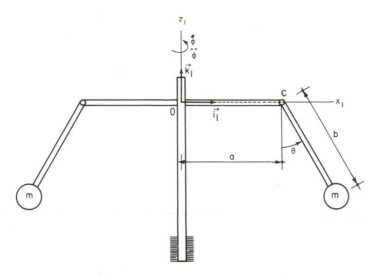

Figure 2.23   Rotating mechanism of Example 2.10.

Let $x_1 y_1 z_1$ be a coordinate system that rotates about the z_1 axis with angular velocity $\vec{\omega} = \dot{\phi}\vec{k}_1$. The position of m relative to this frame of reference is given by

$$\vec{r} = (a + b \sin \theta)\vec{i}_1 - b \cos \theta \vec{k}_1 \tag{2.78}$$

The relative velocity and acceleration are obtained from (2.78) as

$$\dot{\vec{r}} = b\dot{\theta} \cos \theta \vec{i}_1 + b\dot{\theta} \sin \theta \vec{k}_1 \tag{2.79}$$

$$\ddot{\vec{r}} = (b\ddot{\theta} \cos \theta - b\dot{\theta}^2 \sin \theta)\vec{i}_1 + (b\ddot{\theta} \sin \theta + b\dot{\theta}^2 \cos \theta)\vec{k}_1 \tag{2.80}$$

Noting that the origin O has zero velocity, (2.64) yields the absolute velocity of m as

$$\begin{aligned} \vec{v} &= \dot{\vec{r}} + \vec{\omega} \times \vec{r} \\ &= b\dot{\theta} \cos \theta \vec{i}_1 + b\dot{\theta} \sin \theta \vec{k}_1 + \dot{\phi}\vec{k}_1 \times [(a + b \sin \theta)\vec{i}_1 - b \cos \theta \vec{k}_1] \\ &= b\dot{\theta} \cos \theta \vec{i}_1 + \dot{\phi}(a + b \sin \theta)\vec{j}_1 + b\dot{\theta} \sin \theta \vec{k}_1 \end{aligned}$$

Since the acceleration of origin O is zero, (2.65) yields the absolute acceleration of m as

$$\vec{a} = \ddot{\vec{r}} + 2\vec{\omega} \times \dot{\vec{r}} + \dot{\vec{\omega}} \times \vec{r} + \vec{\omega} \times (\vec{\omega} \times \vec{r}) \tag{2.81}$$

Substituting from (2.78), (2.79), and (2.80) in (2.81) and simplifying the result, we obtain

$$\begin{aligned} \vec{a} = {}& [b\ddot{\theta} \cos \theta - b\dot{\theta}^2 \sin \theta - \dot{\phi}^2(a + b \sin \theta)]\vec{i}_1 \\ & + [2b\dot{\phi}\dot{\theta} \cos \theta + \ddot{\phi}(a + b \sin \theta)]\vec{j}_1 + [b\ddot{\theta} \sin \theta + b\dot{\theta}^2 \cos \theta]\vec{k}_1 \end{aligned}$$

2.9 MOTION RELATIVE TO THE ROTATING EARTH

In many applications, we employ a coordinate system whose origin is attached to a point on the surface of the earth. The earth rotates about its axis and its center revolves around the sun, and hence this coordinate system is not inertial.

However, in some applications the additional acceleration terms introduced by considering the rotation and translation of the earth are negligibly small compared to the relative acceleration of a body, including the acceleration due to gravity. In such cases, we may assume as inertial a coordinate system whose origin is attached to a point on the surface of the earth and which does not rotate relative to the earth. Of course, there are other cases where such a system has to be considered as noninertial.

The acceleration caused by the rotation of the earth is much larger than that due to the translation of the earth's center. Considering the translation of the earth as a secondary effect, let us assume as inertial a coordinate system xyz which is attached to the center O of the earth and whose orientation is fixed in space as shown in Fig. 2.24. The z axis is pointing in the direction of the earth's rotation and the xy plane is the equatorial plane.

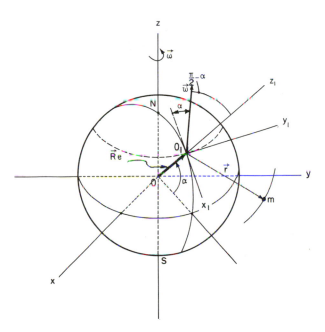

Figure 2.24   The rotating earth.

The coordinate system $x_1 y_1 z_1$ is attached to a point O_1 on the surface of the earth and rotates along with the earth at the same angular velocity $\vec{\omega}$ relative to the xyz frame. The origin O_1 is located at a latitude α as seen from Fig. 2.24. The x_1 axis is tangent to the meridian circle pointing south, y_1 is tangent to the parallel pointing east, and z_1 is in the direction of the local vertical.

The position of mass m relative to the $x_1 y_1 z_1$ coordinate system is denoted by $\vec{r} = x_1 \vec{i}_1 + y_1 \vec{j}_1 + z_1 \vec{k}_1$. Assuming that the earth is a perfect sphere, the vector OO_1 is denoted by $\vec{R}_e = R_e \vec{k}_1$, where R_e is the radius of the earth. Employing (2.65), the acceleration of mass m may be expressed as

$$\vec{a} = \vec{a}_1 + \ddot{\vec{r}} + 2\vec{\omega} \times \dot{\vec{r}} + \dot{\vec{\omega}} \times \vec{r} + \vec{\omega} \times (\vec{\omega} \times \vec{r}) \qquad (2.82)$$

The angular velocity $\vec{\omega}$ of the earth is expressed as

$$\vec{\omega} = -(\omega \cos \alpha)\vec{i}_1 + (\omega \sin \alpha)\vec{k}_1 \qquad (2.83)$$

where it is assumed that ω is a constant and $\omega = 7.27 \times 10^{-5}$ rad/s, which corresponds to one rotation per day. Hence, $\dot{\omega} = 0$. In (2.82), the term $\vec{a}_1$ is the acceleration of the origin O_1 and is given by

$$\vec{a}_1 = \vec{\omega} \times (\vec{\omega} \times \vec{R}_e) \qquad (2.84)$$

Equation (2.82) may now be written as

$$\vec{a} = \vec{\omega} \times (\vec{\omega} \times \vec{R}_e) + \ddot{\vec{r}} + 2\vec{\omega} \times \dot{\vec{r}} + \vec{\omega} \times (\vec{\omega} \times \vec{r}) \qquad (2.85)$$

Substituting the various expressions in (2.85) and carrying out the vector cross products, the components of the acceleration are expressed as

$$a_{x1} = -R_e\omega^2 \sin \alpha \cos \alpha + \ddot{x}_1 - 2\omega\dot{y}_1 \sin \alpha - \omega^2 x_1 \sin^2 \alpha$$
$$\qquad -\omega^2 z_1 \sin \alpha \cos \alpha$$
$$a_{y1} = \ddot{y}_1 + 2\omega\dot{x}_1 \sin \alpha + 2\omega\dot{z}_1 \cos \alpha - \omega^2 y_1$$
$$a_{z1} = -R_e\omega^2 \cos^2 \alpha + \ddot{z}_1 - 2\omega\dot{y}_1 \cos \alpha - \omega^2 x_1 \sin \alpha \cos \alpha$$
$$\qquad -\omega^2 z_1 \cos^2 \alpha \qquad (2.86)$$

The radius of the earth is given by $R_e = 6.37 \times 10^6$ m (3960 miles) and $\omega = 7.27 \times 10^{-5}$ rad/s. Hence, in the first equation of (2.86) we get $R_e\omega^2 \sin \alpha \cos \alpha = 0.0337 \sin \alpha \cos \alpha < 0.0337$ m/s². If errors in the second digit after the decimal point are neglected compared to the acceleration of gravity, which is equal to $g = 9.81$ m/s², the term $R_e\omega^2 \sin \alpha \cos \alpha$ can be dropped from the equation. If the displacements and velocities are sufficiently small such that $2\omega\dot{y}_1 \ll 1$, $\omega^2 x_1 \ll 1$, and $\omega^2 z_1 \ll 1$, then the only significant term in the first equation (2.86) is $\ddot{x}_1$. Under these conditions, we obtain

$$a_{x1} \approx \ddot{x}_1$$
$$a_{y1} \approx \ddot{y}_1 \qquad (2.87)$$
$$a_{z1} \approx \ddot{z}_1$$

Hence, in this case, a coordinate system fixed to a point on the surface of the earth and rotating with the earth may be considered as inertial and this will be implied in many applications that we consider. Of course, there are other applications where these restrictions are not satisfied and (2.86) must be employed. The effect of the Coriolis components in (2.86) can be observed in the manner in which water spirals when draining out of a sink and wind spirals toward a zone of low pressure.

2.10 SUMMARY

This chapter has dealt with the kinematics of motion without considering the cause of motion, which will be covered in the following chapters dealing with kinetics. The expressions for the acceleration has a simple form when an inertial frame of reference is employed. However, in some applications, for one reason or another, it is more advantageous to employ a noninertial coordinate system. For this reason, various coordinate systems, including tangential and normal, polar and cylindrical, and translating and rotating rectangular coordinates have been discussed. Expressions for the velocity and acceleration have been obtained in terms of different coordinate systems. These results will be employed in later chapters dealing with the kinetics of motion.

PROBLEMS

2.1. As observed from the deck of a ship traveling due north at a speed of 10 km/h, the wind appears to form an angle of 30° east of north. When the speed of the ship is increased to 20 km/h, the wind appears to form an angle of 20° east of north. Assume that during the period of observation, the wind velocity is constant and the ship travels in a straight line. Determine the magnitude and direction of true wind velocity.

2.2. The position vector of a particle measured with respect to Cartesian inertial coordinate system is given by $\vec{r} = x\vec{i} + y\vec{j} + z\vec{k}$. Express in terms of x, y, and z and their first and second derivatives:
(a) The tangential component of the acceleration of the particle.
(b) The normal component of its acceleration.
(c) The radius of curvature of the path described by the particle.

2.3. The crank OB of an engine has a constant counter clockwise angular velocity of ω_0 rad/s (Fig. P2.3). As a function of angle θ, determine:
(a) The angular velocity and acceleration of connecting rod BP.
(b) The velocity and acceleration of piston P.
Give your answers in terms of components along the inertial axes $Oxyz$.

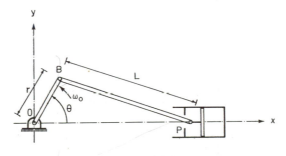

Figure P2.3

2.4. A turbine rotor of radius R is rotating at a constant angular speed ω_o about a fixed axis (Fig. P2.4). A straight vane of length L is welded rigidly to the rotor and the angle between the vane and the radial line is θ. A fluid particle slides outward along the vane tip at a relative speed u which is constant. Determine the velocity $\vec{v}_p$ and acceleration $\vec{a}_p$ of the fluid particle as it leaves the vane. Use rotating coordinate system $Oxyz$.

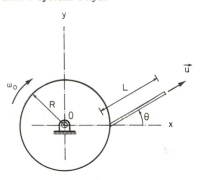

Figure P2.4

2.5. The turret on a tank is rotating about the vertical axis at angular speed $\dot{\phi}$ and the barrel is being raised at an angular speed $\dot{\theta}$ (Fig. P2.5). Both $\dot{\theta}$ and $\dot{\phi}$ are constants. The tank has a constant forward speed of V_t. If a cannon is fired with a muzzle velocity $\dot{s}$ and acceleration $\ddot{s}$ relative to the barrel, determine the velocity $\vec{v}_c$ and acceleration $\vec{a}_c$ of the cannon as it leaves the barrel. Employ $Oxyz$ coordinate system rotating with the turret at $\vec{\omega} = \dot{\phi}\vec{j}$.

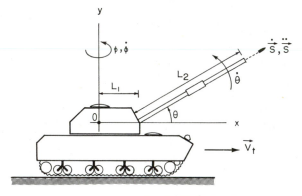

Figure P2.5

2.6. A particle P is moving across a disk in a straight line AB with a constant speed V_0 relative to the disk (Fig. P2.6). The coordinate system xyz is rotating with the disk at angular velocity $\omega_o\vec{k}$ and angular acceleration $\alpha_o\vec{k}$. Determine the velocity and acceleration of P in terms of the xyz coordinate system.

2.7. A radar antenna rotates about a fixed vertical axis at a constant angular velocity $\omega_o\vec{j}$ (Fig. P2.7). The angle θ oscillates as $\theta = a_o + a_1 \sin \omega_1 t$. Determine the velocity and acceleration of probe P using the rotating coordinate system xyz attached to the vertical shaft.

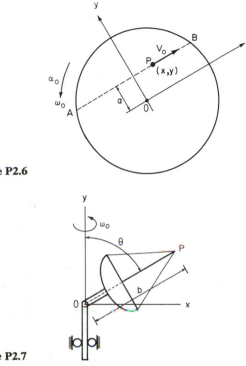

Figure P2.6

Figure P2.7

2.8. Water flows through a sprinkler arm OAB with a velocity $\vec{v}_o$ relative to the arm (Fig. P2.8). The arm rotates counterclockwise at a constant angular speed ω_o. Determine the acceleration of a particle of water as it leaves the arm at B. Employ rotating coordinate system $Oxyz$.

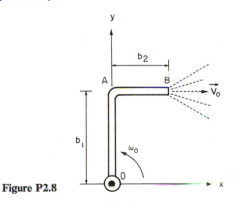

Figure P2.8

2.9. An automobile is traveling due north at a constant speed of 80 km/h along a straight road (Fig. P2.9). It is in the northern hemisphere at 40° latitude. Determine the acceleration of the vehicle in terms of north, east, and local vertical components.

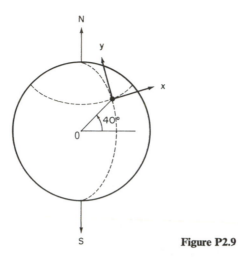

Figure P2.9

REFERENCES

1. Meirovitch, L., *Methods of Analytical Dynamics*, McGraw-Hill Book Company, New York, 1970.
2. Kane, T. R., *Dynamics*, Holt, Rinehart and Winston, New York, 1968.
3. McCuskey, S. W., *An Introduction to Advanced Dynamics*, Addison-Wesley Publishing Company, Inc., Reading, Mass., 1959.
4. Halfman, R. L., *Dynamics*, Vol. 1, Addison-Wesley Publishing Company, Inc., Reading, Mass., 1962.
5. Beer, F. P., and Johnston, E. R., *Vector Mechanics for Engineers, Dynamics*, 3rd ed., McGraw-Hill Book Company, New York, 1977.

3

DYNAMICS OF PARTICLES: NEWTON'S LAW, ENERGY, AND MOMENTUM METHODS

3.1 INTRODUCTION

The early part of this chapter is concerned with the derivation of the equations of motion for a system of particles by direct application of Newton's second law. Physical coordinate systems such as Cartesian, tangential and normal, and polar coordinates are employed to express the equations of motion. Some of the coordinates may not be independent but related to the others by kinematic constraints which are employed simultaneously with the equations of motion. An alternative method of deriving the equations of motion based on Lagrangian techniques and employing generalized coordinates is covered in Chapter 5.

It is recalled from Chapter 1 that a particle is defined as a body of any size or shape that only translates without rotation. This implies that the resultant moment acting on a particle is zero. When a body only translates without rotation, all points of the body have the same velocity and the same acceleration at any instant of time. Hence, a particle may be considered as a point mass.

The latter part of this chapter covers the energy and momentum methods based on some principles of dynamics. The advantage of employing these principles is that answers to some simple problems can be obtained directly without formulating the equations of motion and obtaining their solution. Furthermore, impact between particles is best studied by employing these principles. The principle of work and energy relates directly the force, mass, velocity, and displacement, while the principle of impulse and momentum relates the force, mass, velocity, and time.

The two-body central force motion is discussed in the final part of the chapter. It is concerned with two particles that move in space under the influence of forces exerted by the particles on each other along the line joining them. The two-body problem, together with Newton's law of gravitation, is then employed to study satellite dynamics and orbital mechanics.

3.2 EQUATIONS OF MOTION OF A PARTICLE

From Newton's second law, it is seen that when the resultant force acting on a particle is not zero, the particle moves so that the resultant force vector is equal to the time rate of change of the linear momentum vector; that is,

$$\sum \vec{F} = \frac{d}{dt}(m\vec{v}) \tag{3.1}$$

where $\sum \vec{F}$ is the resultant force, m the mass, $\vec{v}$ the velocity vector measured with respect to inertial frame of reference, and $m\vec{v}$ is the linear momentum. Considering a particle that does not gain or lose mass (e.g., a rocket can lose mass due to depletion of fuel), and restricting the velocities to values that are much smaller than the velocity of light, the mass becomes independent of time and (3.1) may be written as

$$\sum \vec{F} = m\frac{d\vec{v}}{dt} = m\vec{a} \tag{3.2}$$

where the acceleration vector $\vec{a}$ is measured with respect to the inertial coordinate system (i.e., the acceleration is "absolute"). Employing an inertial Cartesian coordinate system and letting $\vec{r}$ denote the position vector of the particle from the origin, (3.2) becomes

$$\sum \vec{F} = m\ddot{\vec{r}} \tag{3.3}$$

and its three components are given by

$$\sum F_x = m\ddot{x}$$
$$\sum F_y = m\ddot{y} \tag{3.4}$$
$$\sum F_z = m\ddot{z}$$

The acceleration has the simplest form when the coordinate system is inertial. Integration of these equations yields the position $\vec{r}(t)$ of the particle at any time instant t. An unconstrained particle has three degrees of freedom in the x, y, and z directions, respectively. However, one or more of the degrees of freedom may be constrained. For example, if the motion of the particle is constrained to the xy plane, then the third equation of (3.4) becomes $\sum F_z = 0$.

Sometimes it is convenient to employ a noninertial coordinate system $O_1 x_1 y_1 z_1$ of Fig. 2.18, rotating at angular velocity $\vec{\omega}$ and whose origin O_1 has an acceleration $\vec{a}_1$. Denoting the position of the particle from the origin O_1 by $\vec{r}$ and expressing all the vectors with respect to the $x_1 y_1 z_1$ coordinate system, we note from (2.65) that (3.2) may be expressed as

$$\sum \vec{F} = m[\vec{a}_1 + \ddot{\vec{r}} + 2\vec{\omega} \times \dot{\vec{r}} + \dot{\vec{\omega}} \times r + \vec{\omega} \times (\vec{\omega} \times \vec{r})] \qquad (3.5)$$

In case tangential and normal coordinates are employed, then from (2.13) the tangential and normal components of the equation of motion are expressed as

$$\sum F_t = m \frac{dv}{dt}$$
$$\sum F_n = m \frac{v^2}{\rho} \qquad (3.6)$$

If polar coordinates are selected to represent plane motion, then from (2.20) the radial and transverse components of the equation of motion are given by

$$\sum F_r = m(\ddot{r} - r\dot{\theta}^2)$$
$$\sum F_\theta = m(r\ddot{\theta} + 2\dot{r}\dot{\theta}) \qquad (3.7)$$

Equations (3.7) will be employed later for the study of two-body central force motion. The choice of the coordinate system is very important and the right choice can simplify the solution of the equations of motion.

3.2.1 State-Variable Formulation

Several methods of analysis of dynamic systems are based on the state-space representation. In this approach, the equations of motion are represented as a set of first-order coupled differential equations described by

$$\{\dot{x}\} = \{f(x_1, \ldots, x_n, Q_1, \ldots, Q_m, t)\}$$

where the elements of the vector (i.e., the column matrix $\{x\}$) are called the state variables and $Q_1, \ldots, Q_m$ are input forces. The right-hand sides of the state equations are in general nonlinear functions of the state variables, forces, and time. The derivatives of the state variables do not appear on the right-hand sides of the state equations. Depending on the number of particles and the total degrees of freedom, the correct number of variables must be chosen so that the equations of motion can be represented in this form. The choice of state variables, however, is not unique.

The future behavior of a dynamic system may be specified in terms of initial conditions at any instant of time and the inputs from that time onward. The knowledge of past inputs is not required to determine the future behavior. The n numbers required to specify the future behavior of a dynamic system represent the initial state of the system and the variables used to represent these numbers at each instant of time are called state variables. Hence, the state-variable vector $\{x(t)\}$ consists of time functions whose values at any specified time represent the state of the dynamic system at that time. The n-dimensional space with the state variables as coordinates is called the state space as discussed later in Chapter 6. The state-variable formulation will be clarified by several examples that are given later.

Example 3.1

A particle of mass m slides down the surface of a smooth spherical radome of radius R (Fig. 3.1). It starts at the top where $\theta = 0$ with a small angular velocity $\dot{\theta}_0$. Neglecting friction, determine the angle θ_m at which it loses contact with the surface.

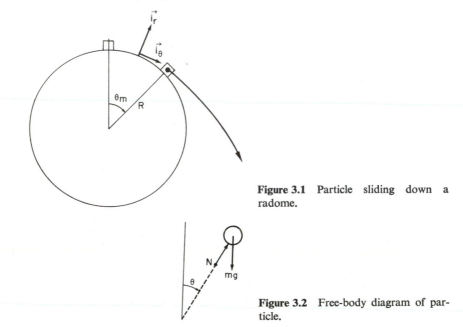

Figure 3.1 Particle sliding down a radome.

Figure 3.2 Free-body diagram of particle.

This problem is solved here employing the equations of motion in polar coordinates. The free-body diagram of the particle for any angle $\theta < \theta_m$ is shown in Fig. 3.2, where N is the normal force. Employing (3.7), we note that $r = R$, constant, so that $\dot{r} = \ddot{r} = 0$ and we get

$$\sum F_r = N - mg \cos \theta = -mR\dot{\theta}^2 \tag{3.8}$$

$$\sum F_\theta = mg \sin \theta = mR\ddot{\theta}$$

The first of these two equations is merely a constraint equation that yields an expression for N, whereas the second equation is a differential equation for $\theta(t)$. The condition for the particle to leave the surface is that $N = 0$. Hence, from the first equation of (3.8), we obtain

$$\dot{\theta}_m^2 = \frac{g}{R} \cos \theta_m \tag{3.9}$$

The second equation of (3.8) can be integrated once to obtain a relationship between $\dot{\theta}$ and θ as follows. Since

$$\ddot{\theta} = \frac{d}{dt}\dot{\theta} = \frac{d\dot{\theta}}{d\theta}\frac{d\theta}{dt} = \dot{\theta}\frac{d\dot{\theta}}{d\theta}$$

from the second equation of (3.8), we obtain

$$\int_{\dot\theta_o}^{\dot\theta_m} \dot\theta\, d\dot\theta = \int_0^{\theta_m} \frac{g}{R} \sin\theta\, d\theta$$

or

$$\dot\theta_m^2 = \dot\theta_o^2 + \frac{2g}{R}(1 - \cos\theta_m) \tag{3.10}$$

Equating the right-hand sides of (3.9) and (3.10), we get

$$\frac{g}{R}\cos\theta_m = \dot\theta_o^2 + \frac{2g}{R}(1 - \cos\theta_m)$$

or

$$\cos\theta_m = \frac{R\dot\theta_o^2}{3g} + \frac{2}{3} \tag{3.11}$$

The initial velocity $\dot\theta_o$ must be small enough so that the right-hand side of (3.11) is less than unity. The equations of motion can also be formulated by using the tangential and normal coordinates. From (3.6), after noting that $v = R\dot\theta$ and $\rho = R$, we obtain

$$\sum F_t = mg\sin\theta = mR\ddot\theta$$

$$\sum F_n = -N + mg\cos\theta = mR\dot\theta^2$$

which are the same equations as (3.8). It will be seen later that the answer to this simple problem can be more easily obtained from the work–energy principle without formulating the equations of motion and obtaining their solution.

Coulomb friction could be included to oppose the sliding motion. Coulomb friction is a constant force that opposes the motion. In the θ direction, the frictional force becomes $F_f = -\mu N \operatorname{sgn}\dot\theta$, where the signum function is defined by $\operatorname{sgn}\dot\theta = +1$ if $\dot\theta > 0$, $\operatorname{sgn}\dot\theta = -1$ if $\dot\theta < 0$, and $-1 \le \operatorname{sgn}\dot\theta \le 1$ for $\dot\theta = 0$ as shown in Fig. 3.3. After substituting for N from the first equation of (3.8), the differential equation of motion in the θ direction becomes

$$mR\ddot\theta + \mu(mg\cos\theta - mR\dot\theta^2)\operatorname{sgn}\dot\theta - mg\sin\theta = 0, \qquad \theta \le \theta_m \tag{3.12}$$

The foregoing equation is nonlinear and numerical integration can be employed to obtain a solution. Numerical integration techniques are discussed in Chapter 7.

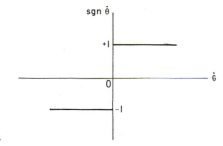

Figure 3.3   Signum function.

Example 3.2

A rigid shaft is rotating at constant angular velocity ω_o about a vertical axis as shown in Fig. 3.4. A mass m can slide with Coulomb friction on the shaft OD and is restrained by a linear spring of stiffness k and unstrained length L. Obtain the equations of motion of mass m.

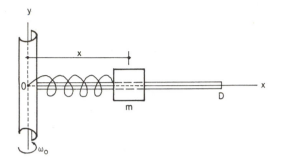

Figure 3.4   Sliding mass.

We employ a rotating coordinate system with origin at O and rotating with the angular velocity $\vec{\omega} = \omega_0 \vec{j}$ as shown in Fig. 3.4. The free-body diagram of the mass is shown in Fig. 3.5, where $k(x - L)$ is the spring force, F_f the friction force, mg the weight, and N_y and N_z are the components of the reaction along the y and z axes. Referring to (3.5) and letting $\vec{a}_1$ be the acceleration of the origin and $\vec{r}$ the position of the mass with respect to this coordinate system, we note that

$$\vec{a}_1 = 0, \qquad \dot{\vec{\omega}} \times \vec{r} = 0$$
$$\vec{r} = x\vec{i}, \, 2\vec{\omega} \times \dot{\vec{r}} = 2\omega_0\vec{j} \times \dot{x}\vec{i} = -2\omega_0\dot{x}\vec{k}$$
$$\vec{\omega} \times (\vec{\omega} \times \vec{r}) = \omega_0\vec{j} \times (\omega_0\vec{j} \times x\vec{i}) = -\omega_0^2 x\vec{i}$$

Hence, the absolute acceleration vector becomes

$$\vec{a} = \ddot{x}\vec{i} - 2\omega_0\dot{x}\vec{k} - \omega_0^2 x\vec{i}$$

From the free-body diagram of Fig. 3.5, the equations of motion can be written as follows:

$$x \text{ axis:} \quad -k(x - L) - F_f = m(\ddot{x} - \omega_0^2 x)$$
$$y \text{ axis:} \quad N_y - mg = 0$$
$$z \text{ axis:} \quad N_z = -2m\omega_0\dot{x}$$

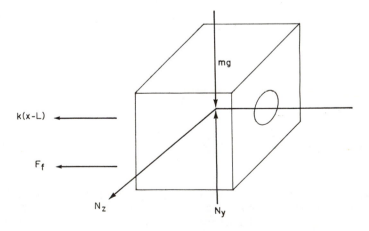

Figure 3.5   Free-body diagram of mass.

The second and third of these equations yield expressions for components of the reaction, whereas the first equation is the differential equation of motion. Now,

$$F_f = \mu N \operatorname{sgn} \dot{x} = \mu[(2m\omega_o\dot{x})^2 + (mg)^2]^{1/2} \operatorname{sgn} \dot{x}$$

Hence, the equation of motion becomes

$$m\ddot{x} + \mu[(2m\omega_o\dot{x})^2 + (mg)^2]^{1/2} \operatorname{sgn} \dot{x} + k(x - L) - m\omega_o^2 x = 0 \qquad (3.13)$$

Sometimes it is desirable for the purpose of analysis, as discussed earlier, to express the equations of motion as a set of first-order equations. The variables chosen to represent the equations in this form are known as state variables and the equations are known as state equations. Letting $x_1 = x$ and $x_2 = \dot{x}$, (3.13) may be expressed as

$$\dot{x}_1 = x_2$$

$$\dot{x}_2 = -\frac{k}{m} x_1 + \omega_o^2 x_1 - \frac{\mu}{m}[(2m\omega_o x_2)^2 + (mg)^2]^{1/2} \operatorname{sgn} x_2 + \frac{kL}{m} \qquad (3.14)$$

The first of these equations is merely a definition, whereas the second is obtained from the equation of motion (3.13).

Example 3.3

A ball of mass m is made to resolve in a horizontal circle at a constant angular velocity ω_o as shown in Fig. 3.6. If the maximum allowable tension in the cord is $T_{\max}$, determine the maximum allowable velocity ω_o and the corresponding value of angle $\theta_{\max}$.

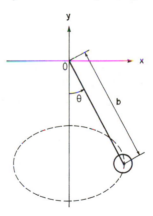

Figure 3.6   Ball revolving in circle.

We employ a rotating coordinate system $Oxyz$ with fixed origin O and angular velocity $\vec{\omega} = \omega_o\vec{j}$ as shown in Fig. 3.6. In (3.5), the only nonzero acceleration term is given by

$$\vec{a} = \vec{\omega} \times (\vec{\omega} \times \vec{r}) = \omega_o\vec{j} \times [\omega_o\vec{j} \times (b \sin\theta\vec{i} - b\cos\theta\vec{j})] = -\omega_o^2 b \sin\theta\vec{i}$$

The free-body diagram of the ball is shown in Fig. 3.7, where T is the tension in the cord. Hence, we obtain the following equations:

$$x \text{ axis:} \quad -T\sin\theta = -m\omega_o^2 b \sin\theta \qquad (3.15)$$

$$y \text{ axis:} \quad T\cos\theta - mg = 0 \qquad (3.16)$$

Figure 3.7   Free-body diagram of ball.

From (3.15), we get max $\omega_o^2 = T_{max}/mb$ and from (3.16) it follows that $\theta_{max} = \cos^{-1}(mg/T_{max})$.

3.3 EQUATIONS OF MOTION OF A SYSTEM OF PARTICLES

Newton's second law has been stated for a single particle but it can be easily extended to study the motion of a system of particles. In the first method, which is convenient when the particles are constrained because they are connected by massless linkages, cables, and other devices, we employ the free-body diagram for each individual particle and obtain the equations of motion for each particle. The constraint forces now appear in the equations of motion and have to be eliminated. It should be noted that for each constraint force there is an equal and opposite constraint force according to Newton's third law. This method is illustrated in Example 3.4.

In the second method, which is convenient for a system of free particles, we consider the motion of the mass center of the system. Consider a system of particles as shown in Fig. 3.8. The forces acting on each particle are separated into two parts. Let $\vec{F}_i$ be the resultant of all external forces acting on ith particle and $\sum\limits_{j=1}^{n} \vec{f}_{ij}$ be the resultant of the internal forces exerted on the ith particle by the other particles. It is noted that $\vec{f}_{ii} = 0$ since there are no interacting forces between a particle and itself. The internal forces may be caused by a central force law such as Newton's law of gravitation or Coulomb's law describing the forces among electrically charged particles. Letting δ_{ij}^* denote the complementary Kronecker delta function (i.e., $\delta_{ij}^* = 1 - \delta_{ij}$, which is 0 for $i = j$ and 1 for $i \neq j$), and employing Newton's second law to the ith particle, we obtain

$$\vec{F}_i + \sum_{j=1}^{n} \delta_{ij}^* \vec{f}_{ij} = m_i \vec{a}_i \tag{3.17}$$

Summing up over the entire system of particles, we get

$$\sum_{i=1}^{n} \vec{F}_i + \sum_{i=1}^{n} \sum_{j=1}^{n} \delta_{ij}^* \vec{f}_{ij} = \sum_{i=1}^{n} m_i \vec{a}_i \tag{3.18}$$

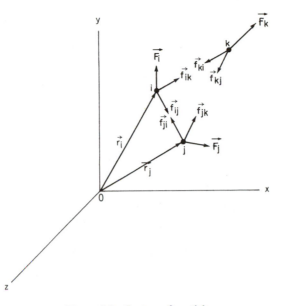

Figure 3.8 System of particles.

The internal forces $\vec{f}_{ij}$ and $\vec{f}_{ji}$ are equal and opposite and hence $\vec{f}_{ij} + \vec{f}_{ji} = 0$. It follows that in (3.18), we get

$$\sum_{i=1}^{n} \sum_{j=1}^{n} \delta_{ij}^{*} \vec{f}_{ij} = 0$$

and that equation becomes

$$\sum_{i=1}^{n} \vec{F}_{i} = \sum_{i=1}^{n} m_{i} \vec{a}_{i} \tag{3.19}$$

For the system of particles as a whole, the sum of the internal forces reduces to zero. However, it does not imply that the internal forces have no effect on the particles. The gravitational forces that the sun and the planets exert on each other sum up to zero for the solar system but cause the motion of the planets around the sun. The constraint forces, if applied by themselves, will not cause the system to move.

The mass center of the system of particles is defined by the position vector $\vec{r}_{c}$, which satisfies the relationship

$$m\vec{r}_{c} = \sum_{i=1}^{n} m_{i} \vec{r}_{i} \tag{3.20a}$$

where m is defined as the total system mass, that is,

$$m = \sum_{i=1}^{n} m_{i}$$

Employing inertial coordinate system and differentiating both sides of (3.20a) with respect to time, the velocity $\vec{v}_c$ and acceleration $\vec{a}_c$ of the center of mass are given by

$$m\vec{v}_c = \sum_{i=1}^{n} m_i \vec{v}_i$$

$$m\vec{a}_c = \sum_{i=1}^{n} m_i \vec{a}_i$$

(3.20b)

From (3.19) and (3.20), we obtain

$$\sum_{i=1}^{n} \vec{F}_i = m\vec{a}_c$$

(3.21)

This equation states that the mass center of a system of particles moves as if the entire mass of the system were concentrated at that point and all the external forces were applied there. The quantity $m\vec{v}_c$ is the linear momentum of the system of particles. If no external force acts on a system of particles, the left-hand side of (3.21) is zero and the linear momentum is conserved (i.e., $m\vec{v}_c$ = constant).

Example 3.4

A particle of mass m_1 is free to slide on a horizontal bar with Coulomb friction under the action of a force P. Mass m_2 is pivoted from m_1 by a massless rigid link of length b. Obtain the equations of motion for this system of particles shown in Fig. 3.9.

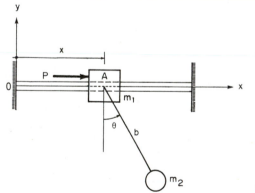

Figure 3.9   System of two particles.

This problem is solved by drawing the free-body diagram for each particle as shown in Fig. 3.10, where the constraint force F is the tension in the rod. Cartesian inertial coordinates are employed to designate the position of m_1 and polar coordinates with origin at the moving point A for the position of m_2.

From (3.3) the acceleration of m_1 is given by $\vec{a}_1 = \ddot{x}\vec{i}$. The acceleration of m_2 is obtained from (3.7) after noting that $r = b$, $\dot{r} = \ddot{r} = 0$, and the origin has acceleration $\ddot{x}\vec{i}$. Hence, the acceleration of m_2 becomes

$$\vec{a}_2 = \ddot{x}\vec{i} - b\dot{\theta}^2\vec{i}_r + b\ddot{\theta}\vec{i}_\theta$$

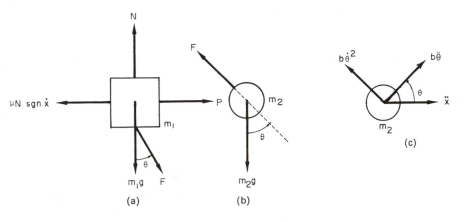

Figure 3.10   Free-body diagram of the two masses.

We get the following equations for mass m_1:

$$x \text{ axis:}\quad P - \mu N \,\text{sgn}\, \dot{x} + F \sin\theta = m_1 \ddot{x} \tag{3.22}$$

$$y \text{ axis:}\quad N - m_1 g - F \cos\theta = 0 \tag{3.23}$$

The equations for mass m_2 are as follows:

$$r \text{ axis:}\quad m_2 g \cos\theta - F = m_2 \ddot{x} \sin\theta - m_2 b \dot{\theta}^2 \tag{3.24}$$

$$\theta \text{ axis:}\quad -m_2 g \sin\theta = m_2 \ddot{x} \cos\theta + m_2 b \ddot{\theta} \tag{3.25}$$

Now, the constraint force F is eliminated by substituting for it from (3.24) in (3.23) to obtain

$$N = m_1 g + (m_2 g \cos\theta - m_2 \ddot{x} \sin\theta + m_2 b \dot{\theta}^2) \cos\theta$$

The foregoing equation is employed to eliminate the constraint force N from (3.22). The two coupled equations of motion are now given by

$$m_1 \ddot{x} + m_2 \ddot{x} \sin^2\theta - m_2 g \sin\theta \cos\theta - m_2 b \dot{\theta}^2 \sin\theta$$
$$+ \mu[m_1 g + (m_2 b \dot{\theta}^2 + m_2 g \cos\theta - m_2 \ddot{x} \sin\theta) \cos\theta] \,\text{sgn}\, \dot{x} = P \tag{3.26}$$
$$m_2 \ddot{x} \cos\theta + m_2 b \ddot{\theta} + m_2 g \sin\theta = 0 \tag{3.27}$$

In order to express these equations as a set of first-order equations, we choose the displacements and velocities as the state variables. Let $x_1 = x$, $x_2 = \theta$, $x_3 = \dot{x}$, and $x_4 = \dot{\theta}$. Now (3.26) and (3.27) become

$$m_1 \dot{x}_3 + m_2 \dot{x}_3 \sin^2 x_2 - m_2 g \sin x_2 \cos x_2 - m b x_4^2 \sin x_2$$
$$+ \mu[m_1 g + (m_2 b x_4^2 + m_2 g \cos x_2 - m_2 \dot{x}_3 \sin x_2) \cos x_2] \,\text{sgn}\, \dot{x} = P$$

and

$$m_2 \dot{x}_3 \cos x_2 + m_2 b \dot{x}_4 + m_2 g \sin x_2 = 0$$

These equations can be expressed as

$$\dot{x}_1 = x_3$$
$$\dot{x}_2 = x_4$$
$$\dot{x}_3 = f_3(x_1, x_2, x_3, x_4, P)$$
$$\dot{x}_4 = f_4(x_1, x_2, x_3, x_4, P)$$

where the first two equations are obtained from the definition of the state variables and the last two from the equations of motion (3.26) and (3.27).

Example 3.5

A projectile of mass m has a velocity $v_o\vec{i}$ and altitude $h_o\vec{j}$ at the instant when an explosion breaks the projectile into two parts of masses m_1 and m_2, respectively. The coordinate system is shown in Fig. 3.11.

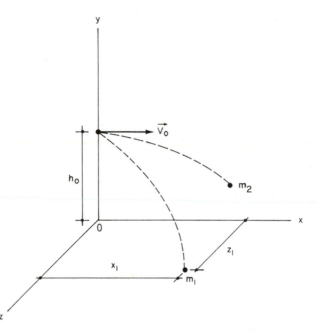

Figure 3.11   Trajectory of an exploding projectile.

The part of mass m_1 strikes the ground t_1 seconds later at a location $x_1\vec{i} + z_1\vec{k}$ from the origin. Determine the position $x_2\vec{i} + y_2\vec{j} + z_2\vec{k}$ of mass m_2 at that instant. Neglect the aerodynamic drag.

This problem is resolved by considering the motion of the mass center. Neglecting aerodynamic drag, the equations of motion for the mass center of the two parts are

$$\ddot{x}_c = 0, \qquad \dot{x}_c(0) = v_0, \qquad x_c(0) = 0$$
$$\ddot{y}_c = -g, \qquad \dot{y}_c(0) = 0, \qquad y_c(0) = h_o$$
$$\ddot{z}_c = 0, \qquad \dot{z}_c(0) = 0, \qquad z_c(0) = 0$$

The solution of these equations of motion is given by

$$x_c = v_0 t$$
$$y_c = h_0 - \tfrac{1}{2}gt^2$$
$$z_c = 0$$

At time t_1, it follows that

$$x_c(t_1) = v_o t_1$$
$$y_c(t_1) = h_o - \tfrac{1}{2}gt_1^2$$
$$z_c(t_1) = 0$$

We also have the relationship

$$m\vec{r}_c(t_1) = m_1\vec{r}_1(t_1) + m_2\vec{r}_2(t_1) \qquad \text{where } m = m_1 + m_2$$

Hence, we obtain

$$m[v_o t_1 \vec{i} + (h_o - \tfrac{1}{2}gt_1^2)\vec{j}] = m_1[x_1\vec{i} + z_1\vec{k}] + m_2[x_2\vec{i} + y_2\vec{j} + z_2\vec{k}]$$

It follows that

$$x_2 = \frac{mv_o t_1 - m_1 x_1}{m_2}$$

$$y_2 = \frac{m}{m_2}(h_o - \tfrac{1}{2}gt_1^2)$$

$$z_2 = -\frac{m_1}{m_2}z_1$$

3.4 ANGULAR MOMENTUM OF A SYSTEM OF PARTICLES

First, we consider a single particle of mass m acted upon by a resultant force $\vec{F}$. The particle has velocity $\vec{v}$ measured with respect to an inertial coordinate system $Oxyz$ as shown in Fig. 3.12. The linear momentum of the particle is $m\vec{v}$. The moment of the linear momentum vector about the fixed point O is $\vec{r} \times m\vec{v}$. This is referred to as the angular momentum vector $\vec{H}_o$ of the particle about point O and is given by

$$\vec{H}_o = \vec{r} \times m\vec{v} \tag{3.28}$$

The vector $\vec{H}_o$ is perpendicular to the plane containing $\vec{r}$ and $m\vec{v}$ and has magnitude $H_o = rmv \sin\theta$, where θ is the angle between $\vec{r}$ and $m\vec{v}$ as shown in Fig. 3.12. The sense of $\vec{H}_o$ is given by the right-hand rule. Resolving the vectors $\vec{r}$ and $m\vec{v}$ into components, we can write

$$\vec{H}_o = \begin{vmatrix} \vec{i} & \vec{j} & \vec{k} \\ x & y & z \\ mv_x & mv_y & mv_z \end{vmatrix} \tag{3.29}$$

Next, we compute the derivative with respect to time of $\vec{H}_o$. From (3.28) we obtain

$$\dot{\vec{H}}_o = \dot{\vec{r}} \times m\vec{v} + \vec{r} \times m\dot{\vec{v}} \tag{3.30}$$

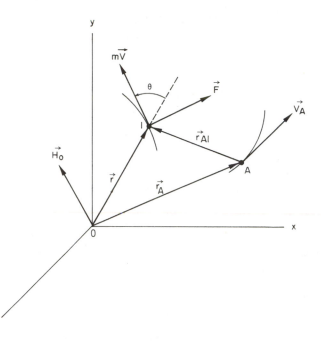

Figure 3.12   Angular momentum of a particle.

Since the inertial coordinate system is employed, the first term on the right-hand side of (3.30) becomes $\vec{v} \times m\vec{v} = 0$ and the second term $\vec{r} \times m\vec{v} = \vec{r} \times m\vec{a} = \vec{r} \times \vec{F}$, where $\vec{F}$ is the resultant force on the particle. Hence, we get

$$\dot{\vec{H}}_o = \vec{r} \times \vec{F}$$
$$= \vec{M}_o \tag{3.31}$$

which states that the resultant moment of the forces about O is equal to the rate of change of angular momentum about O. We now consider the moment of the momentum about a moving point A as shown in Fig. 3.12. We obtain

$$\vec{H}_A = \vec{r}_{A1} \times m\vec{v}$$

and its derivative with respect to time becomes

$$\dot{\vec{H}}_A = \dot{\vec{r}}_{A1} \times m\vec{v} + \vec{r}_{A1} \times m\dot{\vec{v}} \tag{3.32}$$

Now, $m\dot{\vec{v}} = \vec{F}$ and denoting $\vec{r}_{A1} \times \vec{F}$ by $\vec{M}_A$, the foregoing equation can be written as

$$\dot{\vec{H}}_A = \dot{\vec{r}}_{A1} \times m\vec{v} + \vec{M}_A \tag{3.33}$$

This equation states that the resultant moment of the forces about a moving point A is in general not equal to the rate of change of angular momentum about A. We get $\dot{\vec{H}}_A = \vec{M}_A$ only when $\dot{\vec{r}}_{A1} \times \vec{v} = 0$. When A coincides with the fixed point O, $\dot{\vec{r}}_{A1} \times \vec{v} = \vec{v} \times \vec{v} = 0$ and we obtain the result of (3.31). Also, $\dot{\vec{H}}_A = \vec{M}_A = \vec{0}$ when $\vec{r}_{A1} = 0$ and point A coincides with the particle.

Now,

$$\vec{r}_{A1} \times m\vec{v} = (-\vec{r}_A + \vec{r}) \times m\vec{v}$$
$$= (-\vec{r}_A + \vec{v}) \times m\vec{v}$$
$$= -\vec{r}_A \times m\vec{v}$$
$$= -\vec{v}_A \times m\vec{v}$$

and (3.33) can also be written as

$$\dot{\vec{H}}_A = -\vec{v}_A \times m\vec{v} + \vec{M}_A \tag{3.34}$$

It can be seen that when A is any fixed point, we have $\vec{v}_A = 0$ and $\dot{\vec{H}}_A = \vec{M}_A$.

The foregoing results are now generalized to a system of n particles with masses $m_1, \ldots, m_n$ each of which is acted upon by a resultant external force $\vec{F}_i$, $i = 1, 2, \ldots, n$. Each particle has velocity $\vec{v}_i$ with respect to an inertial coordinate system $Oxyz$ as shown in Fig. 3.13. The linear momentum of the system of particles is $m_1\vec{v}_1 + \cdots + m_n\vec{v}_n = m\vec{v}_c$, where $m = \sum_{i=1}^{n} m_i$ and $\vec{v}_c$ is the velocity of the center of mass as seen from (3.20). The moment of the momentum of the system of particles about a moving point A is given by

$$\vec{H}_A = \sum_{i=1}^{n} \vec{r}_{Ai} \times m_i\vec{v}_i \tag{3.35}$$

The derivative with respect to time of this equation yields

$$\dot{\vec{H}}_A = \sum_{i=1}^{n} \dot{\vec{r}}_{Ai} \times m_i\vec{v}_i + \sum_{i=1}^{n} \vec{r}_{Ai} \times m_i\dot{\vec{v}}_i \tag{3.36}$$

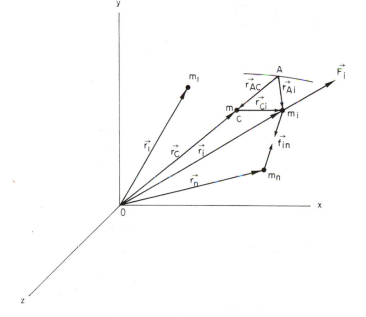

Figure 3.13 Angular momentum of a system of particles.

From Fig. 3.13, it is seen that

$$\vec{r}_{Ai} = \vec{r}_{AC} + \vec{r}_{Ci} \quad \text{and} \quad \vec{v}_i = \vec{r}_c + \vec{r}_{Ci}$$

where C is the mass center. Also,

$$\sum_{i=1}^{n} \vec{r}_{Ai} \times m_i \vec{v}_i = \sum_{i=1}^{n} \vec{r}_{Ai} \times \vec{F}_i$$

since

$$\sum_{i=1}^{n} \sum_{j=1}^{n} (\vec{r}_{Ai} \times \vec{f}_{ij}) = 0$$

because the internal forces $\vec{f}_{ij}$ and $\vec{f}_{ji}$ are equal and opposite. Substituting these results in (3.36), we obtain

$$\dot{\vec{H}}_A = \sum_{i=1}^{n} (\vec{r}_{AC} + \vec{r}_{Ci}) \times m_i (\vec{r}_c + \vec{r}_{Ci}) + \sum_{i=1}^{n} \vec{r}_{Ai} \times \vec{F}_i \qquad (3.37)$$

Now, $\sum_{i=1}^{n} m_i \vec{r}_{Ci} = 0$ since C is the center of mass, $\vec{r}_c = \vec{v}_c$, and $\sum_{i=1}^{n} \vec{r}_{Ai} \times \vec{F}_i = \vec{M}_A$. Hence, (3.37) becomes

$$\dot{\vec{H}}_A = \vec{r}_{AC} \times m\vec{v}_c + \vec{M}_A \qquad (3.38)$$

which is the generalization of (3.33) to a system of particles. Again, it is seen that the resultant moment of the forces about a moving point A is in general not equal to the rate of change of angular momentum about A. When point A coincides with the fixed origin O, $\vec{r}_{AC} = \vec{r}_c = \vec{v}_e$ and (3.38) becomes

$$\dot{\vec{H}}_o = \vec{M}_o \qquad (3.39)$$

When the moving point A coincides with the moving center of mass C, then $\vec{r}_{AC} = 0$ and (3.38) reduces to

$$\dot{\vec{H}}_c = \vec{M}_c \qquad (3.40)$$

Hence, it can be seen that if it is necessary to take moments about a moving point, it is advantageous to choose the moving center of mass as that point.

Example 3.6

In this example, we again consider the two particles of Example 3.4 shown in Fig. 3.9. For the moving point A, we choose the location of the particle of mass m_1. We have the following equations to describe the positions and velocities.

$$\vec{r}_1 = x\vec{i}, \qquad \vec{r}_2 = (x + b \sin \theta)\vec{i} - b \cos \theta \vec{j}$$
$$\vec{v}_1 = \dot{x}\vec{i}, \qquad \vec{v}_2 = (\dot{x} + b\dot{\theta} \cos \theta)\vec{i} + b\dot{\theta} \sin \theta \vec{j}$$
$$(m_1 + m_2)\vec{r}_c = m_1 x\vec{i} + m_2[(x + b \sin \theta)\vec{i} - b \cos \theta \vec{j}]$$
$$(m_1 + m_2)\vec{v}_c = m_1 \dot{x}\vec{i} + m_2[(\dot{x} + b\dot{\theta} \cos \theta)\vec{i} + b\dot{\theta} \sin \theta \vec{j}]$$

Since the point A is located at the particle of mass m_1, only $m_2 \vec{v}_2$ has a nonzero moment about A and we get

$$\dot{\vec{H}}_A = (b \sin \theta \vec{i} - b \cos \theta \vec{j}) \times m_2[(\dot{x} + b\dot{\theta} \cos \theta)\vec{i} + b\dot{\theta} \sin \theta \vec{j}]$$
$$= (m_2 b^2 \dot{\theta} + m_2 b\dot{x} \cos \theta)\vec{k} \qquad (3.41)$$
$$\dot{\vec{H}}_A = (m_2 b^2 \ddot{\theta} + m_2 b\ddot{x} \cos \theta - m_2 b\dot{x}\dot{\theta} \sin \theta)\vec{k}$$

The only force which has nonzero moment at A is $-m_2 g \vec{j}$ and we get

$$\vec{M}_A = (b \sin \theta \vec{i} - b \cos \theta \vec{j}) \times -m_2 g \vec{j}$$
$$= -m_2 g b \sin \theta \vec{k} \qquad (3.42)$$

Now,

$$\vec{r}_{AC} = \vec{r}_c - x \vec{i} \qquad \text{and} \qquad \dot{\vec{r}}_{AC} = \vec{v}_c - \dot{x} \vec{i}$$

Hence,

$$\dot{\vec{r}}_{AC} \times m \vec{v}_c = (\vec{v}_c - \dot{x} \vec{i}) \times (m_1 + m_2) \vec{v}_c$$
$$= -\dot{x} \vec{i} \times (m_1 + m_2) \vec{v}_c$$
$$= -\dot{x} \vec{i} \times [m_1 \dot{x} \vec{i} + m_2 (\dot{x} + b \dot{\theta} \cos \theta) \vec{i} + m_2 b \dot{\theta} \sin \theta \vec{j}]$$
$$= -m_2 b \dot{x} \dot{\theta} \sin \theta \vec{k} \qquad (3.43)$$

Substitution from (3.41), (3.42), and (3.43) in (3.38) yields

$$m_2 b^2 \ddot{\theta} + m_2 b \ddot{x} \cos \theta - m_2 b \dot{x} \dot{\theta} \sin \theta = -m_2 b \dot{x} \dot{\theta} \sin \theta - m_2 g b \sin \theta$$

or

$$m_2 b \ddot{\theta} + m_2 \ddot{x} \cos \theta = -m_2 g \sin \theta \qquad (3.44)$$

which is the same as the equation of motion (3.25).

3.5 PRINCIPLE OF WORK AND ENERGY

We now consider the principle of work and energy, which is derived from the first integral of the equation of motion obtained from Newton's second law. This principle provides quick answers to simple problems without the formulation of the equations of motion and their solution. Besides, work and energy are useful concepts in formulating the equations of motion by the Lagrange method. There are several advantages in using the principle. Both work and energy are scalars. The forces of constraints and other forces that do no work can be ignored. A system of particles can be considered as a whole.

Consider a particle of mass m which is acted upon by a resultant force $\vec{F}$. Let $\vec{r}$ denote the position of the particle as shown in Fig. 3.14. Let the particle move from position $\vec{r}$ to $\vec{r} + d\vec{r}$. The vector $d\vec{r}$ is called the displacement of the particle. The work of the force $\vec{F}$ corresponding to the displacement $d\vec{r}$ is defined as

$$dW = \vec{F} \cdot d\vec{r} \qquad (3.45)$$

Assuming that the coordinate system of Fig. 3.14 is inertial and applying Newton's second law, we get $\vec{F} = m\ddot{\vec{r}}$. Hence, (3.45) becomes

$$dW = \vec{F} \cdot d\vec{r} = m\ddot{\vec{r}} \cdot d\vec{r}$$
$$= m\ddot{\vec{r}} \cdot d\dot{\vec{r}} = d(\tfrac{1}{2} m \dot{\vec{r}} \cdot \dot{\vec{r}})$$
$$= dT \qquad (3.46)$$

where the kinetic energy T of the particle is defined by $T = \tfrac{1}{2} m \dot{\vec{r}} \cdot \dot{\vec{r}}$ and $\dot{\vec{r}}$ is the particle velocity. When the particle moves from position $\vec{r}_1$ to $\vec{r}_2$ under the

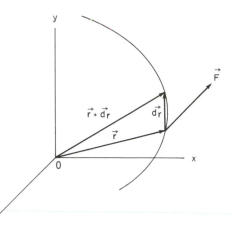

Figure 3.14 Work done by a force.

action of the resultant force $\vec{F}$, integrating (3.46), it follows that

$$\int_{\vec{r}_1}^{\vec{r}_2} \vec{F} \cdot d\vec{r} = T_2 - T_1 \tag{3.47}$$

where the subscripts 1 and 2 denote the kinetic energy corresponding to positions $\vec{r}_1$ and $\vec{r}_2$, respectively. The foregoing equation states that when a particle moves from position $\vec{r}_1$ to $\vec{r}_2$ under the action of a force $\vec{F}$, the work done by the force is equal to the change in the kinetic energy of the particle. This is called the principle of work and energy. The work done may be positive or negative depending on the direction of the displacement and the direction of the force. Expressing the kinetic energy as $T = \frac{1}{2}m\vec{v} \cdot \vec{v}$, it should be noted from the derivation that the velocity used to determine the kinetic energy should be measured with respect to an inertial system of coordinates. If the position $\vec{r}$ is expressed in terms of a noninertial coordinate system whose angular velocity is $\vec{\omega}$ and whose origin has velocity $\vec{v}_o$, then it is recalled from Chapter 2 that the absolute velocity is given by

$$\vec{v} = \vec{v}_o + \dot{\vec{r}} + \vec{\omega} \times \vec{r} \tag{3.48}$$

and it is employed in computing the kinetic energy.

The generalization of this principle to a system of n particles is straightforward. The quantity T now represents the kinetic energy of the entire system: that is,

$$T = \frac{1}{2} \sum_{i=1}^{n} m_i \vec{v}_i \cdot \vec{v}_i$$

The work is the sum of the work of all the forces acting on the particles of the system; that is,

$$\sum_{i=1}^{n} \int_{\vec{r}_{i,1}}^{\vec{r}_{i,2}} \left(\vec{F}_i + \sum_{j=1}^{n} \vec{f}_{ij} \right) \cdot d\vec{r}_i$$

It should be noted that, while the internal forces $\vec{f}_{ij}$ and $\vec{f}_{ji}$ are equal and opposite the sum of the work done by the internal forces may not add to zero because the particles on which they act undergo different displacements. Hence, in computing the work done, both internal and external forces should be considered. Those forces that do no work are ignored.

Example 3.7

A system of two masses m_1 and m_2 shown in Fig. 3.15 is at rest when a constant force P is applied to mass m_1. The coefficient of friction between each mass and the horizontal plane is μ. Determine the velocity of mass m_1 after it has moved a distance d.

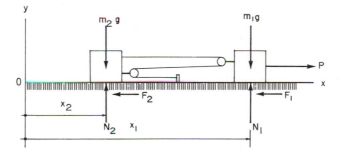

Figure 3.15   System of two masses.

Masses m_1 and m_2 are constrained kinematically by the relationship $2x_1 - 3x_2 = 0$. Hence, $v_2 = \frac{2}{3} v_1$.

The forces acting on each of the masses are shown in Fig. 3.15. The tension in the cable is a constraint force and does no work. Also, forces $m_1 g$, $m_2 g$, N_1, and N_2 do no work and are ignored. When mass m_1 has moved a distance d, mass m_2 moves a distance $\frac{2}{3} d$. Hence, the total work done is given by:

$$\text{work} = Pd - F_1 d - F_2 \tfrac{2}{3} d$$

But $F_1 = \mu N_1$, $F_2 = \mu N_2$, $N_1 = m_1 g$, and $N_2 = m_2 g$. Hence, the work done becomes:

$$\text{work} = Pd - \mu m_1 g d - \tfrac{2}{3} \mu m_2 g d$$

Since the system starts from rest, $T_1 = 0$ and T_2 becomes

$$T_2 = \tfrac{1}{2} m_1 v_1^2 + \tfrac{1}{2} m_2 v_2^2$$
$$= \tfrac{1}{2} m_1 v_1^2 + \tfrac{1}{2} \cdot \tfrac{4}{9} m_2 v_1^2$$

Hence,

$$v_1^2 = \frac{Pd - \mu m_1 g d - \tfrac{2}{3} \mu m_2 g d}{\tfrac{1}{2} m_1 + \tfrac{4}{18} m_2}$$

Example 3.8

The magnitude of the velocity of the projectile of Example 3.5 is v_o when it explodes into two parts of masses m_1 and m_2. Immediately after the explosion, the magnitudes of the velocities of m_1 and m_2 are observed to be v_1 and v_2, respectively. Assuming that the explosion is instantaneous, determine the work done by the internal forces during the explosion.

The expressions for the kinetic energy just before and after the explosion are obtained as

$$T_1 = \tfrac{1}{2}(m_1 + m_2)v_o^2$$
$$T_2 = \tfrac{1}{2}m_1 v_1^2 + \tfrac{1}{2}m_2 v_2^2$$

Since the explosion is instantaneous, the displacement of the external forces during the explosion is zero and hence their work is also zero. Only internal forces do work, which is given by:

$$\text{work by internal forces} = T_2 - T_1 = \tfrac{1}{2}m_1 v_1^2 + \tfrac{1}{2}m_2 v_2^2 - \tfrac{1}{2}(m_1 + m_2)v_o^2$$

3.5.1 Conservative Forces and the Principle of Conservation of Energy

A force is called conservative if the work done by the force in a closed path is zero; that is,

$$\oint \vec{F} \cdot d\vec{r} = 0 \tag{3.49}$$

This line integral can be converted to surface integral by Stokes's theorem, and we obtain

$$\iint \vec{\nabla} \times \vec{F} \cdot d\vec{n} = 0$$

where $\vec{\nabla}$ is the differential operator del or nabla, and $\vec{n}$ is the unit vector normal to the surface. This leads to the condition that $\vec{\nabla} \times \vec{F} = 0$ for a force to be conservative. But the curl of a vector vanishes if and only if the vector is the gradient of a scalar. We denote this scalar function associated with vector $\vec{F}$ by $-U$, where U is called the potential energy and write

$$\vec{F} = -\vec{\nabla} U \tag{3.50}$$

The negative sign in the foregoing equation is employed so that if the work is positive, the potential energy decreases, and vice versa. The potential energy U is a function of position only. From (3.45), the infinitesimal work done by the force in a displacement $d\vec{r}$ is given by

$$dW = \vec{F} \cdot dr = -\vec{\nabla} U \cdot d\vec{r}$$
$$= -dU$$

where in Cartesian coordinates we have

$$-dU = -\left(\frac{\partial U}{\partial x} dx + \frac{\partial U}{\partial y} dy + \frac{\partial U}{\partial z} dz\right)$$

Hence, when a particle moves from position $\vec{r}_1$ to $\vec{r}_2$ under the action of a conservative force $\vec{F}$, the work done can be obtained from the change in potential energy as

$$\int_{\vec{r}_1}^{\vec{r}_2} \vec{F} \cdot d\vec{r} = U_1(\vec{r}_1) - U_2(\vec{r}_2) \qquad (3.51)$$

where the right-hand side denotes that U_1 and U_2 are functions of $\vec{r}_1$ and $\vec{r}_2$, respectively. Now, if *all* the forces acting on a system of particles are conservative, then in the work–energy principle (3.47) we substitute from (3.51) for the work done to obtain

$$U_1 - U_2 = T_2 - T_1$$

or

$$U_1 + T_1 = U_2 + T_2 = E \qquad (3.52)$$

where E is a constant and is the total mechanical energy of a system of particles. The foregoing equation states that when all the forces are conservative, the total mechanical energy, which is the sum of kinetic and potential energies, is conserved. This is called the principle of conservation of energy. It should be noted that the only area of applications, where all the forces are conservative and mechanical energy is conserved, is the area of orbital mechanics, which is discussed later. In almost all other applications, we encounter some friction, drag, or dissipative mechanisms. In such cases, total mechanical energy is not conserved and (3.52) provides only an approximation when the effect of dissipative mechanisms is negligible. Usually, the conservative forces in dynamics are due to spring or elasticity and Newton's law of gravitation.

Example 3.9

A mechanism for shooting a plunger is shown in Fig. 3.16. The mass of the plunger is m and the undeformed length of the spring is ℓ. It is compressed to a length ℓ_1 by a force P and then released when it expands to length ℓ_2. Determine the velocity of the plunger as it leaves the mechanism, assuming that $\ell > \ell_2 > \ell_1$. Neglect friction.

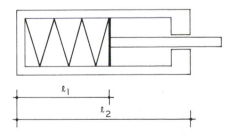

Figure 3.16   Mechanism for shooting plunger.

The spring force is conservative. If the spring is stretched by an amount δ from its unstretched length, the spring force is $F = -k\delta$, where k is the spring constant and it is assumed that the spring is linear. Now, if the spring is given a displacement $d\delta$, the increment of work becomes

$$dW = F\,d\delta = -k\delta\,d\delta$$

and the work in stretching it from δ_1 to δ_2 is

$$W = \int F \, d\delta = -\int_{\delta_1}^{\delta_2} k\delta \, d\delta$$

$$= \tfrac{1}{2}k\delta_1^2 - \tfrac{1}{2}k\delta_2^2$$

It follows that the potential energy of the spring is given by $U = \tfrac{1}{2}k\delta^2$. Neglecting friction, the only force that does work is the spring force and hence the total mechanical energy is conserved. We then obtain

$$T_1 + U_1 = T_2 + U_2$$

where $T_1 = 0$, $U_1 = \tfrac{1}{2}k(\ell - \ell_1)^2$, $T_2 = \tfrac{1}{2}mv_2^2$, and $U_2 = \tfrac{1}{2}k(\ell - \ell_2)^2$. Hence,

$$\tfrac{1}{2}k(\ell - \ell_1)^2 = \tfrac{1}{2}mv_2^2 + \tfrac{1}{2}k(\ell - \ell_2)^2$$

or

$$v_2^2 = \frac{k}{m}[(\ell - \ell_1)^2 - (\ell - \ell_2)^2]$$

Also, $P = k(\ell - \ell_1)$; that is, $k = P/(\ell - \ell_1)$. It follows that

$$v_2^2 = \frac{P}{m(\ell - \ell_1)}[(\ell - \ell_1)^2 - (\ell - \ell_2)^2]$$

3.6 PRINCIPLE OF IMPULSE AND MOMENTUM

The principle of impulse and momentum is another principle of dynamics. It is also derived from the first integral of the equations of motion obtained from Newton's second law. Consider a particle of mass m acted upon by a resultant force $\vec{F}$. From Newton's second law, we get

$$\vec{F} = \frac{d}{dt}(m\vec{v}) \tag{3.53}$$

where $m\vec{v}$ is the linear momentum. This equation can be written as

$$\vec{F} \, dt = d(m\vec{v})$$

Integrating with respect to time from t_1 to t_2, we obtain

$$\int_{t_1}^{t_2} \vec{F} \, dt = \int d(m\vec{v})$$

$$= m\vec{v}_2 - m\vec{v}_1 \tag{3.54}$$

where the subscripts 1 and 2 on the right-hand side designate the velocities at times t_1 and t_2, respectively. The left-hand side of this equation is called the linear impulse and the right-hand side is the change in linear momentum. This is called the principle of impulse and momentum. It states that the linear impulse is equal to the change in linear momentum. It should be noted that unlike work and energy, which are scalars, impulse and momentum are vector quantities. The advantage of employing this principle is that when the left-hand side of (3.54) can be integrated, answers to certain problems can be obtained quickly without integrating the equations of motion. If the resultant force acting on a particle is zero (i.e., $\vec{F} = 0$), then (3.54) becomes

$$m\vec{v}_2 = m\vec{v}_1 = m\vec{v} = \text{constant} \tag{3.55}$$

which is a statement of the conservation of linear momentum.

For a system of n particles, the linear impulse momentum principle can be stated as

$$\sum \int_{t_1}^{t_2} \vec{F}\, dt = \left(\sum_{i=1}^{n} m_i \vec{v}_i\right)_2 - \left(\sum_{i=1}^{n} m_i \vec{v}_i\right)_1 \tag{3.56}$$

Also, integrating (3.39) with respect to time from t_1 to t_2, we obtain

$$\sum \int_{t_1}^{t_2} \vec{M}_o\, dt = (\vec{H}_o)_2 - (\vec{H}_o)_1 \tag{3.57}$$

This equation states that the sum of the angular impulses of the external forces about the origin is equal to the change in angular momentum of the system. If no external force acts on the particles of the system, then the left-hand side of (3.57) is zero and we get

$$(\vec{H}_o)_2 = (\vec{H}_o)_1 = \vec{H}_o, \qquad \text{constant}$$

which is a statement of the conservation of angular momentum about the origin.

Example 3.10

A collar of mass m slides on a rod with Coulomb friction under the action of a force P shown in Fig. 3.17. The coefficient of friction is μ. Determine the time at which the collar comes to rest again.

The forces acting on the collar are shown in Fig. 3.17. The linear momentum in the y and z directions is zero at all times. In the x direction, we obtain

$$\int_0^{t_2} (P - F)dt = mv_2 - mv_1$$

where t_2 is the time at which the collar comes to rest again. Since $v_1 = v_2 = 0$, we have

$$\int_0^{t_2} P\, dt = \int_0^{t_2} F\, dt$$

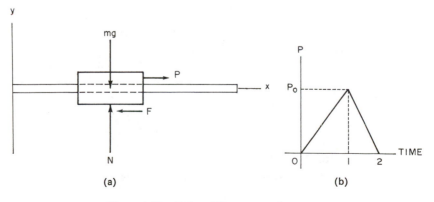

Figure 3.17   Collar sliding on a rod.

The value of the integral on the left-hand side is P_o, as seen from Fig. 3.17, and $F = \mu N = \mu mg$. Hence, we obtain

$$P_o = \mu mg t_2$$

or

$$t_2 = \frac{P_o}{\mu mg}$$

3.7 TWO-BODY CENTRAL FORCE MOTION

When the resultant force acting on a particle always passes through a fixed point O, the particle is said to be moving under a central force, and the point O is referred to as the center of force. The force may be directed toward or away from the center of force. Since the force passes through O as shown in Fig. 3.18, we have $\sum \vec{M}_o = 0$ and from (3.31) it follows that $\dot{\vec{H}}_o = 0$; that is, $\vec{H}_o = $ constant for all time. Hence, under a central force field, the angular momentum is a constant in both magnitude and direction.

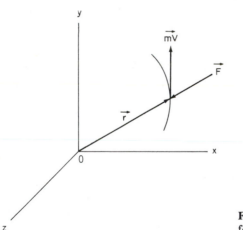

Figure 3.18 Motion under a central force.

Since $\vec{H}_o = \vec{r} \times m\vec{v} = $ constant, the motion takes place in a plane defined by some initial position vector $\vec{r}$ and initial velocity vector $\vec{v}$. The constant vector $\vec{H}_o$ is of course perpendicular to this plane. Employing polar coordinates to represent this plane motion, we have $\vec{r} = r\vec{i}_r$, and $\vec{v} = \dot{r}\vec{i}_r + r\dot{\theta}\vec{i}_\theta$, as shown in Chapter 2. Hence, it follows that

$$\vec{H}_o = r\vec{i}_r \times m(\dot{r}\vec{i}_r + r\dot{\theta}\vec{i}_\theta)$$
$$= mr^2\dot{\theta}\vec{k} = \text{constant} \qquad (3.58)$$

This equation may also be given a geometric interpretation. As shown in Fig. 3.19, an element of area in polar coordinates has the expression

$$dA = \tfrac{1}{2}r^2\,d\theta$$

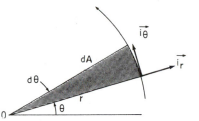

Figure 3.19 Motion under central force
in polar coordinates.

Differentiating both sides of this equation, we obtain

$$\dot{A} = \tfrac{1}{2}r^2\dot{\theta} \tag{3.59}$$

and from (3.58) it is seen that $\dot{A}$ = constant. This is the statement of Kepler's
second law for planetary motion, which is stated later in this chapter. The
quantity $\dot{A}$ is called the areal velocity.

We now consider two-body central force motion, where two particles are
free to move in space under the influence of forces exerted by the particles on
each other along the line joining them. In the following, it is shown that this
problem can be reduced to that of a single particle moving under a central
force, the center of force being the other moving particle. Let m_1 and m_2 be the
masses of the two particles and $\vec{r}_1$ and $\vec{r}_2$ be their position vectors, respectively,
with respect to an inertial coordinate system $Oxyz$ shown in Fig. 3.20. The

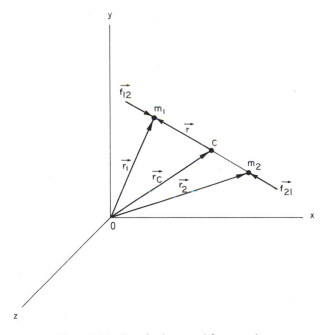

Figure 3.20 Two-body central force motion.

position of the mass center is defined in the usual manner as

$$m_1 \vec{r}_1 + m_2 \vec{r}_2 = (m_1 + m_2) \vec{r}_c$$

Let $\vec{r}$ be the radius vector from m_2 to m_1. From Fig. 3.20, it is seen that

$$\vec{r}_1 = \vec{r}_c + \frac{m_2}{m_1 + m_2} \vec{r} \tag{3.60}$$

$$\vec{r}_2 = \vec{r}_c - \frac{m_1}{m_1 + m_2} \vec{r} \tag{3.61}$$

Let $\vec{f}_{ij}$ be the internal force on m_i due to m_j, there being no external forces. Force $\vec{f}_{ij}$ may be an attractive or repulsive force. The equations of motion for each of the two particles are

$$\vec{f}_{12} = m_1 \ddot{\vec{r}}_1 = m_1 \ddot{\vec{r}}_c + \frac{m_1 m_2}{m_1 + m_2} \ddot{\vec{r}} \tag{3.62}$$

$$\vec{f}_{21} = m_2 \ddot{\vec{r}}_2 = m_2 \ddot{\vec{r}}_c - \frac{m_1 m_2}{m_1 + m_2} \ddot{\vec{r}} \tag{3.63}$$

where substitution from (3.60) and (3.61) has been employed. Since the internal forces add up to zero, adding (3.62) and (3.63), we obtain

$$(m_1 + m_2) \ddot{\vec{r}}_c = 0 \tag{3.64}$$

It is concluded that $\ddot{\vec{r}}_c = 0$ (i.e., the center of mass is unaccelerated). It follows from (3.62) and (3.63) that

$$\vec{f}_{12} = \frac{m_1 m_2}{m_1 + m_2} \ddot{\vec{r}} \tag{3.65}$$

$$\vec{f}_{21} = -\frac{m_1 m_2}{m_1 + m_2} \ddot{\vec{r}} \tag{3.66}$$

Equation (3.65) may be given the following interpretation. As seen from Fig. 3.20, $\vec{r}$ is the position of m_1 relative to m_2 and hence $\ddot{\vec{r}}$ is the acceleration of m_1 relative to m_2. We can now select the position of m_2 as the origin of the coordinate system and in (3.65) treat $\ddot{\vec{r}}$ as the absolute acceleration, provided that the mass m_1 is replaced by an equivalent mass $m_1 m_2/(m_1 + m_2)$. The motion of m_1 can then be considered as that of a single particle under the action of a central force. The center of force is the position of m_2. Alternatively, we can employ (3.66) and study the central force motion of m_2, the center of force being the position of m_1. In this case, the mass m_2 would be replaced by the equivalent mass $m_1 m_2/(m_1 + m_2)$.

In order to study the orbit of m_1 around m_2, we select the position of m_2 as the origin and for simplicity of notation let $m = m_1 m_2/(m_1 + m_2)$. Employing polar coordinates to represent this plane motion and making use of (2.20) the equations of motion of m_1 are represented by

$$m(\ddot{r} - r\dot{\theta}^2) = f \tag{3.67}$$

$$m(r\ddot{\theta} + 2\dot{r}\dot{\theta}) = 0 \tag{3.68}$$

where m is the equivalent mass of m_1 and $f = f_{12}$ is assumed to be a repulsive central force. By differentiating (3.58) with respect to time, it can be verified that we obtain (3.68). Equations (3.67) and (3.68) are valid for any central force and have applications in atomic and nuclear physics. Here, we restrict ourselves only to the study of orbital mechanics in the next section.

3.8 ORBITS OF PLANETS AND SATELLITES

The orbit of one body around another is governed by Newton's law of gravitation. In this case, the central force is attractive and is given by

$$f = -\frac{Gm_1m_2}{r^2}$$

Hence, the equations of motion (3.67) and (3.68) become

$$\frac{m_1m_2}{m_1 + m_2}(\ddot{r} - r\dot{\theta}^2) = -\frac{Gm_1m_2}{r^2} \tag{3.69}$$

$$\frac{m_1m_2}{m_1 + m_2}(r\ddot{\theta} + 2\dot{r}\dot{\theta}) = 0 \tag{3.70}$$

Equation (3.70) can be integrated once with respect to time resulting in $r^2\dot{\theta} = h$, where h is a constant and from (3.58) it is seen that h is the angular momentum per unit mass. Hence, (3.69) and (3.70) become

$$\ddot{r} - r\dot{\theta}^2 = -\frac{G(m_1 + m_2)}{r^2} \tag{3.71}$$

$$r^2\dot{\theta} = h \tag{3.72}$$

These nonlinear equations can be integrated in a closed form by making a substitution for r. Let $1/r = u$. Eliminating the time dependence from (3.71), we get

$$\dot{r} = \frac{dr}{d\theta}\frac{d\theta}{dt} = \frac{h}{r^2}\frac{dr}{d\theta} = -h\frac{d}{d\theta}\left(\frac{1}{r}\right) = -h\frac{du}{d\theta} \tag{3.73}$$

$$\ddot{r} = \frac{d\dot{r}}{d\theta}\frac{d\theta}{dt} = \frac{h}{r^2}\frac{d}{d\theta}(\dot{r}) = -h^2u^2\frac{d^2u}{d\theta^2} \tag{3.74}$$

where we have substituted for $\dot{\theta}$ from (3.72). Employing (3.72) and (3.74) in (3.71), we obtain the linear equation

$$\frac{d^2u}{d\theta^2} + u = \frac{G(m_1 + m_2)}{h^2} \tag{3.75}$$

where the right-hand side forcing function is a constant. Its solution is obtained by adding the particular solution $G(m_1 + m_2)/h^2$ to the complementary solution $C\cos(\theta - \theta_o)$ to obtain

$$\frac{1}{r} = u = \frac{G(m_1 + m_2)}{h^2} + C\cos(\theta - \theta_o) \tag{3.76}$$

Equation (3.75) being of second order, its solution (3.76) contains two constants of integration, C and θ_o. Defining a new constant e as $e = Ch^2/G(m_1 + m_2)$, (3.76) can be expressed as

$$\frac{1}{r} = \frac{G(m_1 + m_2)}{h^2}[1 + e\cos(\theta - \theta_o)] \tag{3.77}$$

The constant θ_o can be eliminated by choosing the polar axis so that $\theta_o = 0$ and the constant e can be evaluated from the total mechanical energy of the body. Equation (3.77) is the equation of a conic section in polar coordinates r and θ and may represent a hyperbola, parabola, ellipse, or circle, depending on the value of e, which is known as the eccentricity of the conic section. The constant e is evaluated as follows. It can be verified by employing (3.49) that the force due to Newton's law of gravitation is conservative. Hence, it has associated with it a potential energy U. From (3.50), we get

$$f = -\frac{Gm_1m_2}{r^2} = -\frac{\partial U}{\partial r} \tag{3.78}$$

Choosing the reference position for the potential energy at infinity so that $U(\infty) = 0$ and integrating (3.78), we obtain

$$U(r) = \int_{\infty}^{r} \frac{Gm_1m_2}{r^2}\,dr = -\frac{Gm_1m_2}{r}$$

Using the equivalent mass, the potential energy per unit mass becomes

$$U(r) = -\frac{G(m_1 + m_2)}{r}$$

The kinetic energy per unit mass is given by

$$T = \tfrac{1}{2}v^2 = \tfrac{1}{2}(\dot{r}^2 + r^2\dot{\theta}^2)$$

so that the total mechanical energy per unit mass becomes

$$E = T + U = \tfrac{1}{2}(\dot{r}^2 + r^2\dot{\theta}^2) - \frac{G(m_1 + m_2)}{r} \tag{3.79}$$

In (3.79), letting $r^2\dot{\theta}^2 = h^2/r^2$ and substituting for r from (3.77) and for $\dot{r}$ obtained by differentiating (3.77), it can be verified that

$$e = \left\{1 + \frac{2Eh^2}{[G(m_1 + m_2)]^2}\right\}^{1/2} \tag{3.80}$$

It should be noted that the central force is the only force in this case and its being a conservative force, the total mechanical energy (3.79) is conserved. A point of the orbit where $dr/d\theta = 0$ is called an apsis. For an open orbit such as a hyperbola or parabola, there is only one apsis but for an ellipse there exist two apsides. The shortest distance from the force center to one of the apsides is called the pericentron and the longer one is called the apocentron. Measuring θ from the pericentron, we set $\theta_o = 0$ in (3.77).

We now distinguish the following four orbits:

Case 1: $e > 1$, $E > 0$. The orbit is a hyperbola that is an open orbit. The particle comes from infinity as shown in Fig. 3.21, reaches the minimum distance at the apsis where the potential energy has a minimum and, hence, the kinetic energy a maximum, and escapes to infinity.

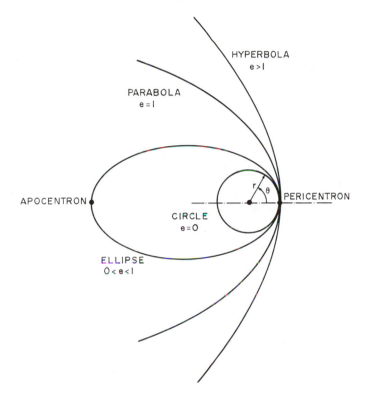

Figure 3.21 Four possible orbits.

Case 2: $e = 1$, $E = 0$. The orbit is a parabola that is an open orbit with the lowest energy. At the pericentron r_p, $\dot{r} = 0$ and using (3.79), we get

$$E = \tfrac{1}{2}r_p^2\dot{\theta}^2 - \frac{G(m_1 + m_2)}{r_p} = 0$$

Letting $v_e = r_p\dot{\theta}$ in the foregoing equation, it is seen that

$$v_e = \left[\frac{2G(m_1 + m_2)}{r_p}\right]^{1/2} \tag{3.81}$$

This velocity v_e, which is called the escape velocity, is the minimum velocity for which an open orbit is obtained.

Case 3: $0 < e < 1$, $-[G(m_1 + m_2)]^2/2h^2 < E < 0$. The orbit is an ellipse. From (3.77), the pericentron and apocentron are obtained as

$$r_p = \frac{h^2}{G(m_1 + m_2)}\left(\frac{1}{1 + e}\right) \tag{3.82}$$

$$r_a = \frac{h^2}{G(m_1 + m_2)}\left(\frac{1}{1 - e}\right) \tag{3.83}$$

Case 4: $e = 0$, $E = -[G(m_1 + m_2)]^2/2h^2$. The orbit is a circle which is a special case of the elliptic orbit. For a circular orbit, the radial velocity $\dot{r} = 0$ and there is a balance between the centrifugal and gravitational forces. Denoting the velocity $r\dot{\theta} = v_c$ and noting that in (3.77) we have $e = 0$ and $h = r^2\dot{\theta} = rv_c$, we get

$$v_c = \left[\frac{G(m_1 + m_2)}{r}\right]^{1/2} \tag{3.84}$$

On comparing (3.81) and (3.84), it is seen that the escape velocity is $\sqrt{2}\,v_c$.

Some of the results obtained in the foregoing had been discovered by Kepler purely from observations of the orbits of planets around the sun before Newton had formulated his law of gravitation. Kepler's three laws of planetary motion may be stated as follows:

1. Each planet describes an ellipse, with the sun located at one of its foci.
2. The radius vector from the sun to a planet sweeps equal areas in equal times.
3. The squares of the periodic times of the planets are proportional to the cubes of the semimajor axes of their orbits.

The first law states a special case of our results: Case 3, where $0 < e < 1$. The second law is proved by (3.59) and the third law can be easily verified for elliptic orbits.

Example 3.11

A satellite is projected into space from the earth with a velocity v_o at a distance r_o from the center of the earth by the last stage of its launching rocket (Fig. 3.22). The velocity v_o was designed to send the satellite into a circular orbit of radius r_0. However, owing

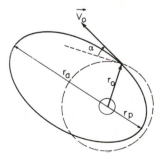

Figure 3.22 Orbit of satellite.

to a malfunction of the control, the satellite is not projected horizontally but at an angle α to the horizontal and, as a result, is projected into an elliptic orbit. Find the perigee and apogee of the orbit. (For an earth-centered orbit, the pericentron and apocentron are called perigee and apogee, respectively.)

First, we consider the circular orbit for which

$$E = \frac{1}{2} v_o^2 - \frac{G(m_1 + m_2)}{r_o} \qquad \text{and} \qquad v_o^2 = \frac{G(m_1 + m_2)}{r_o}$$

where m_1 and m_2 are the masses of the satellite and earth, respectively. Since $m_1 \ll m_2$, $m_1 + m_2 \simeq m_2$ and the equivalent mass of the satellite becomes $m_1 m_2/(m_1 + m_2) \simeq m_1$, which is its actual mass. Eliminating v_o for the circular orbit, we obtain

$$E = -\frac{1}{2} \frac{G(m_1 + m_2)}{r_o} \tag{3.85}$$

Now consider the elliptic orbit. Let subscript A denote the apsis. Since at the apsis $\dot{r} = 0$, we get

$$E = \frac{1}{2} v_A^2 - \frac{G(m_1 + m_2)}{r_A} \tag{3.86}$$

But according to the data, the energy from (3.85) is equal to that of (3.86). Hence,

$$\frac{1}{2} v_A^2 - \frac{G(m_1 + m_2)}{r_A} = -\frac{1}{2} \frac{G(m_1 + m_2)}{r_o} \tag{3.87}$$

Since the elliptic orbit conserves the angular momentum,

$$(v_o \cos \alpha) r_o = v_A r_A$$

or

$$v_A^2 = \frac{v_o^2 r_o^2 \cos^2 \alpha}{r_A^2} \qquad \text{where} \quad v_o^2 = \frac{G(m_1 + m_2)}{r_o}$$

Hence,

$$v_A^2 = G(m_1 + m_2) \frac{r_o \cos^2 \alpha}{r_A^2} \tag{3.88}$$

Substituting for v_A^2 in (3.87) from (3.88) and simplifying the resultant expression, we get

$$r_A^2 - 2r_o r_A + r_o^2 \cos^2 \alpha = 0$$

The two roots of this equation are

$$r_A = r_o \pm r_o(1 - \cos^2 \alpha)^{1/2}$$
$$= r_o(1 \pm \sin \alpha)$$

Hence, the perigee and apogee of the orbit are

$$r_p = r_o(1 - \sin \alpha)$$
$$r_a = r_o(1 + \sin \alpha)$$

Example 3.12

A spacecraft is in a circular orbit of radius r_o with speed v_o around a body whose mass center is at 0 (Fig. 3.23). Its engine is fired, thus increasing the speed of the spacecraft from v_o to αv_o, where $1 < \alpha^2 < 2$. Show that the maximum distance r_m from O reached

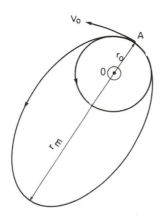

Figure 3.23   Orbit of a satellite.

by the spacecraft depends only upon r_o and α, and express the ratio r_m/r_o as a function of α.

For the circular orbit, we have

$$v_o^2 = \frac{G(m_1 + m_2)}{r_o} \tag{3.89}$$

For the elliptic orbit, we measure θ from the pericentron A and let $\theta_o = 0$ in (3.77). Hence, at A, $\theta = 0$ and from (3.77) we get

$$\frac{1}{r_o} = \frac{G(m_1 + m_2)}{h^2}(1 + e) \tag{3.90}$$

At the apocentron, $\theta = 180°$ and it follows that

$$\frac{1}{r_m} = \frac{G(m_1 + m_2)}{h^2}(1 - e) \tag{3.91}$$

Adding (3.90) and (3.91), we obtain

$$\frac{1}{r_m} + \frac{1}{r_o} = \frac{2G(m_1 + m_2)}{h^2} \tag{3.92}$$

But $G(m_1 + m_2) = v_o^2 r_o$ from (3.89) and $h = r_o(v_A)_e = r_o\alpha v_o$. Then (3.92) yields

$$\frac{1}{r_m} + \frac{1}{r_o} = \frac{2}{r_o\alpha^2}$$

Hence,

$$\frac{1}{r_m} = \frac{1}{r_o}\left(\frac{2}{\alpha^2} - 1\right)$$

$$\frac{r_m}{r_o} = \frac{\alpha^2}{2 - \alpha^2}$$

We get a real, positive solution for r_m only if $\alpha^2 < 2$. If $\alpha^2 = 2$, $r_m \longrightarrow \infty$ and hence for $\alpha^2 \geq 2$, there is no closed orbit and the escape velocity is reached.

Example 3.13

A spacecraft describes a circular orbit at a radius r_o from the center of the earth. Preparatory to reentry, its speed is reduced to a value v_o so that it is placed in an elliptic trajectory that intersects the earth's surface. Determine the angle θ where splashdown

will occur. The radius of the earth is R. Neglect the drag after the spacecraft enters the earth's atmosphere.

First, it should be noted that there is drag on the satellite after it enters the earth's atmosphere. The subatmospheric flight is therefore no longer a central force motion and the results of our analysis are not valid. An approximate answer can be obtained by neglecting drag and it can serve as a guideline for more accurate formulation of the subatmospheric motion.

Choosing θ as shown in Fig. 3.24, we set $\theta_o = 0$ in (3.77) and since at point A of the elliptic orbit $\theta = 180°$, we get

$$\frac{1}{r_o} = \frac{G(m_1 + m_2)}{h^2}(1 - e)$$

where $h^2 = r_o^2 v_o^2$ for the elliptic orbit. Solving for e, we obtain

$$e = \frac{G(m_1 + m_2) - r_o v_o^2}{G(m_1 + m_2)}$$

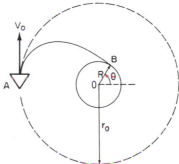

Figure 3.24   Splashdown of satellite.

Substituting this value of e and $r = R$ for point B of splashdown in (3.77), it follows that

$$\frac{1}{R} = \frac{G(m_1 + m_2)}{r_o^2 v_o^2} + \frac{G(m_1 + m_2) - r_o v_o^2}{r_o^2 v_o^2}\cos\theta$$

Hence,

$$\cos\theta = \frac{\dfrac{1}{R} - \dfrac{G(m_1 + m_2)}{r_o^2 v_o^2}}{\dfrac{G(m_1 + m_2)}{(r_o v_o)^2} - \dfrac{1}{r_o}}$$

Since $m_1 \ll m_2$ where m_1 and m_2 are the masses of the satellite and earth, respectively, it follows that $G(m_1 + m_2) \simeq Gm_2$. Now, on the surface of the earth the weight of m_1 becomes

$$m_1 g = \frac{Gm_1 m_2}{R^2}$$

that is, $Gm_2 = gR^2$, where g is the acceleration of gravity. Then the angle θ is obtained from

$$\cos\theta = \frac{\dfrac{1}{R} - \dfrac{gR^2}{(r_o v_o)^2}}{\dfrac{gR^2}{(r_o v_o)^2} - \dfrac{1}{r_o}}$$

3.9 SUMMARY

The first part of this chapter dealt with the formulation of the equations of motion by the application of Newton's second law. Both a single particle and a system of particles have been considered and different coordinate systems have been employed depending on their convenience. Next the principles of work–energy and impulse–momentum have been studied. They provide quick answers to many simple problems. Finally, the two-body central force motion has been studied and orbits of planets and satellites are considered. The two-body problem admits a closed-form solution and many systems can be adequately represented by such a model. The motion of the earth around the sun may be adequately studied by neglecting the effect of other planets. However, in certain applications the motion of a system of n bodies has to be considered. The n-body problem in general does not admit a closed-form solution and computer simulation becomes necessary.

PROBLEMS

3.1. A bead of mass m slides under gravity along a wire bent in the form of a parabola $y = 1 + cx^2$ (Fig. P3.1). The coefficient of friction between bead and wire is μ. Friction force opposing the motion is μN, where N is the normal force. At the same time, the wire rotates about y axis at constant angular speed ω_o. Obtain the equation of motion of the bead by Newton's law, employing the coordinate x shown in Fig. P3.1.

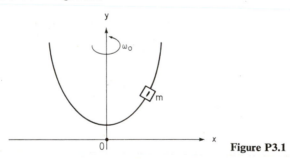

Figure P3.1

3.2. A box of mass m_2 rests on a box of mass m_1 (Fig. P3.2). The coefficient of static friction between the two boxes is μ_2 and the coefficient of sliding friction between box 1 and the ground is μ_1. A horizontal force P is applied to box 1. If box 2 is not to slip on box 1, determine the maximum value of P.

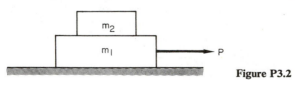

Figure P3.2

3.3. A barge of mass m_1 with an automobile of mass m_2 on its deck is initially at rest (Fig. P3.3). The automobile is now driven forward at a constant speed v_o relative to the barge. Neglecting the resistance of the water, determine the velocity of the barge and the distance it moves when the automobile has moved a distance d.

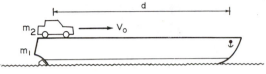

Figure P3.3

3.4. A rod is rotating freely at speed ω_0 about a vertical axis (Fig. P3.4). At time $t = 0$, two sliders are released from rest at $r = a$. The mass of the rod is M, and its moment of inertia about the centroidal axis is $\frac{1}{12}ML^2$. The mass of each slider is m.
   (a) Find the angular velocity of the system after the sliders come to rest at each end of the rod.
   (b) Find the loss of kinetic energy.
   (c) Where did the lost kinetic energy go?
   (d) Set up the equation for finding the radial position of the sliders as a function of time before they hit the stops.

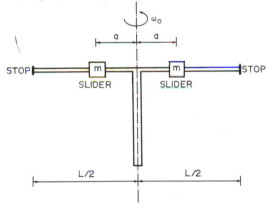

Figure P3.4

3.5. A ball of mass m moves on a frictionless table (Fig. P3.5). The ball is attached to a rubber band that goes through a hole in the middle of the table and is fastened

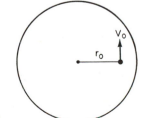

Figure P3.5

to the floor. The pull of the rubber band on the ball is proportional to the distance between the ball and the hole. The ball is now held at a point r_o away from the hole and released at time $t = 0$ with a velocity v_o perpendicular to the rubber band.
(a) Obtain the subsequent motion of the ball and find its trajectory.
(b) Is there an escape velocity $v_o = v_e$?

3.6. Prove Kepler's third law of planetary motion.

3.7. Show that the values v_1 and v_2 of the speed of an earth satellite at the perigee A and apogee B of an elliptic orbit are defined by

$$v_1^2 = \frac{2gR_e^2}{r_1 + r_2}\frac{r_2}{r_1} \qquad \text{and} \qquad v_2^2 = \frac{2gR_e^2}{r_1 + r_2}\frac{r_1}{r_2}$$

where g is the acceleration of gravity and R_e is the radius of the earth (Fig. P3.7).

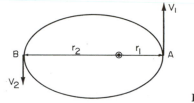

Figure P3.7

3.8. A space vehicle approaches Saturn along a hyperbolic trajectory of eccentricity $e = 2$ (Fig. P3.8). As the vehicle reaches a distance r_A closest to Saturn, retro rockets are fired to slow the vehicle and place it in a circular orbit. Show that the velocities of the vehicle just before and after firing of retro rockets are given by

$$v_A = \left(\frac{3GM}{r_A}\right)^{1/2} \qquad \text{and} \qquad v_c = \left(\frac{GM}{r_A}\right)^{1/2}$$

where M is the mass of Saturn.

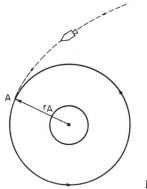

Figure P3.8

3.9. A satellite in circular orbit around the earth at an altitude of 1130 km is to be given a new orbit (Fig. P3.9). The engines are aligned radially, imparting an additional velocity of 4 km/s to the satellite outward. Determine the eccentricity e of the orbit. Is the orbit open or closed?

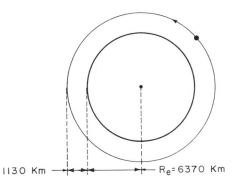

Figure P3.9   1130 Km ⟶|◀─▶|◀──────▶|← R_e= 6370 Km

REFERENCES

1. Meirovitch, L., *Methods of Analytical Dynamics*, McGraw-Hill Book Company, New York, 1970.

2. Kane, T. R., *Dynamics*, Holt, Rinehart and Winston, New York, 1968.

3. Beer, F. P., and Johnston, E. R., *Vector Mechanics for Engineers, Dynamics*, 3rd ed., McGraw-Hill Book Company, New York, 1977.

4. Halfman, R. L., *Dynamics*, Vol. 1, Addison-Wesley Publishing Company, Inc., Reading, Mass., 1962.

4

DYNAMICS OF RIGID BODIES: NEWTON'S LAW, ENERGY, AND MOMENTUM METHODS

4.1 INTRODUCTION

This chapter is devoted to the study of dynamics of rigid bodies by the direct application of Newton's second law. A flexible body may be regarded as being composed of an infinite number of particles and it has an infinite number of degrees of freedom. By assumption, a rigid body does not deform and, hence, the distance between any two of its particles is a constant. As a result, an unconstrained rigid body has only six degrees of freedom: three translational and three rotational.

Out of the six coordinates required to describe the motion of an unconstrained rigid body, three can be chosen as the components of the position vector of a reference point of the body. The remaining three, however, cannot be chosen as the components of its angular displacement because, as seen from Chapter 2, components of a finite angular displacement do not constitute a vector. In the next chapter, dealing with Lagrangian dynamics, we employ three Euler angles to describe the rotational motion of a rigid body. In this chapter the six coordinates selected to describe the motion of an unconstrained rigid body are the three components of the position vector of a reference point of the body and the three components of its angular velocity vector. It will be seen that the rotational motion of a single rigid body is uncoupled from the translational motion by selecting the reference point as the center of mass of the body.

We begin this chapter with a brief discussion of kinematics and then develop expressions for the linear and angular momentum of a rigid body. The principal axes and principal moments of inertia are discussed. The equations of motion of a rigid body undergoing translation and rotation are derived. Euler equations and modified Euler equations of motion are then studied. In the latter part of this chapter, the principle of work and energy and the principle of impulse and momentum are employed to analyze the motion of rigid bodies. The motion of a gyroscope is then studied. Finally, the motion of a system of connected rigid bodies is discussed.

4.2 KINEMATICS OF A RIGID BODY

The velocities of all points in a rigid body are found by knowing angular velocity vector $\vec{\omega}$ of the body and linear velocity of any point in the body.

We consider a rigid body which has angular velocity vector $\vec{\omega}$ with respect to an inertial frame XYZ (Fig. 4.1). The coordinate system, $Oxyz$, has its origin at a reference point, O, of the body and it rotates at the same angular velocity $\vec{\omega}$ as the body. Such a coordinate system, $Oxyz$, is called a body coordinate system. Letting P be an arbitrary point in the body, its position from O is defined by $\vec{r}$, which when decomposed in the xyz coordinate system becomes

$$\vec{r} = x\vec{i} + y\vec{j} + z\vec{k} \tag{4.1}$$

The position vector $\vec{R}$ of the point P with respect to the inertial coordinate system XYZ is the vector sum of $\vec{R}_o$ and $\vec{r}$:

$$\vec{R} = \vec{R}_o + \vec{r} \tag{4.2}$$

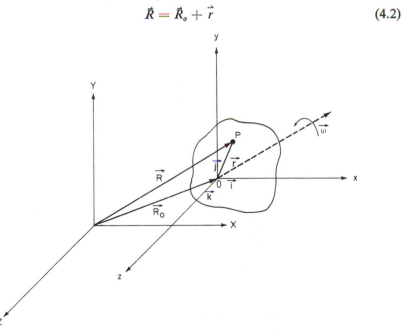

Figure 4.1 Rigid body.

The absolute velocity of P is

$$\vec{v} = \frac{d\vec{R}}{dt} = \frac{d\vec{R}_o}{dt} + \frac{d\vec{r}}{dt} \tag{4.3}$$

After noting that $\vec{r}$ has been expressed in terms of a rotating coordinate system in (4.1), we get

$$\frac{d\vec{r}}{dt} = \frac{dx}{dt}\vec{i} + \frac{dy}{dt}\vec{j} + \frac{dz}{dt}\vec{k} + x\frac{d\vec{i}}{dt} + y\frac{d\vec{j}}{dt} + z\frac{d\vec{k}}{dt} \tag{4.4}$$

In the foregoing equation, the first three terms vanish, since the distance between any two points O and P of a rigid body is constant. As shown by (2.58), the last three terms can be expressed as $\vec{\omega} \times \vec{r}$. Denoting $d\vec{R}_o/dt$ by $\vec{v}_o$, which is the linear velocity of the origin of the xyz coordinate system, (4.3) becomes

$$\vec{v} = \vec{v}_o + \vec{\omega} \times \vec{r} \tag{4.5}$$

This equation could have been written directly from (2.64) by replacing $\vec{v}_1$ by $\vec{v}_o$ and letting $\vec{r} = 0$. The acceleration $\vec{a}$ of P with respect to the inertial coordinate system is

$$\vec{a} = \frac{d\vec{v}}{dt} = \frac{d\vec{v}_o}{dt} + \frac{d}{dt}(\vec{\omega} \times \vec{r})$$

$$= \vec{a}_o + \frac{d\vec{\omega}}{dt} \times \vec{r} + \vec{\omega} \times \frac{d\vec{r}}{dt} \tag{4.6}$$

where $\vec{a}_o$ denotes the acceleration of the origin of the coordinate system xyz and $d\vec{\omega}/dt = \dot{\vec{\omega}}$ is the angular acceleration of the body. Substituting $\vec{\omega} \times \vec{r}$ for $d\vec{r}/dt$ in (4.6), we obtain

$$\vec{a} = \vec{a}_o + \dot{\vec{\omega}} \times \vec{r} + \vec{\omega} \times (\vec{\omega} \times \vec{r}) \tag{4.7}$$

Again, the foregoing equation could be written directly from (2.65) by replacing $\vec{a}_1$ by $\vec{a}_o$ and letting $\vec{r} = \dot{\vec{r}} = 0$.

The reason for selecting a body coordinate system is that the inertia matrix to be discussed later is a constant with respect to such axes. In some cases, a rigid body possesses axial symmetry with the result that two of its principal moments of inertia are equal. In such cases, it is possible to choose a coordinate system which has an angular velocity $\vec{\omega}$ that is different from the angular velocity $\vec{\Omega}$ of the body, and yet the inertia matrix remains constant with respect to such axes as discussed later. Then, (4.5) and (4.6) are not valid for determining the velocity and acceleration of a point of the body. Letting $\vec{\omega}_{B/F}$ denote the angular velocity of the body, with respect to the reference frame, the angular velocity of the body is related to that of the coordinate system by

$$\vec{\Omega} = \vec{\omega} + \vec{\omega}_{B/F} \tag{4.8}$$

From (2.64), the velocity of a point of the body may be represented by

$$\vec{v} = \vec{v}_o + \dot{\vec{r}} + \vec{\omega} \times \vec{r}$$

In this equation, the relative velocity $\dot{\vec{r}} = \vec{\omega}_{B/F} \times \vec{r}$. Hence, we obtain

$$\vec{v} = \vec{v}_o + \vec{\omega}_{B/F} \times \vec{r} + \vec{\omega} \times \vec{r} \tag{4.9}$$

The acceleration of a point of the body is obtained from (2.65) as

$$\vec{a} = \vec{a}_o + \ddot{\vec{r}} + 2\vec{\omega} \times \dot{\vec{r}} + \dot{\vec{\omega}} \times \vec{r} + \vec{\omega} \times (\vec{\omega} \times \vec{r})$$

In this equation, the relative acceleration $\ddot{\vec{r}}$ becomes

$$\ddot{\vec{r}} = \dot{\vec{\omega}}_{B/F} \times \vec{r} + \vec{\omega}_{B/F} \times (\vec{\omega}_{B/F} \times \vec{r})$$
$$2\vec{\omega} \times \dot{\vec{r}} = 2\vec{\omega} \times (\vec{\omega}_{B/F} \times \vec{r})$$

Hence, it follows that

$$\vec{a} = \vec{a}_o + \dot{\vec{\omega}}_{B/F} \times \vec{r} + \vec{\omega}_{B/F} \times (\vec{\omega}_{B/F} \times \vec{r}) + 2\vec{\omega} \times (\vec{\omega}_{B/F} \times \vec{r})$$
$$+ \dot{\vec{\omega}} \times \vec{r} + \vec{\omega} \times (\vec{\omega} \times \vec{r}) \tag{4.10}$$

Example 4.1

A circular disk of radius R is rotating about a vertical axis with angular velocity $\vec{\omega}_1$ and angular acceleration $\dot{\vec{\omega}}_1$. Determine the velocity and acceleration of a point P at the rim of the disk shown in Fig. 4.2.

Let xyz be a body coordinate system whose origin is at the center of the disk C and whose angular velocity $\vec{\omega}_1 = \omega_1 \vec{j}$, the body angular velocity. To determine the velocity of P, in (4.3) we let $\vec{v}_o - \vec{v}_c = 0$ since the point C is a fixed point. Hence, we obtain

$$\vec{v} = \vec{\omega} \times \vec{r} = \omega_1 \vec{j} \times R\vec{i} = -\omega_1 R\vec{k}$$

The acceleration of P is obtained from (4.7) by letting $\vec{a}_o = \vec{a}_c = 0$ since C is a fixed point. Hence, we get

$$\vec{a} = \dot{\omega}_1 \vec{j} \times R\vec{i} + \omega_1 \vec{j} \times (\omega_1 \vec{j} \times R\vec{i})$$
$$= -\dot{\omega}_1 R\vec{k} - \omega_1^2 R\vec{i}$$

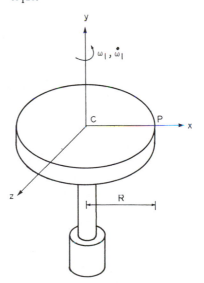

Figure 4.2   Rotating disk.

Example 4.2

A rigid cone with apex half-angle α rolls steadily without slip on a horizontal surface so that it precesses about the vertical Z axis at a constant angular rate ω_o. The height of the cone is h and its base radius is R. Determine (a) the velocity and acceleration of point P at the base of the cone shown in Fig. 4.3, and (b) the velocity and acceleration of C, the mass center of the cone.

The coordinate system $OXYZ$ is inertial with the Z axis being vertical. The cone rolls on the horizontal XY plane. In Fig. 4.3, xyz is a coordinate system with origin at the fixed point O and rotating at constant angular velocity ω_o about the vertical Z axis. Hence, the angular velocity of this coordinate system is written as $\vec{\omega} = \omega_o \sin \alpha \vec{i} + \omega_o \cos \alpha \vec{k}$.

(a) In order to determine the angular velocity $\vec{\Omega}$ of the body, we first determine the spin of the cone about the x axis. Letting this spin be $\omega_1 \vec{i}$, since the cone rolls without slipping and hence point A has zero velocity, we obtain

$$v_A = \frac{h}{\cos \alpha} \omega_o + R\omega_1 = 0$$

or

$$\omega_1 = -\frac{h}{R} \frac{\omega_o}{\cos \alpha} = -\frac{\omega_o}{\sin \alpha} = -\omega_o \csc \alpha$$

The angular velocity of the body is then obtained as

$$\vec{\Omega} = (\omega_o \sin \alpha - \omega_o \csc \alpha)\vec{i} + \omega_o \cos \alpha \vec{k}$$

and the angular velocity of the body relative to the rotating frame becomes

$$\vec{\omega}_{B/F} = -\omega_o \csc \alpha \vec{i}$$

In order to determine the velocity and acceleration of point P, we note that its position vector $\vec{r} = h\vec{i} + R\vec{k}$. Hence, we obtain

$$\vec{\omega}_{B/F} \times \vec{r} = \omega_o R \csc \alpha \vec{j}$$
$$\vec{\omega}_{B/F} \times (\vec{\omega}_{B/F} \times \vec{r}) = -\omega_o^2 R \csc^2 \alpha \vec{k}$$
$$2\vec{\omega} \times (\vec{\omega}_{B/F} \times \vec{r}) = 2\omega_o^2 R\vec{k} - 2\omega_o^2 R \cot \alpha \vec{i}$$
$$\vec{\omega} \times (\vec{\omega} \times \vec{r}) = -\omega_o^2 \cos \alpha(h \cos \alpha - R \sin \alpha)\vec{i}$$
$$\vec{\omega} \times \vec{r} = \omega_o(h \cos \alpha - R \sin \alpha)\vec{j}$$
$$\vec{v}_o = \vec{a}_o = \dot{\vec{\omega}}_{B/F} = \dot{\vec{\omega}} = 0$$

Substituting these results in (4.9) and (4.10) and simplifying the expressions, the velocity and acceleration of point P are obtained as

$$\vec{v}_p = \omega_o[R \csc \alpha + h \cos \alpha - R \sin \alpha]\vec{j}$$
$$\vec{a}_p = \omega_o^2 \cos \alpha[-2R \csc \alpha - h \cos \alpha + R \sin \alpha]\vec{i}$$
$$+ \omega_o^2[-R \csc^2 \alpha + 2R + h \sin \alpha \cos \alpha - R \sin^2 \alpha]\vec{k}$$

(b) In order to obtain the velocity and acceleration of center of mass C, we note that its position vector is $\vec{r} = \frac{3}{4}h\vec{i}$. The velocity and acceleration are obtained by setting

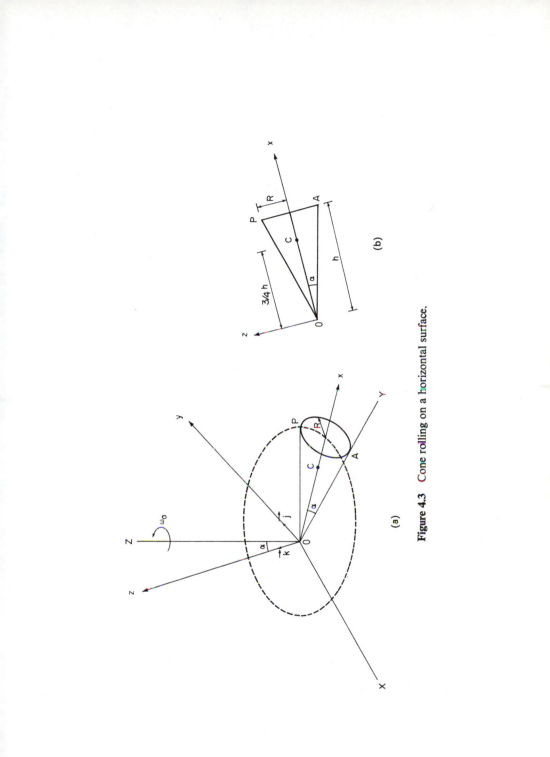

Figure 4.3 Cone rolling on a horizontal surface.

$R = 0$ and replacing h by $\frac{3}{4}h$ in the equations expressing the velocity and acceleration of point P. Hence, we get

$$\vec{v}_c = \frac{3}{4}h\omega_o \cos \alpha \vec{j}$$

$$\vec{a}_c = -\frac{3}{4}h\omega_o^2 \cos^2 \alpha \vec{i} + \frac{3}{4}h\omega_o^2 \sin \alpha \cos \alpha \vec{k}$$

4.3 LINEAR AND ANGULAR MOMENTUM OF A RIGID BODY

4.3.1 Linear Momentum

We consider a rigid body as shown in Fig. 4.4. The mass center of the body is C, and O is a given reference point. The total mass m of the rigid body is written as

$$m = \int_m dm \tag{4.11}$$

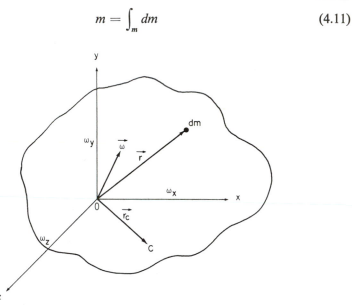

Figure 4.4 Rigid body.

Coordinate system xyz has its origin at the reference point O of the body and it is rotating with respect to an inertial coordinate system at angular velocity $\vec{\omega}$, which is the angular velocity of the body. Hence, xyz is a body coordinate system. The radius vector from the origin O to the center of mass C is defined by

$$\vec{r}_c = \frac{1}{m} \int_{body} \vec{r} \, dm \tag{4.12}$$

Hence, it follows that if the origin O coincides with center of mass C, then $\vec{r}_c = 0$. The linear momentum $\vec{L}$ of a rigid body is the vector sum of the linear momenta of the individual particles that make up the rigid body. We consider a mass particle dm with position vector $\vec{r}$ as shown in Fig. 4.4. From (4.5), the velocity of dm is given by

$$\vec{v} = \vec{v}_o + \vec{\omega} \times \vec{r}$$

Hence, the linear momentum of the body becomes

$$\vec{L} = \vec{v}_o \int dm + \vec{\omega} \times \int \vec{r}\, dm$$

and after employing (4.11) and (4.12), we obtain

$$\vec{L} = m(\vec{v}_o + \vec{\omega} \times \vec{r}_c)$$
$$= m\vec{v}_c \qquad\qquad (4.13)$$

The linear momentum of a rigid body therefore is the product of the total mass and the velocity of its mass center. If the origin O coincides with the center of mass C, then $\vec{r}_c = 0$ and $\vec{L} = m\vec{v}_o$.

4.3.2 Angular Momentum

The angular momentum of a mass particle dm of Fig. 4.4 is the moment of its linear momentum about the origin of the coordinate system and is given by $\vec{r} \times (\vec{v}_o + \vec{\omega} \times \vec{r})\, dm$. The angular momentum of the rigid body about the origin O is obtained as

$$\vec{H}_o = \int \vec{r} \times (\vec{v}_o + \vec{\omega} \times \vec{r})\, dm$$

$$= -\vec{v}_o \times \int \vec{r}\, dm + \int \vec{r} \times (\vec{\omega} \times \vec{r})\, dm \qquad (4.14)$$

If the reference point O of the body is a fixed point in inertial space, then $\vec{v}_o = 0$ and we obtain

$$\vec{H}_o = \int \vec{r} \times (\vec{\omega} \times \vec{r})\, dm \qquad \text{with } O \text{ fixed} \qquad (4.15)$$

If the origin of the coordinate system is located at its center of mass C (i.e., if O coincides with C), we have $\int \vec{r}\, dm = 0$ and it follows that

$$\vec{H}_c = \int \vec{r} \times (\vec{\omega} \times \vec{r})\, dm \qquad (4.16)$$

where $\vec{r}$ is the position of mass particle dm from the center of mass. Equation (4.16) is valid even though the moving center of mass has a velocity $\vec{v}_c$. The motivation for choosing the origin of the coordinate system either at its center of mass or at a fixed point of the body, if such a point exists, becomes obvious. The rotational equations of motion are uncoupled from the translational. Henceforth, we shall assume, unless mentioned otherwise, that the origin of the coordinate system to describe the motion of a rigid body has been selected judiciously in this manner.

It can be easily verified that

$$\vec{r} \times (\vec{\omega} \times \vec{r}) = \vec{i}[y(\omega_x y - \omega_y x) - z(\omega_z x - \omega_x z)]$$
$$+ \vec{j}[z(\omega_y z - \omega_z y) - x(\omega_x y - \omega_y x)]$$
$$+ \vec{k}[x(\omega_z x - \omega_x z) - y(\omega_y z - \omega_z y)] \qquad (4.17)$$

Substituting (4.17) into (4.15), we obtain

$$\vec{H}_o = \vec{i} H_x + \vec{j} H_y + \vec{k} H_z \qquad (4.18)$$

where

$$H_x = \omega_x \int_m (y^2 + z^2)\, dm - \omega_y \int_m xy\, dm - \omega_z \int_m xz\, dm \qquad (4.19a)$$

$$H_y = -\omega_x \int_m xy\, dm + \omega_y \int_m (x^2 + z^2)\, dm - \omega_z \int_m yz\, dm \qquad (4.19b)$$

$$H_z = -\omega_x \int_m xz\, dm - \omega_y \int_m yz\, dm + \omega_z \int_m (x^2 + y^2)\, dm \qquad (4.19c)$$

Since

$$I_x = \int_m (y^2 + z^2)\, dm, \qquad I_y = \int_m (x^2 + z^2)\, dm, \qquad I_z = \int_m (x^2 + y^2)\, dm$$

$$I_{xy} = I_{yx} = -\int xy\, dm$$

$$I_{yz} = I_{zy} = -\int_m yz\, dm$$

$$I_{xz} = I_{zx} = -\int_m xz\, dm$$

we can rewrite (4.19a), (4.19b), and (4.19c) in the form

$$\begin{Bmatrix} H_x \\ H_y \\ H_z \end{Bmatrix} = \begin{bmatrix} I_x & I_{xy} & I_{xz} \\ I_{xy} & I_y & I_{yz} \\ I_{xz} & I_{yz} & I_z \end{bmatrix} \begin{Bmatrix} \omega_x \\ \omega_y \\ \omega_z \end{Bmatrix} \qquad (4.20)$$

Equation (4.20) may be written compactly as a single matrix equation

$$\{H\}_o = [I]_o\{\omega\}_o \qquad (4.21)$$

where it should be noted that point O either coincides with the moving center of mass C or is fixed in inertial space.

The column matrix $\{H\}_o$ contains the components of the vector $\vec{H}_o$, whereas the matrix $\{\omega\}_o$ includes the components of the angular velocity vector $\vec{\omega}$. It may be noted that both $\vec{H}_o$ and $\vec{\omega}$ are independent of the orientation of the xyz coordinate frame having its origin at O. However, the elements of the matrices $\{H\}_o$ and $\{\omega\}_o$ depend on the orientation of the coordinate system. If a different coordinate frame $Ox'y'z'$ is used, the vectors $\vec{H}_o$ and $\vec{\omega}$ will remain unchanged but the column matrices $\{H'\}_o$ and $\{\omega'\}_o$ will have different elements.

The matrix $[I]_o$ contains elements that are the moment and product of inertias of the body with respect to a particular coordinate system. If a different frame $Ox'y'z'$ is selected, the resulting matrix $[I']_o$ will have different elements. The matrix $[I]_o$ is called the *inertia matrix* and its transformation law is similar to that of a stress or strain tensor at a point.

In the foregoing development, it was assumed that a body coordinate system has been employed so that the angular velocity $\vec{\omega}$ of the coordinate system is the same as the angular velocity of the body. However, it was discussed earlier that when a rigid body possesses axial symmetry it is possible to choose a coordinate system which has an angular velocity $\vec{\omega}$ which is different from the angular velocity $\vec{\Omega}$ of the body. In this case, from (4.9) the angular momentum about the origin of the coordinate system of a mass particle dm is given by $\vec{r} \times (\vec{v}_o + \vec{\omega}_{B/F} \times \vec{r} + \vec{\omega} \times \vec{r})\,dm$. Since we have $\vec{\Omega} = \vec{\omega}_{B/F} + \vec{\omega}$, following the foregoing procedure we find that (4.21) is modified as

$$\{H\}_o = [I]_o\{\Omega\}_o \qquad (4.22)$$

Example 4.3

For the rigid cone of Example 4.2, find the linear and angular momentum of the cone with the origin of the coordinate system located at (a) the fixed point O, and (b) the moving center of mass C.

(a) First, we consider the rotating coordinate system xyz of Fig. 4.3 with origin at the fixed point O. From Example 4.2 we know that the angular velocities of the coordinate system and the body are given, respectively, by

$$\vec{\omega} = \omega_o \sin \alpha\, \vec{i} + \omega_o \cos \alpha\, \vec{k}$$

and

$$\vec{\Omega} = (\omega_o \sin \alpha - \omega_o \operatorname{cosec} \alpha)\vec{i} + \omega_o \cos \alpha\, \vec{k}$$

The velocity of the center of mass is given by

$$\vec{v}_c = \tfrac{3}{4}h\omega_o \cos \alpha\, \vec{j}$$

From (4.13), the linear momentum of the cone becomes

$$\vec{L} = m\vec{v}_c = \tfrac{3}{4}mh\omega_o \cos \alpha\, \vec{j}$$

It can be verified that for the coordinate system employed, the products of inertia terms vanish and the diagonal inertia matrix becomes

$$[I]_o = \begin{bmatrix} I_x & 0 & 0 \\ 0 & I_y & 0 \\ 0 & 0 & I_z \end{bmatrix}_o$$

$$= \begin{bmatrix} \tfrac{3}{10}mR^2 & 0 & 0 \\ 0 & m(\tfrac{3}{20}R^2 + \tfrac{3}{5}h^2) & 0 \\ 0 & 0 & m(\tfrac{3}{20}R^2 + \tfrac{3}{5}h^2) \end{bmatrix}$$

Substituting these results in (4.22), we obtain

$$\{H\}_o = \begin{Bmatrix} \tfrac{3}{10}mR^2\omega_o(\sin \alpha - \operatorname{cosec} \alpha) \\ 0 \\ m(\tfrac{3}{20}R^2 + \tfrac{3}{5}h^2)\omega_o \cos \alpha \end{Bmatrix}$$

(b) The linear and angular momentum of the cone are now obtained by employing a coordinate system whose origin is at the mass center C and whose angular velocity is again

$$\vec{\omega} = \omega_o \sin \alpha \vec{i} + \omega_o \cos \alpha \vec{k}$$

as shown in Fig. 4.5.

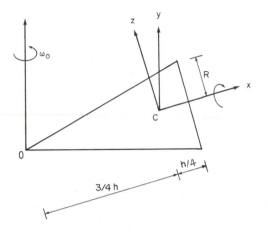

Figure 4.5 Cone rolling on horizontal surface.

This coordinate system remains parallel to the $Oxyz$ considered previously with origin at O. The expressions for the angular velocity $\vec{\Omega}$ of the cone and for the velocity of center of mass C remain unchanged. Hence, the linear momentum of the cone is still given by

$$\vec{L} = m\vec{v}_c = \tfrac{3}{4}mh\omega_o \cos \alpha \vec{j}$$

From (4.22) the expression for the angular momentum becomes

$$\{H\}_c = [I]_c\{\Omega\}_c$$

It can be verified that the products of inertia terms vanish and the diagonal inertia matrix $[I]_c$ is given by

$$[I]_c = \begin{bmatrix} \tfrac{3}{10}mR^2 & 0 & 0 \\ 0 & m(\tfrac{3}{20}R^2 + \tfrac{3}{80}h^2) & 0 \\ 0 & 0 & m(\tfrac{3}{20}R^2 + \tfrac{3}{80}h^2) \end{bmatrix}$$

Hence, the angular moment becomes

$$\{H\}_c = [I]_c\{\Omega\}_c = \begin{Bmatrix} \tfrac{3}{10}mR^2\omega_o(\sin \alpha - \operatorname{cosec} \alpha) \\ 0 \\ m(\tfrac{3}{20}R^2 + \tfrac{3}{80}h^2)\omega_o \cos \alpha \end{Bmatrix}$$

4.3.3 Parallel Axes Theorem of Inertia Matrix

There is an inertia matrix associated with every point of a rigid body. Our interest is in obtaining the relationships between components of the inertia matrix corresponding to different points of a rigid body but referred to parallel

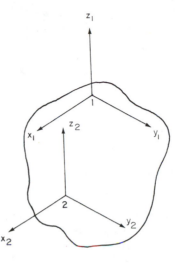

Figure 4.6 Parallel axes theorem.

coordinate systems. Two parallel coordinate systems $x_1y_1z_1$ and $x_2y_2z_2$ are shown in Fig. 4.6 with origins at points 1 and 2 of a rigid body.

Let $x_2 = x_1 + a$, $y_2 = y_1 + b$, and $z_2 = z_1 + c$. When the inertia matrix $[I]_1$ is known, the problem is to determine the inertia matrix $[I]_2$. We get

$$I_{x_2x_2} = \int (y_2^2 + z_2^2)\, dm$$

$$= \int [(y_1 + b)^2 + (z_1 + c)^2]\, dm$$

$$= I_{x_1x_1} + 2b \int y_1\, dm + 2c \int z_1\, dm + (b^2 + c^2)m$$

$$= I_{x_1x_1} + 2m(by_c + cz_c) + m(b^2 + c^2)$$

where x_c, y_c, and z_c denote the position coordinates of the center of mass C from the point 1. Also,

$$I_{y_2z_2} = -\int y_2z_2\, dm$$

$$= -\int (y_1 + b)(z_1 + c)\, dm$$

$$= I_{y_1z_1} - m(cy_c + bz_c) - mbc$$

Similarly, expressions can be obtained for the other elements. Now, if point 1 of the body coincides with the center of mass C, then $x_c - y_c = z_c = 0$ and we obtain

$$[I]_2 = [I]_c + m \begin{bmatrix} b^2 + c^2 & -ab & -ac \\ -ab & c^2 + a^2 & -bc \\ -ac & -bc & a^2 + b^2 \end{bmatrix} \tag{4.23}$$

This result is known as the parallel axes theorem. We note again that the origin of the coordinate system for the matrix $[I]_c$ is at the center of mass. Otherwise, (4.23) will be modified by inclusion of terms containing x_c, y_c, and z_c.

Example 4.4

Consider the problem of Example 4.3, where the two parallel axes are located at the center of mass C and the fixed point O, respectively. Here, we have $a^2 = (\tfrac{3}{4}h)^2$, $b = c = 0$. It can be verified that matrices $[I]_o$ and $[I]_c$ listed in Example 4.3 do indeed satisfy (4.23). Hence, one of the inertia matrices could be obtained by knowing the other matrix and employing (4.23).

4.3.4 Translation Theorem for the Angular Momentum

The angular momentum of a body about any point P can be expressed in terms of the linear momentum of the body and its angular momentum about its center of mass. The angular momentum of the body about any point P shown in Fig. 4.7 can be expressed as

$$\vec{H}_p = \int \vec{r}_p \times \vec{v}\, dm$$

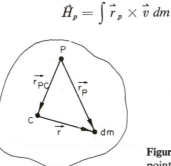

Figure 4.7 Angular momentum about point P.

If $\vec{\omega}$ is the angular velocity of the body, the velocity of dm is given by $\vec{v} = \vec{v}_c + \vec{\omega} \times \vec{r}$. Hence,

$$\vec{H}_p = \int \vec{r}_p \times (\vec{v}_c + \vec{\omega} \times \vec{r})\, dm$$

$$= \int (\vec{r}_{pc} + \vec{r}) \times (\vec{v}_c + \vec{\omega} \times \vec{r})\, dm$$

Now, $(\int \vec{r}\, dm) \times \vec{v}_c = \vec{r}_{pc} \times \vec{\omega} \times \int \vec{r}\, dm = 0$ since the integral vanishes. It follows that

$$\vec{H}_p = \vec{r}_{pc} \times m\vec{v}_c + \int \vec{r} \times (\vec{\omega} \times \vec{r})\, dm \qquad (4.24)$$

Here, $m\vec{v}_c = \vec{L}$, the linear momentum of the body, and the second term on the right-hand side of (4.24) is the angular momentum of the body about the center of mass. We therefore obtain

$$\vec{H}_p = \vec{r}_{pc} \times \vec{L} + \vec{H}_c \qquad (4.25)$$

The foregoing result is called the translation theorem for angular momentum.

Example 4.5

We consider the rolling cone of Example 4.3. Obtain $\{H\}_o$, knowing $\{H\}_c$.

We have

$$\{H\}_o = \vec{r}_{oc} \times \vec{L} + \{H\}_c$$

where $\vec{r}_{oc} = \frac{3}{4}h\vec{i}$, $\vec{L} = m\vec{v}_c = \frac{3}{4}mh\omega_o \cos\alpha\vec{j}$, and $\{H\}_c$ has been obtained in Example 4.3. Since $\vec{r}_{oc} \times \vec{L} = (\frac{3}{4}h)^2 m\omega_o \cos\alpha\vec{k}$, we obtain

$$\{H\}_o = \left\{ \begin{array}{c} \frac{3}{10}mR^2\omega_o(\sin\alpha - \csc\alpha) \\ 0 \\ m(\frac{3}{20}R^2 + \frac{3}{80}h^2)\omega_o \cos\alpha + (\frac{3}{4}h)^2 m\omega_o \cos\alpha \end{array} \right\}$$

It can be verified that this expression is identical to the one obtained for $\{H\}_o$ in Example 4.3 by direct method.

4.4 PRINCIPAL AXES

It is often convenient to deal with rigid-body dynamic problems using the coordinate system fixed in the body for which all products of inertias are zero simultaneously; that is, the inertia matrix $[I]$ is *diagonal*. The three mutually orthogonal coordinate axes are known as *principal axes* and the corresponding moments of inertia are referred to as the *principal moments of inertia*. The three planes formed by the principal axes are called *principal planes*.

For principal axes, (4.21) assumes the simple form

$$\vec{H}_o = \vec{i}\,\omega_x I_x + \vec{j}\,\omega_y I_y + \vec{k}\,\omega_z I_z \tag{4.26}$$

and the scalar components of the angular momentum vector $\vec{H}_o$ become

$$H_x = I_x\omega_x, \qquad H_y = I_y\omega_y, \qquad H_z = I_z\omega_z \tag{4.27}$$

In order to determine the principal axes and the principal moments of inertia, we consider the rotational transformation of the coordinate system. Consider two systems of coordinates xyz and $x_1y_1z_1$ which have the same common origin O but are rotated with respect to each other as shown in Fig. 4.8. Let the angular momentum when expressed with respect to the xyz and $x_1y_1z_1$ coordinate systems be noted by $\{H\}$ and $\{H\}_1$, respectively. Then $\{H\}$ and $\{H\}_1$ are related by the rotational transformation matrix $[C]$ discussed in Section 2.6. We have

$$\{H\}_1 = [C]\{H\} \tag{4.28}$$

where, as shown in Chapter 2, the rotation transformation matrix $[C]$ is made up of direction cosines between the $x_1y_1z_1$ and xyz axes. Substituting $\{H\}_1 = [I]_1\{\omega\}_1$ and $\{H\} = [I]\{\omega\}$ in (4.28), we obtain

$$[I]_1\{\omega\}_1 = [C][I]\{\omega\}$$
$$= [C][I][C]^T[C]\{\omega\} \tag{4.29}$$

where $[C]^T$ is the transpose of matrix $[C]$ and $[C]^T[C]$ is an identity matrix.

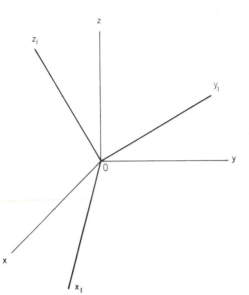

Figure 4.8 Rotated coordinate systems.

Letting $[C]\{\omega\} = \{\omega\}_1$ in (4.29), it is seen that

$$[I]_1 = [C][I][C]^T \tag{4.30}$$

We seek an orthogonal transformation matrix $[C]$ that diagonalizes a general inertia matrix $[I]$. Then from (4.30) we obtain

$$[C]^T \begin{bmatrix} I_1 & 0 & 0 \\ 0 & I_2 & 0 \\ 0 & 0 & I_3 \end{bmatrix} = [I][C]^T$$

Equating the corresponding columns, we get the eigenvalue equation

$$I \begin{Bmatrix} C_{11} \\ C_{12} \\ C_{13} \end{Bmatrix} = [I] \begin{Bmatrix} C_{11} \\ C_{12} \\ C_{13} \end{Bmatrix}$$

where I is I_1, I_2, or I_3. This equation has a nontrivial solution only if $\det |[I] - I| = 0$. This yields the characteristic equation

$$\begin{vmatrix} I_x - I & I_{xy} & I_{xz} \\ I_{xy} & I_y - I & I_{yz} \\ I_{xz} & I_{yz} & I_z - I \end{vmatrix} = 0 \tag{4.31}$$

This is a cubic equation in I which always has three real roots I_1, I_2, and I_3. These roots are the principal moments of inertia. Since $[I]$ is a square-symmetric real matrix, it can be shown that the eigenvalues of such a matrix are always real. It will be shown later that the rotational part of the kinetic energy is given by $\frac{1}{2}\{\omega\}^T[I]\{\omega\} \geq 0$. Hence, matrix $[I]$ is positive semidefinite. It can be shown that the eigenvalues of a positive-semidefinite matrix are nonnegative. Hence, (4.31) always has three real, nonnegative eigenvalues. If all three eigenvalues are unequal, the directions of the three eigenvectors are mutually perpendicular and are uniquely determined. In case the eigenvalues are not all unequal, the principal directions are not uniquely determined. For example, if $I_1 = I_2 \neq I_3$, the direction of principal axis associated with I_3 is uniquely determined but any direction perpendicular to direction of I_3 is a principal axis. The principal axis associated with I_3 is an axis of inertial symmetry. In case $I_1 = I_2 = I_3$, any three mutually perpendicular vectors form a set of principal axes.

The directions of the principal axes can be obtained by determining the three eigenvectors corresponding to I_1, I_2, and I_3, respectively. Normalization of the eigenvectors, so that the length of each eigenvector is unity, then yields the direction cosines between each of the principal axes and the axes xyz. The principal axes are then the body axes for which the inertia matrix is diagonal. When the angular velocity $\vec{\omega}$ of a rigid body is directed along a principal axis of inertia, the angular momentum vector $\vec{H}_o$ and the $\vec{\omega}$ angular velocity vector have the same direction. Otherwise, they have different directions, as seen from (4.27). The diagonalization of matrices is covered in Chapter 6 and hence the details are not given here.

Many rigid bodies have a plane of symmetry. For example, for the rigid body shown in Fig. 4.9, the xy plane is a plane of symmetry; that is, for every mass particle dm with coordinates (x, y, z), there exists a mass dm with coordinates $(x, y, -z)$. Hence, we find that the products of inertia terms

$$I_{yz} = I_{zy} = -\int_{body} yz \, dm = 0$$

$$I_{xz} = I_{zx} = -\int_{body} zx \, dm = 0$$

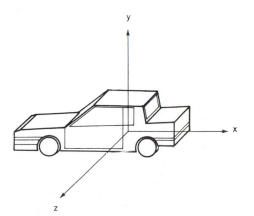

Figure 4.9 Rigid body with plane of symmetry.

The inertia matrix with respect to the xyz axes becomes

$$[I] = \begin{bmatrix} I_x & I_{xy} & 0 \\ I_{xy} & I_y & 0 \\ 0 & 0 & I_3 \end{bmatrix} \tag{4.32}$$

It is noted that here the z axis is a principal axis with principal moment of inertia $I_z = I_3$. The other two principal axes, x_1 and y_1, are obtained by rotation through angle θ about the z axis as shown in Fig. 4.10. From Chapter 2, the rotation transformation matrix becomes

$$[C] = \begin{bmatrix} \cos\theta & \sin\theta & 0 \\ -\sin\theta & \cos\theta & 0 \\ 0 & 0 & 1 \end{bmatrix} \tag{4.33}$$

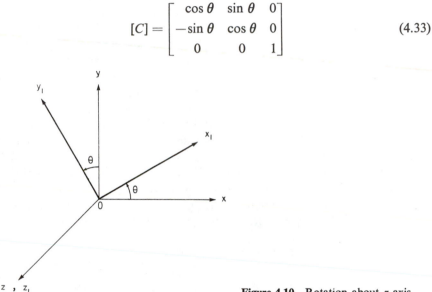

Figure 4.10   Rotation about z axis.

From (4.30), we have

$$\begin{bmatrix} I_1 & 0 & 0 \\ 0 & I_2 & 0 \\ 0 & 0 & I_3 \end{bmatrix} = [C][I][C]^T \tag{4.34}$$

where $[C]$ is given by (4.33). The third column and third row of the matrices on the left- and right-hand sides of (4.34) are identical. Equating the corresponding elements of matrices on both sides of (4.34), we obtain the following three equations:

$$I_1 = I_x \cos^2\theta + I_y \sin^2\theta + 2I_{xy}\sin\theta\cos\theta \tag{4.35}$$

$$0 = I_{xy}(\cos^2\theta - \sin^2\theta) + (I_y - I_x)\sin\theta\cos\theta \tag{4.36}$$

$$I_2 = I_x \sin^2\theta + I_y \cos^2\theta - 2I_{xy}\sin\theta\cos\theta \tag{4.37}$$

Expressing (4.36) in terms of 2θ, we obtain

$$\tan 2\theta = \frac{I_{xy}}{\frac{1}{2}(I_x - I_y)} \tag{4.38}$$

The principal moments of inertia I_1 and I_2 are then evaluated by substituting this result in (4.35) and (4.37) respectively.

4.5 EQUATIONS OF MOTION FOR A RIGID BODY

As mentioned earlier, an unconstrained rigid body has six degrees of freedom, and six equations of motion are needed to specify its configuration. Three equations may be chosen to represent the translation of the mass center, and three equations for the rotation about the mass center. Let xyz represent body axes with origin at the center of mass C as shown in Fig. 4.11. The angular velocity $\vec{\omega}$ of this coordinate system is the same as the angular velocity of the body. Let m be the mass of the body, $\vec{F}$ the resultant of the external forces acting on the body, and $\vec{M}_c$ be the resultant moment of external forces and couples about the mass center C. The equations of motion for the rigid body may be written by direct application of Newton's second law as

$$\frac{d}{dt}\vec{L} = \frac{d}{dt}(m\vec{v}_c) = m\frac{d}{dt}\vec{v}_c = \vec{F} \tag{4.39}$$

$$\frac{d}{dt}\vec{H}_c = \vec{M}_c \tag{4.40}$$

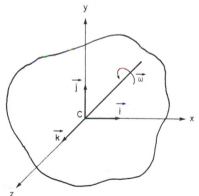

Figure 4.11   Motion of a rigid body.

where $\vec{v}_c$ is the velocity vector of the mass center C and $\vec{H}_c$ is the angular momentum vector given by (4.16). It is seen that by choosing the center of mass C as the origin of the coordinate system, the rotational equations of motion (4.40) are uncoupled from the translational equations of motion (4.39). Since $\vec{v}_c$ and $\vec{H}_c$ have been expressed in terms of a rotating coordinate system xyz with angular velocity $\vec{\omega}$, it is seen from Chapter 2 that

$$\frac{d\vec{v}_c}{dt} = \dot{v}_x\vec{i} + \dot{v}_y\vec{j} + \dot{v}_z\vec{k} + \vec{\omega} \times \vec{v}_c$$

$$= \dot{v}_x\vec{i} + \dot{v}_y\vec{j} + \dot{v}_z\vec{k} + (v_z\omega_y - v_y\omega_z)\vec{i}$$
$$+ (v_x\omega_z - v_z\omega_x)\vec{j} + (v_y\omega_x - v_x\omega_y)\vec{k} \qquad (4.41)$$

$$\frac{d}{dt}\vec{H}_c = \dot{H}_x\vec{i} + \dot{H}_y\vec{j} + \dot{H}_z\vec{k} + \vec{\omega} \times \vec{H}_c \qquad (4.42a)$$

Since the components of $\vec{H}_c$ are given by (4.20), we get

$$\dot{H}_x = I_x\dot{\omega}_x + I_{xy}\dot{\omega}_y + I_{xz}\dot{\omega}_z$$
$$\dot{H}_y = I_{xy}\dot{\omega}_x + I_y\dot{\omega}_y + I_{yz}\dot{\omega}_z \qquad (4.42b)$$
$$\dot{H}_z = I_{xz}\dot{\omega}_x + I_{yz}\dot{\omega}_y + I_z\dot{\omega}_z$$

The inertia terms are constants with time as the axes xyz form a body coordinate system. We also have

$$\vec{\omega} \times \vec{H}_c = (\omega_y H_z - \omega_z H_y)\vec{i} + (\omega_z H_x - \omega_x H_z)\vec{j} + (\omega_x H_y - \omega_y H_x)\vec{k} \qquad (4.43)$$

Substituting from (4.41) in (4.39), the translational equations of motion are given by

$$m(\dot{v}_x + v_z\omega_y - v_y\omega_z) = F_x$$
$$m(\dot{v}_y + v_x\omega_z - v_z\omega_x) = F_y \qquad (4.44)$$
$$m(\dot{v}_z + v_y\omega_x - v_x\omega_y) = F_z$$

Similarly, substitution from (4.42a), (4.42b), and (4.43) in (4.40) yields

$$I_x\dot{\omega}_x + I_{xy}(\dot{\omega}_y - \omega_x\omega_z) + I_{xz}(\dot{\omega}_z + \omega_x\omega_y)$$
$$+ (I_z - I_y)\omega_y\omega_z + I_{yz}(\omega_y^2 - \omega_z^2) = M_x$$
$$I_{xy}(\dot{\omega}_x + \omega_y\omega_z) + I_y\dot{\omega}_y + I_{yz}(\dot{\omega}_z - \omega_x\omega_y)$$
$$+ (I_x - I_z)\omega_x\omega_z + I_{xz}(\omega_z^2 - \omega_x^2) = M_y \qquad (4.45)$$
$$I_{xz}(\dot{\omega}_x - \omega_y\omega_z) + I_{yz}(\dot{\omega}_y + \omega_x\omega_z) + I_z\dot{\omega}_z$$
$$+ (I_y - I_x)\omega_x\omega_y + I_{xy}(\omega_x^2 - \omega_y^2) = M_z$$

Equations (4.44) and (4.45) are the six equations of motion for a rigid body. The moments and inertia terms in (4.45) are with respect to a body axes with the origin at the mass center C. Sometimes, a rigid body may have a point O which is fixed in inertial space. In such cases, the origin of the body coordinate system xyz may be chosen as this fixed point O. It is seen from (4.15) that in this case also the rotational motion is uncoupled from the translational one. Equation (4.39) remains unchanged and it is seen from (4.13) that

$$\vec{v}_c = \vec{v}_o + \vec{\omega} \times \vec{r}_c = \vec{\omega} \times \vec{r}_c \qquad \text{since} \qquad \vec{v}_o = 0$$

and hence

$$\frac{d}{dt}\vec{v}_c = \vec{\omega} \times \vec{r}_c + \vec{\omega} \times (\vec{\omega} \times \vec{r}_c) \qquad (4.46)$$

which is identical to (4.41). In (4.45), the moments and inertia terms are now with respect to a body axes with origin at the fixed point O.

4.6 EULER'S AND MODIFIED EULER'S EQUATIONS OF MOTION

4.6.1 Euler Equations

A considerable simplification can be made in the general rotational equations of motion if the body coordinate axes x, y, and z are selected such that they are the principal axes with origin either at the mass center C, or at a point of the body O fixed in inertial space (in case such a point exists). The following assumptions are made:

1. The origin of the coordinate system is either at the center of mass C or at a point of the body O fixed in inertial space.
2. The coordinate system xyz is a body coordinate system so that its angular velocity $\vec{\omega}$ is the same as the angular velocity of the body.
3. The axes are principal axes.

Assumptions 1 and 2 have been made in the derivation of (4.45). But assumption 3 is an additional assumption. With this choice of body axes, all the product of inertia terms vanish and (4.45) reduces to

$$I_1\dot{\omega}_1 + (I_3 - I_2)\omega_2\omega_3 = M_1$$
$$I_2\dot{\omega}_2 + (I_1 - I_3)\omega_1\omega_3 = M_2 \qquad (4.47)$$
$$I_3\dot{\omega}_3 + (I_2 - I_1)\omega_1\omega_2 = M_3$$

where I_1, I_2, and I_3 are the principal moments of inertias, ω_1, ω_2, and ω_3 are the components of the angular velocity vector $\vec{\omega}$ along the principal axes, and M_1, M_2, and M_3 represent the components of the moment vector $\vec{M}$ along the principal axes.

Equations (4.47) are known as *Euler's equations of motion*. These equations are relatively simple as compared to (4.45) and are often used in describing the rotational motion of a rigid body.

4.6.2 Modified Euler Equations

We consider a rigid body which has at least two of its three principal moments of inertia at the mass center equal (Fig. 4.12). We choose the principal axis 1 as the axis of symmetry and imply that I_2 and I_3 are equal. We designate the moment of inertia about the symmetry axis and about a transverse principal axis through the mass center C by I_α and I_t, respectively. Thus,

$$I_1 = I_\alpha$$
$$I_2 = I_3 = I_t \qquad (4.48)$$

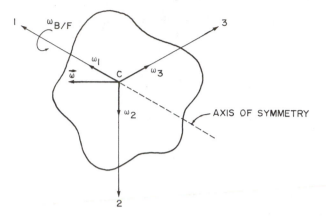

Figure 4.12   Rigid body with an axis of symmetry.

We now discard assumption 2 made in the derivation of Euler's equations of motion and in (4.45) but retain assumptions 1 and 3. However, letting $\vec{\omega} = \omega_1 \vec{i} + \omega_2 \vec{j} + \omega_3 \vec{k}$ be the angular velocity of the coordinate system, the angular velocity $\vec{\Omega}$ of the body is restricted to be

$$\vec{\Omega} = (\omega_1 + \omega_{B/F})\vec{i} + \omega_2 \vec{j} + \omega_3 \vec{k} \tag{4.49}$$

where $\omega_{B/F}\vec{i}$ is the angular velocity of the body with respect to the coordinate frame. The angular momentum vector is now given by (4.22) and the components are

$$\begin{aligned} H_1 &= I_a \Omega_1 = I_a(\omega_1 + \omega_{B/F}) \\ H_2 &= I_t \Omega_2 = I_t \omega_2 \\ H_3 &= I_t \Omega_3 = I_t \omega_3 \end{aligned} \tag{4.50}$$

The rotational equations of motion are given by (4.40), where

$$\frac{d\vec{H}}{dt} = I_a \dot{\Omega}_1 \vec{i} + I_t \dot{\Omega}_2 \vec{j} + I_t \dot{\Omega}_3 \vec{k} + \vec{\omega} \times \vec{H}$$

Hence, the *modified Euler's equations* of motion are obtained as

$$\begin{aligned} I_a(\dot{\omega}_1 + \dot{\omega}_{B/F}) &= M_1 \\ I_t \dot{\omega}_2 + (I_a - I_t)\omega_1 \omega_3 + I_a \omega_{B/F} \omega_3 &= M_2 \\ I_t \dot{\omega}_3 + (I_t - I_a)\omega_1 \omega_2 - I_a \omega_{B/F} \omega_2 &= M_3 \end{aligned} \tag{4.51}$$

Equations (4.51) have an added flexibility of being able to specify the spin $\omega_{B/F}\vec{i}$ in an arbitrary time-varying manner. Even though the axes do not constitute a body coordinate system, the principal moments of inertia still remain time invariant because axis 1 is an axis of symmetry. Any two orthogonal axes in the plane perpendicular to axis 1 constitute principal axes. In some cases, all three principal moments of inertia are equal, as, for example, when the body is a sphere or cube with the origin of axes at the center of mass. In such cases, we can discard restriction (4.49) and specify the angular velocity $\vec{\Omega}$ of the body quite arbitrarily and write the equations as $\dot{\vec{H}} + \vec{\omega} \times \vec{H} = \vec{M}$.

4.6.3 State-Variable Formulation of the Equations of Motion

Since the rotational equations of motion are uncoupled from the translational ones, the Euler's equations of motion (4.47) may be expressed in the state-variable form as

$$\dot{\omega}_1 = -\frac{1}{I_1}(I_3 - I_2)\omega_2\omega_3 + \frac{1}{I_1}M_1$$

$$\dot{\omega}_2 = -\frac{1}{I_2}(I_1 - I_3)\omega_1\omega_3 + \frac{1}{I_2}M_2 \qquad (4.52)$$

$$\dot{\omega}_3 = -\frac{1}{I_3}(I_2 - I_1)\omega_1\omega_2 + \frac{1}{I_3}M_3$$

where ω_1, ω_2, and ω_3 are the state variables and M_1, M_2, and M_3 are the inputs. The translational equations of motion (4.44) may also be expressed in the state-variable form as

$$\dot{v}_x = v_y\omega_3 - v_z\omega_2 + \frac{1}{m}F_x$$

$$\dot{v}_y = v_z\omega_1 - v_x\omega_3 + \frac{1}{m}F_y \qquad (4.53)$$

$$\dot{v}_z = v_x\omega_2 - v_y\omega_1 + \frac{1}{m}F_z$$

where v_x, v_y, and v_z are three additional state variables and F_x, F_y and F_z are inputs. Equations (4.52) and (4.53) may also be combined in the form

$$\{\dot{x}\} = \{f(x_1, \ldots, x_6, u_1, \ldots, u_6)\} \qquad (4.54)$$

where $\{x\}$ is a six-dimensional column matrix defined as

$$\{x\} = \lfloor \omega_1, \omega_2, \omega_3, v_x, v_y, v_z \rfloor^T$$

and $u_1, \ldots, u_6$ represent the inputs. In order to express the general rotational equations of motion (4.45) in state-variable form, they can be rearranged as

$$\begin{bmatrix} I_x & I_{xy} & I_{xz} \\ I_{xy} & I_y & I_{yz} \\ I_{xz} & I_{yz} & I_z \end{bmatrix} \begin{Bmatrix} \dot{\omega}_x \\ \dot{\omega}_y \\ \dot{\omega}_z \end{Bmatrix} = \begin{Bmatrix} g_1(\omega_x, \omega_y, \omega_z, M_x) \\ g_2(\omega_x, \omega_y, \omega_z, M_y) \\ g_3(\omega_x, \omega_y, \omega_z, M_z) \end{Bmatrix}$$

where g_1, g_2, and g_3 are nonlinear functions of their arguments and can be defined from (4.45). Inverting the inertia matrix, the state-variable equations are obtained as

$$\begin{Bmatrix} \dot{\omega}_x \\ \dot{\omega}_y \\ \dot{\omega}_z \end{Bmatrix} = [I]^{-1} \begin{Bmatrix} g_1 \\ g_2 \\ g_3 \end{Bmatrix} = \begin{Bmatrix} f_1 \\ f_2 \\ f_3 \end{Bmatrix} \qquad (4.55)$$

Example 4.6

Obtain the equations of motion for the rigid cone of Example 4.2 which is rolling steadily without slip on a horizontal surface.

The coordinate system xyz has its origin at the fixed point O as shown in Fig. 4.3 and its angular velocity $\vec{\omega}$ is $\vec{\omega} = \omega_o \sin \alpha\, \vec{i} + \omega_o \cos \alpha \vec{k}$. The angular velocity of the body is given by $\vec{\Omega} = (\omega_o \sin \alpha - \omega_o \operatorname{cosec} \alpha)\vec{i} + \omega_o \cos \alpha \vec{k}$. It is noted that the x axis is a principal axis and also an axis of symmetry. It is recognized that $-\omega_o \operatorname{cosec} \alpha\, \vec{i}$ is a constant spin about the x axis. In Example 4.2, the velocity of the center of mass was obtained as $\vec{v}_c = \frac{3}{4}h\omega_o \cos \alpha\, \vec{j}$. Employing (4.44) for this example, we have

$$v_x = v_z = \dot{v}_x = \dot{v}_z = \dot{v}_y = 0, \qquad v_y = \tfrac{3}{4}h\omega_o \cos \alpha, \qquad \omega_x = \omega_o \sin \alpha,$$

$$\omega_z = \omega_o \cos \alpha, \qquad \omega_y = 0$$

Substituting these values in (4.44), we obtain

$$-\tfrac{3}{4}m\omega_o^2 h \cos^2 \alpha = F_x$$

$$0 = F_y$$

$$\tfrac{3}{4}m\omega_o^2 h \sin \alpha \cos \alpha = F_z$$

The forces that must be applied to maintain this motion are given by the foregoing equations. The angular velocity vector $\vec{H}_o$ for this problem was obtained in Example 4.3. It is noted that the coordinate system consists of the principal axes and that the angular velocity of the body satisfies the restriction (4.49). Hence, the equations employed to describe the rotational motion are the modified Euler's equations. Employing (4.51) for this example, we have

$$I_\alpha = I_x = \tfrac{3}{10}mR^2, \qquad I_y = I_z = I_t = m(\tfrac{3}{20}R^2 + \tfrac{3}{5}h^2)$$

$$\dot{\omega}_x = \dot{\omega}_y = \dot{\omega}_z = \dot{\omega}_{B/F} = 0, \qquad \omega_{B/F} = -\omega_o \operatorname{cosec} \alpha$$

Substituting these values in (4.51), we obtain

$$0 = M_x$$

$$m(\tfrac{3}{20}R^2 - \tfrac{3}{5}h^2)\,\omega_o^2 \sin \alpha \cos \alpha - \tfrac{3}{10}mR^2\omega_o^2 \operatorname{cosec} \alpha \cos \alpha = M_y$$

$$0 = M_z$$

The moments that must be applied to maintain this motion are given by the foregoing equations.

Example 4.7

A rotor shown in Fig. 4.13 rotates about axis x at a constant angular velocity ω_o. The rotor is dynamically unbalanced so that its principal axis x_1 is displaced at an angle θ to the x axis. Determine the reactions at the bearings, which are a distance b apart, due to the unbalance.

We assume that the rotor is statically balanced so that its center of mass C lies on the bearing axis x. When the rotor is statically balanced but the axis of rotation is not a principal axis, the rotor is said to be dynamically unbalanced. Coordinate system xyz has its origin at the center of mass C and its angular velocity $\vec{\omega} = \omega_o \vec{i}$. Axes system $x_1 y_1 z_1$ consists of principal axes. We solve this problem first by employing the coordinate system xyz and the general rotational equations of motion (4.45). This problem

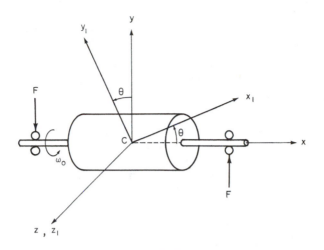

Figure 4.13   Dynamically unbalanced rotor.

is then solved again by employing the principal axes coordinate system and axes $x_1 y_1 z_1$.

Employing the xyz coordinate system, we note, however, that z is a principal axis and hence $I_{xz} = I_{yz} = 0$ but $I_{xy} \neq 0$. Now, we have $\omega_y = \omega_z = \dot{\omega}_x = \dot{\omega}_y = \dot{\omega}_z = 0$ and $\omega_x = \omega_o$. Therefore, from (4.45) we obtain

$$0 = M_x$$
$$0 = M_y$$
$$I_{xy}\omega_o^2 = M_z$$

This moment M_z is supplied by a pair of forces of magnitude $F = (1/b)\, I_{xy}\omega_o^2$ on the shaft at the bearings, which are a distance b apart. These reactions act parallel to the y axis, as shown in Fig. 4.13. They retain a fixed orientation with respect to the body coordinates xyz and thus rotate with the shaft at angular velocity $\omega_o \vec{i}$. The reactions of the shaft on the bearings are opposite in sign to those of the bearings on the shaft. In addition, there are vertical reactions at the bearings due to the weight of rotor.

This problem may also be solved by employing the Euler's equation of motion (4.47). Coordinate system $x_1 y_1 z_1$ consists of principal axes and its angular velocity $\vec{\omega}$ is the same as the angular velocity of the body; that is,

$$\vec{\omega} = \omega_o \cos\theta\, \vec{i}_1 + \omega_o \sin\theta\, \vec{j}_1$$

Employing (4.47), we note that

$$I_{x1} = I_1, \qquad I_{y1} = I_{z1} = I_2, \qquad \omega_3 = \dot{\omega}_1 = \dot{\omega}_2 = \dot{\omega}_3 = 0,$$
$$\omega_1 = \omega_o \cos\theta, \qquad \omega_2 = \omega_o \sin\theta$$

Hence, (4.47) yields

$$0 = M_1$$
$$0 = M_2$$
$$(I_2 - I_1)\omega_o^2 \sin\theta \cos\theta = M_3$$

This moment M_3 is supplied by the reactions of the bearings on the shaft. The reactions have magnitude $(1/b)(I_2 - I_1)\omega_o^2 \sin \theta \cos \theta$. Since the z and z_1 axes are identical, it is expected that

$$I_{xy} = (I_2 - I_1) \sin \theta \cos \theta$$

This equality can in fact be proved by employing (4.35), (4.36), and (4.37).

Example 4.8

The essential structure of a certain type of aircraft turn indicator is shown in Fig. 4.14. A rotor spinning at an angular velocity of ω_r rad/s (constant) counterclockwise as viewed from the right is supported on two springs AC and BD, a distance b apart. The plane executes a horizontal turn at angular velocity ω_o (constant) clockwise as viewed from the top. The coordinate system xyz rotates with the plane. (a) Obtain the modified Euler equations, and (b) determine the change in length of each spring from the equilibrium position. Let k_s be the spring constant of each spring.

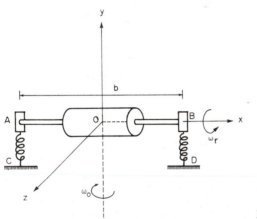

Figure 4.14 Aircraft turn indicator.

(a) The coordinate system xyz has its origin at the center of mass of the rotor and its angular velocity $\vec{\omega} = -\omega_o \vec{j}$. The angular velocity of the body is given by $\vec{\Omega} = -\omega_o \vec{j} + \omega_r \vec{i}$. It is noted that the coordinate system xyz constitutes principal axes and x is an axis of symmetry. Employing the modified Euler's equations, we note that

$$I_\alpha = I_x, \quad I_t = I_y = I_z, \quad \omega_1 = \omega_x = 0, \quad \omega_2 = -\omega_o,$$
$$\omega_3 = \dot{\omega}_1 = \dot{\omega}_2 = \dot{\omega}_3 = 0, \quad \omega_{B/F} = \omega_r, \quad \dot{\omega}_{B/F} = 0$$

Hence, (4.51) yields

$$0 = M_1$$
$$0 = M_2$$
$$I_x \omega_o \omega_r = M_3$$

(b) The moment M_3 is provided by forces $F\vec{j}$ and $-F\vec{j}$ acting on the rotor shaft at B and A, respectively, as shown in Fig. 4.15. Here, $F = (1/b)I_x\omega_o\omega_r$. Hence, spring

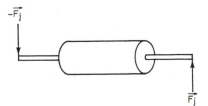

Figure 4.15 Forces on the rotor shaft.

BD is compressed by an amount $(1/bk_s)I_x\omega_o\omega_r$ and spring AC is extended by the same amount.

Example 4.9

A disk of mass m and radius r is attached to the end of a rod OB of negligible weight which is supported by a ball-and-socket joint at O as shown in Fig. 4.16. The rate of precession of the disk about the vertical is observed to be ω_o (constant) and the angle to be β. Determine the rate of spin ω_s of the disk about OB.

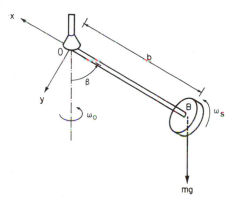

Figure 4.16 Spinning and precessing disk.

    The coordinate system xyz has its origin at the fixed point O and its angular velocity is $\vec{\omega} = \omega_o \cos \beta \vec{i} - \omega_o \sin \beta \vec{j}$.

    The axes xyz constitute principal axis and x axis is an axis of symmetry. The angular velocity of the body is

$$\vec{\Omega} = (\omega_o \cos \beta + \omega_s)\vec{i} - \omega_o \sin \beta \vec{j}$$

In the modified Euler's equations of motion (4.51), we have

$$I_\alpha = I_x, \quad I_t = I_y = I_z, \quad \omega_1 = \omega_o \cos \beta, \quad \omega_{B/F} = \omega_s,$$

$$\omega_2 = -\omega_o \sin \beta, \quad \omega_3 = 0, \quad \dot{\omega}_1 = \dot{\omega}_2 = \dot{\omega}_3 = \dot{\omega}_{B/F} = 0$$

Hence, (4.51) yields

$$0 = M_1$$

$$0 = M_2$$

$$(I_x - I_y)\omega_o^2 \sin \beta \cos \beta + I_x\omega_s\omega_o \sin \beta = M_3$$

Taking moment of forces about O, we obtain

$$\vec{M}_o = -b\vec{i} \times (-mg \cos \beta \vec{i} + mg \sin \beta \vec{j})$$
$$= -bmg \sin \beta \vec{k}$$

Equating the moments and then solving for ω_s, we obtain

$$\omega_s = \frac{I_x - I_y}{I_x} \omega_o \cos \beta - \frac{bmg}{I_x \omega_o}$$

4.7 WORK–ENERGY PRINCIPLE FOR A RIGID BODY

The work–energy principles for a particle derived in Chapter 3 are also valid for a rigid body. The only necessary modification concerns the expressions for the kinetic energy and the work done by external forces. We consider a rigid body and let xyz be a body coordinate system with origin at the mass center C (Fig. 4.17).

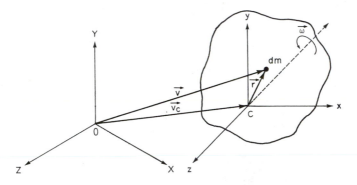

Figure 4.17   Rigid body.

The kinetic energy of the particle with mass dm is defined by

$$dT = \tfrac{1}{2}v^2 \, dm = \tfrac{1}{2}\vec{v} \cdot \vec{v} \, dm \qquad (4.56)$$

but from (4.5) we have

$$\vec{v} = \vec{v}_c + \vec{\omega} \times \vec{r} \qquad (4.57)$$

Substituting (4.57) into (4.56), we get

$$dT = \tfrac{1}{2}\vec{v}_c \cdot \vec{v}_c dm + \vec{v}_c \cdot (\vec{\omega} \times \vec{r}) \, dm + \tfrac{1}{2}(\vec{\omega} \times \vec{r}) \cdot (\vec{\omega} \times \vec{r}) \, dm \qquad (4.58)$$

Integrating (4.58) over the entire mass m of the body we obtain (note that the second term drops out, since $\int_m \vec{r} \, dm = 0$)

$$T = \tfrac{1}{2}m\vec{v}_c \cdot \vec{v}_c + \tfrac{1}{2} \int_m (\vec{\omega} \times \vec{r}) \cdot (\vec{\omega} \times \vec{r}) \, dm \qquad (4.59)$$

We can notice from (4.59) that the kinetic energy of a rigid body consists of two parts: that is,

$$T = T_t + T_r \qquad (4.60)$$

where

$$T_t = \tfrac{1}{2} m \vec{v}_c \cdot \vec{v}_c \qquad (4.61)$$

$$T_r = \tfrac{1}{2} \int_m (\vec{\omega} \times \vec{r}) \cdot (\vec{\omega} \times \vec{r}) \, dm \qquad (4.62)$$

Here, T_t refers to the kinetic energy of translation and T_r is the kinetic energy due to rotation of the rigid body computed in the reference frame translating with mass center C.

From the properties of a triple vector product and (4.16), we have

$$\vec{\omega} \cdot \vec{H}_c = \int_m \vec{\omega} \cdot [\vec{r} \times (\vec{\omega} \times \vec{r})] \, dm = \int_m (\vec{\omega} \times \vec{r}) \cdot (\vec{\omega} \times \vec{r}) \, dm$$

$$= 2T_r \qquad (4.63)$$

or

$$T_r = \tfrac{1}{2} \vec{\omega} \cdot \vec{H}_c \qquad (4.64)$$

We can easily evaluate (4.64) as

$$T_r = \tfrac{1}{2}(I_x \omega_x^2 + I_y \omega_y^2 + I_z \omega_z^2) + I_{xy}\omega_x\omega_y + I_{yz}\omega_y\omega_z + I_{zx}\omega_z\omega_x \qquad (4.65)$$

If the reference frame xyz refers to the principal axes frame with origin at the mass center C, then (4.64) is reduced to the following form:

$$T_r = \tfrac{1}{2}(I_1 \omega_1^2 + I_2 \omega_2^2 + I_3 \omega_3^2) \qquad (4.66)$$

In matrix notation, we get

$$T = \tfrac{1}{2} m \{v_c\}^T \{v_c\} + \tfrac{1}{2} \{\omega\}^T [I]_c \{\omega\} \qquad (4.67)$$

We denote the resultant of all external forces by $\vec{F}$ and $\vec{M}_c$ refers to the resultant moment of the external forces and couples acting on the body about the mass center C. Then,

$$\int_{t_1}^{t_2} \vec{F} \cdot \vec{v}_c \, dt + \int_{t_1}^{t_2} \vec{M}_c \cdot \vec{\omega} \, dt = W_{1,2} \qquad (4.68)$$

Equation (4.68) represents the work done by all external forces and couples in the time interval from t_1 to t_2. We know that work done is equal to the change in kinetic energies T_1 and T_2: that is,

$$W_{1,2} = T_1 - T_2 \qquad (4.69)$$

In (4.68) for the first term we can write

$$\int_{t_1}^{t_2} \vec{F} \cdot \vec{v}_c \, dt = m \int_{v_{c_1}}^{v_{c_2}} \vec{v}_c \cdot d\vec{v}_c = \tfrac{1}{2} m (v_{c_2}^2 - v_{c_1}^2)$$

$$= (T_{2,t} - T_{1,t}) \qquad (4.70)$$

Next, from (4.40), (4.62), and (4.64) we obtain

$$\int_{t_1}^{t_2} \vec{M}_c \cdot \vec{\omega} \, dt = T_{2,r} - T_{1,r} \qquad (4.71)$$

We can notice that the work done by external forces produces a change in the kinetic energy of translation of the body, whereas the work done by the resultant moment of the external forces and couples about C leads to a change in the rotational kinetic energy of the body.

In case all the impressed forces are conservative and their potential is denoted by U, then

$$T + U = E = \text{const.} \tag{4.72}$$

or

$$T_1 + U_1 = T_2 + U_2 \tag{4.73}$$

This is the principle of conservation of mechanical energy. In case the body has a point O fixed in inertial space and the origin of the coordinate system xyz is this point O, then $\vec{v}_o = 0$ and the expression for the kinetic energy becomes

$$T = \tfrac{1}{2}\{\omega\}^T[I]_o\{\omega\} \tag{4.74}$$

so that the kinetic energy may be regarded as due entirely to the rotational motion of the body about the fixed point. In (4.74), the inertia matrix $[I]_o$ is about the fixed point O. As in the case of the modified Euler's equations of motion, if the body has an axis of symmetry, we may let $\vec{\omega}$ be the angular velocity of the coordinate system and $\vec{\Omega}$ be the angular velocity of the body. Following the procedure outlined in the foregoing, it can be shown that the rotational part of the kinetic energy becomes

$$T_r = \tfrac{1}{2}\{\Omega\}^T[I]\{\Omega\} \tag{4.75}$$

Example 4.10

A sphere of mass m and radius r rolls without slipping inside a curved surface of radius R as shown in Fig. 4.18. The sphere is released from rest at $\theta = \pi/2$. Obtain an expression for the velocity of its center of mass as a function of angle θ.

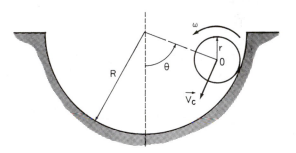

Figure 4.18   Sphere rolling without slipping.

The expression developed for the kinetic energy is given by

$$T = \tfrac{1}{2}m\{v_c\}^T\{v_c\} + \tfrac{1}{2}\{\omega\}^T[I]\{\omega\}$$

We have here a case of plane motion and the second term in the foregoing equation becomes $\tfrac{1}{2}(\tfrac{2}{5}mr^2)\omega^2$. Since the sphere rolls without slipping, $v_c = r\omega$. Also,

$\omega r = -(R - r)\dot{\theta}$. Hence, in terms of θ, the kinetic energy becomes

$$T = \tfrac{1}{2}m(R - r)^2\,\dot{\theta}^2 + \tfrac{1}{2}(\tfrac{2}{5}mr^2)\left(\frac{R - r}{r}\right)^2\dot{\theta}^2$$

$$= \tfrac{7}{10}m(R - r)^2\dot{\theta}^2$$

The expression for the potential energy is given by

$$U = mg(R - r)(1 - \cos\theta)$$

Since the cylinder rolls without slipping, friction does no work and the system is conservative. Hence, $T + V = \text{constant}$. Now, when $\theta = \pi/2$, $T = 0$, and $U = mg(R - r)$. Hence, it follows that $T + U = mg(R - r)$; that is,

$$\tfrac{7}{10}m(R - r)^2\dot{\theta}^2 + mg(R - r)(1 - \cos\theta) = mg(R - r)$$

or

$$\tfrac{7}{10}m(R - r)^2\dot{\theta}^2 = mg(R - r)\cos\theta$$

Hence,

$$\dot{\theta} = \left[\tfrac{10}{7}\frac{g}{R - r}\cos\theta\right]^{1/2}$$

$$\vec{v}_c = -(R - r)\dot{\theta}\vec{i}_\theta = -[\tfrac{10}{7}g(R - r)\cos\theta]^{1/2}\vec{i}_\theta$$

Example 4.11

Obtain the kinetic energy of the precessing and rolling cone of Examples 4.2 and 4.3.

In Example 4.3, the expression obtained for the angular velocity of the body is

$$\vec{\Omega} = (\omega_o \sin\alpha - \omega_o \operatorname{cosec}\alpha)\vec{i} + \omega_o \cos\alpha\vec{k}$$

Employing the coordinate system of Fig. 4.3 with origin at the fixed point O, and noting that $\vec{v}_o = 0$, we get

$$T = \tfrac{1}{2}\{\Omega\}^T[I]_o\{\Omega\}$$

where in Example 4.3, $[I]_o$ is given by

$$[I]_o = \begin{bmatrix} \tfrac{3}{10}mR^2 & 0 & 0 \\ 0 & m(\tfrac{3}{20}R^2 + \tfrac{3}{5}h^2) & 0 \\ 0 & 0 & m(\tfrac{3}{20}R^2 + \tfrac{3}{5}h^2) \end{bmatrix}$$

Hence, we obtain

$$T = \tfrac{1}{2}(\tfrac{3}{10}mR^2)\omega_o^2(\sin\alpha - \operatorname{cosec}\alpha)^2 + \tfrac{1}{2}m(\tfrac{3}{20}R^2 + \tfrac{3}{5}h^2)\omega_o^2\cos^2\alpha$$

Alternatively, employing the $Cxyz$ coordinate system of Fig. 4.5 with origin at the center of mass, we get

$$T = \tfrac{1}{2}m\{v_c\}^T\{v_c\} + \tfrac{1}{2}\{\Omega\}^T[I]_c\{\Omega\}$$

where from Example 4.3, we have

$$\vec{v}_c = \tfrac{3}{4}h\cos\alpha\vec{j}$$

$$[I]_c = \begin{bmatrix} \tfrac{3}{10}mR^2 & 0 & 0 \\ 0 & m(\tfrac{3}{20}R^2 + \tfrac{3}{80}h^2) & 0 \\ 0 & 0 & m(\tfrac{3}{20}R^2 + \tfrac{3}{80}h^2) \end{bmatrix}$$

Hence, we obtain

$$T = \tfrac{1}{2}m\omega_o^2(\tfrac{3}{4}h)^2 \cos^2 \alpha + \tfrac{1}{2}(\tfrac{3}{10}mR^2)\omega_o^2(\sin \alpha - \operatorname{cosec} \alpha)^2$$

$$+ \tfrac{1}{2}m(\tfrac{3}{20}R^2 + \tfrac{3}{80}h^2)\omega_o^2 \cos^2 \alpha$$

It is verified that the expressions obtained for T by both methods are identical.

4.8 IMPULSE–MOMENTUM PRINCIPLE FOR A RIGID BODY

Integration of the force equation (4.39) with respect to time yields the theorem that impulse of a rigid body is equal to the change in momentum; that is,

$$\int_{t_1}^{t_2} \vec{F} \, dt = m[\vec{v}_c(t_2) - \vec{v}_c(t_1)] \tag{4.76}$$

Similarly, integration of the moment equation (4.40) with respect to time yields the theorem that angular impulse for a rigid body in general motion is equal to the change in angular momentum as

$$\int_{t_1}^{t_2} \vec{M}_c \, dt = \vec{H}_c(t_2) - \vec{H}_c(t_1) \tag{4.77}$$

Example 4.12

A cross of mass m is made of two uniform equal rods, each of length $2b$. It is suspended from a ball-and-socket joint at O [Fig. 4.19(a)]. It was at rest when hit by a force of constant magnitude F_o and time duration Δt in the positive z direction at the end A. Determine the angular velocity $\vec{\omega}$ of the cross immediately after impact.

The free-body diagram of the cross is shown in Fig. 4.19(b). Taking moments about the fixed point O, we get

$$\int \vec{M}_o \, dt = (b\vec{i} - b\vec{j}) \times (F_o \Delta t \vec{k}) = -bF_o \Delta t \vec{j} - bF_o \Delta t \vec{i}$$

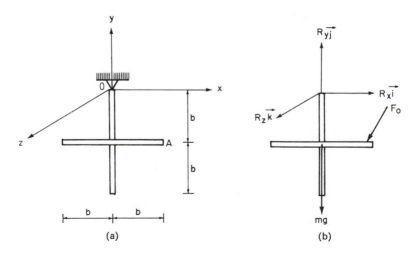

(a)   (b)

Figure 4.19   (a) Body subjected to impulse; (b) free-body diagram.

The coordinate system xyz consists of principal axes. In (4.75) we let $t_1 = 0$ and $t_2 = \Delta t$. Hence,

$$\vec{H}_o(t_2) = I_x \omega_x \vec{i} + I_y \omega_y \vec{j} + I_z \omega_z \vec{k}$$
$$= -bF_o \Delta t\, \vec{i} - bF_o \Delta t\, \vec{j}$$

where

$$I_x = \tfrac{1}{3}\left(\frac{m}{2}\right)(2b)^2 + \left(\frac{m}{2}\right)b^2 = \tfrac{7}{6}mb^2$$

$$I_y = \tfrac{1}{12}\left(\frac{m}{2}\right)(2b)^2 = \tfrac{1}{6}mb^2$$

Hence, immediately after impact, the angular velocity vector is obtained as

$$\begin{Bmatrix} \omega_x \\ \omega_y \\ \omega_z \end{Bmatrix} = \begin{Bmatrix} -\tfrac{6}{7}\dfrac{F_o \Delta t}{mb} \\ -6\dfrac{F_o \Delta t}{mb} \\ 0 \end{Bmatrix}$$

4.9 GYROSCOPE

The term *gyroscope* is applied to any rotating rigid body in which the orientation of its axis of rotation changes. The problem is three-dimensional and can be described by the general principle of angular impulse and momentum for a rigid body with respect to a fixed point.

We consider a rotor of the given diameter 3–3' located in the two gimbals as shown in Fig. 4.20. To define the position of the rotor, we select a fixed reference frame $OXYZ$, with the origin O located at the mass center of the rotor

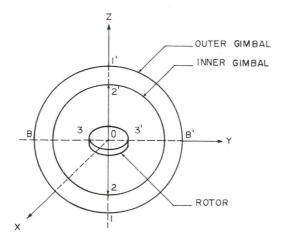

Figure 4.20   Gyroscope.

and the Z axis directed along the line defined by the bearings 1 and 1' of the outer gimbal.

The rotor may attain any arbitrary position by (1) a rotation of the outer gimbal through an angle ϕ about the axis 1–1', (2) a rotation of the inner gimbal through θ about BB', and (3) a rotation of the rotor through ψ about 2–2' as shown in Fig. 4.21. The derivatives $\dot{\phi}$, $\dot{\theta}$, and $\dot{\psi}$ refer to, respectively, the rate of *precession*, the rate of *nutation*, and the rate of *spin* of the gyroscope.

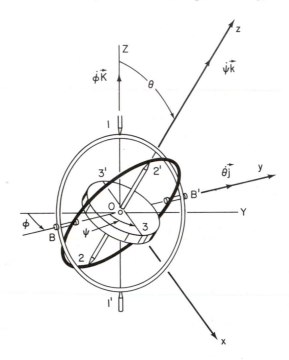

Figure 4.21   Euler angles of gyroscope.

We consider a rotating system of axes $oxyz$ attached to the inner gimbal with axis x along 3–3', axis y along BB', and axis z along 2–2'. We express the angular velocity $\vec{\Omega}$ of the gyroscope with respect to the fixed reference frame $OXYZ$. Thus,

$$\vec{\Omega} = \dot{\phi}\vec{K} + \dot{\theta}\vec{j} + \dot{\psi}\vec{k} \tag{4.78}$$

where $\vec{K}$ is the unit vector along the Z axis and $\vec{i}, \vec{j}$, and $\vec{k}$ refer to the unit vectors along the rotating axes, which are the principal axes for the gyroscope.

We resolve the unit vector $\vec{K}$ into components along the x and z axes:

$$\vec{K} = -\sin\theta\vec{i} + \cos\theta\vec{k} \tag{4.79}$$

Substituting (4.79) into (4.78), we obtain

$$\vec{\Omega} = -\dot{\phi}\sin\theta\vec{i} + \dot{\theta}\vec{j} + (\dot{\psi} + \dot{\phi}\cos\theta)\vec{k} \tag{4.80}$$

The angular momentum $\vec{H}_o$ is obtained by multiplying the components of $\vec{\Omega}$ by principal moments of inertia of the rotor. Let I_α be the moment of inertia of the rotor about its spin axis and I_t its moment of inertia about transverse axes through O. Since $\{H\}_o = [I]\{\Omega\}$, we obtain

$$\vec{H}_o = -I_t\dot{\phi}\sin\theta\,\vec{i} + I_t\dot{\theta}\,\vec{j} + I_\alpha(\dot{\psi} + \dot{\phi}\cos\theta)\vec{k} \tag{4.81}$$

Since the rotating axes are attached to the inner gimbal and do not spin, we express their angular velocity as

$$\vec{\omega} = \dot{\phi}\vec{K} + \dot{\theta}\vec{j} \tag{4.82}$$

Substituting from (4.79) in (4.82), we get

$$\vec{\omega} = -\dot{\phi}\sin\theta\,\vec{i} + \dot{\theta}\,\vec{j} + \dot{\phi}\cos\theta\,\vec{k} \tag{4.83}$$

The rate of change of the angular momentum is given by

$$\frac{d\vec{H}_o}{dt} = I_t\dot{\Omega}_1\,\vec{i} + I_t\dot{\Omega}_2\,\vec{j} + I_\alpha\dot{\Omega}_3\,\vec{k} + \vec{\omega} \times \vec{H}_o$$

Substituting for $\dot{\vec{\Omega}}$, $\vec{H}_o$, and $\vec{\omega}$ in this equation from (4.80), (4.81), and (4.83) and the resulting expression in (4.40), we obtain the three nonlinear differential equations of motion given by

$$M_x = -I_t(\ddot{\phi}\sin\theta + 2\dot{\theta}\dot{\phi}\cos\theta) + I_\alpha\dot{\theta}(\dot{\psi} + \dot{\phi}\cos\theta)$$

$$M_y = I_t(\ddot{\theta} - \dot{\phi}^2\sin\theta\cos\theta) + I_\alpha\dot{\phi}\sin\theta(\dot{\psi} + \dot{\phi}\cos\theta) \tag{4.84}$$

$$M_z = I_\alpha(\ddot{\psi} + \ddot{\phi}\cos\theta - \dot{\theta}\dot{\phi}\sin\theta)$$

We note that the modified Euler equations (4.51) for a body with an axis of symmetry are expressed in terms of angular velocities about orthogonal axis. Equations (4.84) for a body with axis of symmetry are expressed in terms of angular positions, but the rotations are not about three orthogonal axes. The angles ϕ, θ, and ψ are called Euler angles and they can be employed to describe the motion of a body with an axis of symmetry. Another method which is commonly employed for the selection of Euler angles, and which does not depend on the body having an axis of symmetry, is described in Chapter 5.

4.10 SYSTEM OF CONSTRAINED RIGID BODIES

So far we have studied the dynamics of a single rigid body. In some practical applications, we encounter a system of rigid bodies that are connected or coupled to one another in some manner. The connections or couplers eliminate some of the degrees of freedom that a rigid body would have otherwise and the equations of motion of the rigid bodies become coupled. Some examples of a system of connected rigid bodies include rail vehicles forming part of a train and articulated road vehicles such as tractor–semitrailer systems. In the following example, we give the derivation of equations of motion that can be employed to investigate the lateral stability or "jack-knifing" of tractor–semitrailer vehicles.

Example 4.13

Equations of Motion for the Lateral Stability of a Tractor–Semitrailer.

The sprung masses of the tractor and semitrailer are assumed to be rigid bodies. For the study of lateral stability, the bouncing, pitching, and rolling degrees of freedom of both the tractor and semitrailer are neglected. The pitch angles are usually very small and can be neglected, but the effect of roll on tire loading can be introduced at a later stage through semistatic load transfer.

The coordinate system xyz is fixed to the center of mass of the tractor and it translates and yaws with the tractor at its yaw angular velocity ω_1. A coordinate system $x_2 y_2 z_2$ is fixed to the center of mass of the semitrailer and it translates and yaws with the semitrailer at its yaw angular velocity ω_2. The axes z and z_2 are vertical and point downward. These coordinate systems are illustrated in Fig. 4.22. The relative yaw angle between the tractor and semitrailer is denoted by γ. The position of the center of mass of the semitrailer relative to the center of mass of the tractor is determined by the

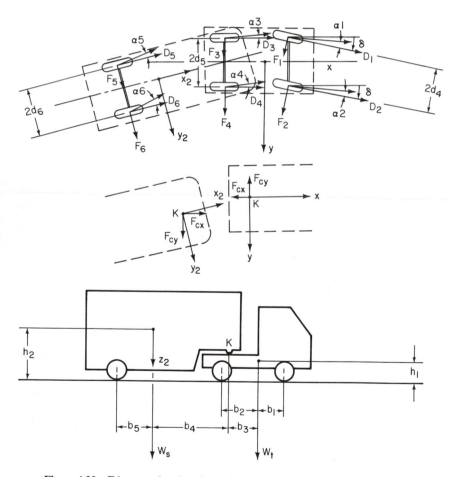

Figure 4.22 Diagram showing the main components of tractor–semitrailer.

fifth-wheel constraint. Hence, the translation of the semitrailer is expressed in terms of the xyz coordinate system.

It is assumed that the suspension is rigid enough that the forces at the wheel–road interface are directly transmitted to the sprung masses. Let F_i and D_i denote the side and driving forces, respectively, acting at the ith wheel. A braking force is obtained by changing the sign of D_i. The slip angle is defined as the angle between the velocity vector of a wheel and the vertical plane of the wheel. Let α_i denote the slip angle of the ith wheel as shown in Fig. 4.22. Tandem axles can be included by modifying the equations. It is assumed that the steering angle δ is the same for both front wheels. The components of the fifth-wheel constraint force in the x and y directions are denoted by F_{cx} and F_{cy}.

Equations of motion of the tractor.   With reference to Fig. 4.22, the translation of the tractor in the x and y directions and its yaw about the z axis are expressed by the following equations in terms of the rotating coordinate system xyz, which has angular velocity $\vec{\omega} = \omega_1 \vec{k}$:

$$m_t(\dot{v}_x - v_y\omega_1) = (D_1 + D_2)\cos\delta - (F_1 + F_2)\sin\delta + D_3 + D_4 - F_{cx} \quad (4.85)$$

$$m_t(\dot{v}_y + v_x\omega_1) = (D_1 + D_2)\sin\delta + (F_1 + F_2)\cos\delta + F_3 + F_4 - F_{cy} \quad (4.86)$$

$$I_t\dot{\omega}_1 = -(F_1 - F_2)d_4\sin\delta + (D_1 - D_2)d_4\cos\delta + (F_1 + F_2)b_1\cos\delta$$
$$+ (D_1 + D_2)b_1\sin\delta + (D_3 - D_4)d_5 - (F_3 + F_4)b_2 + F_{cy}b_3$$

$$(4.87)$$

Equations of motion of the semitrailer.   First, the acceleration of the center of mass of the semitrailer is obtained in terms of the rotating coordinate system xyz as follows. The acceleration of the semitrailer center of mass relative to the fifth wheel becomes

$$\vec{a}_{s/c} = \dot{\omega}_2\vec{k} \times (-b_4\cos\gamma\,\vec{i} + b_4\sin\gamma\,\vec{j})$$
$$+ \omega_2\vec{k} \times [\omega_2\vec{k} \times (-b_4\cos\gamma\,\vec{i} + b_4\sin\gamma\,\vec{j})]$$
$$= (b_4\omega_2^2\cos\gamma - b_4\dot{\omega}_2\sin\gamma)\vec{i} - (b_4\omega_2^2\sin\gamma + b_4\dot{\omega}_2\cos\gamma)\vec{j} \quad (4.88)$$

The acceleration of the fifth wheel is expressed by

$$\vec{a}_c = (\dot{v}_x - v_y\omega_1)\vec{i} + (\dot{v}_y + v_x\omega_1)\vec{j} + \dot{\omega}_1\vec{k} \times -b_3\vec{i}$$
$$+ \omega_1\vec{k} \times (\omega_1\vec{k} \times -b_3\vec{i})$$
$$= (\dot{v}_x - v_y\omega_1 + b_3\omega_1^2)\vec{i} + (\dot{v}_y + v_x\omega_1 - b_3\dot{\omega}_1)\vec{j} \quad (4.89)$$

The acceleration of the semitrailer center of mass is obtained by adding (4.88) and (4.89) as

$$\vec{a}_s = (\dot{v}_x - v_y\omega_1 + b_3\omega_1^2 + b_4\omega_2^2\cos\gamma - b_4\dot{\omega}_2\sin\gamma)\vec{i}$$
$$+ (\dot{v}_y + v_x\omega_1 - b_3\dot{\omega}_1 - b_4\omega_2^2\sin\gamma - b_4\dot{\omega}_2\cos\gamma)\vec{j} \quad (4.90)$$

The translation of the semitrailer in the x and y directions and its yaw about the vertical z_2 axis are now obtained as follows:

$$m_s(\dot{v}_x - v_y\omega_1 + b_3\omega_1^2 + b_4\omega_2^2\cos\gamma - b_4\dot{\omega}_2\sin\gamma)$$
$$= F_{cx} + (D_5 + D_6)\cos\gamma + (F_5 + F_6)\sin\gamma \quad (4.91)$$

$$m_s(\dot{v}_y + v_x\omega_1 - b_3\dot{\omega}_1 - b_4\omega_2^2 \sin\gamma - b_4\dot{\omega}_2 \cos\gamma)$$

$$= F_{cy} - (D_5 + D_6)\sin\gamma + (F_5 + F_6)\cos\gamma \tag{4.92}$$

$$I_s\dot{\omega}_2 = b_4F_{cx}\sin\gamma + b_4F_{cy}\cos\gamma - b_5(F_5 + F_6) + d_6(D_5 - D_6) \tag{4.93}$$

Equations of motion of the tractor-semitrailer.   Adding (4.85) and (4.91), and (4.86) and (4.92), respectively, the translation of the tractor–semitrailer in the x and y directions is expressed by

$$(m_t + m_s)(\dot{v}_x - v_y\omega_1) + m_s(b_3\omega_1^2 + b_4\omega_2^2 \cos\gamma - b_4\dot{\omega}_2 \sin\gamma)$$

$$= (D_1 + D_2)\cos\delta - (F_1 + F_2)\sin\delta + D_3 + D_4 + (D_5 + D_6)\cos\gamma$$

$$+ (F_5 + F_6)\sin\gamma \tag{4.94}$$

$$(m_t + m_s)(\dot{v}_y + v_x\omega_1) + m_s(-b_3\dot{\omega}_1 - b_4\omega_2^2 \sin\gamma - b_4\dot{\omega}_2 \cos\gamma)$$

$$= (D_1 + D_2)\sin\delta + (F_1 + F_2)\cos\delta + F_3 + F_4 - (D_5 + D_6)\sin\gamma$$

$$+ (F_5 + F_6)\cos\gamma \tag{4.95}$$

Substituting for F_{cy} in (4.87) from (4.86), the tractor yaw equation becomes

$$I_t\dot{\omega}_1 + m_tb_3(\dot{v}_y + v_x\omega_1) = (b_1 + b_3)(D_1 + D_2)\sin\delta$$

$$+ (b_1 + b_3)(F_1 + F_2)\cos\delta + (b_3 - b_2)(F_3 + F_4)$$

$$- d_4(F_1 - F_2)\sin\delta + d_4(D_1 - D_2)\cos\delta$$

$$+ d_5(D_3 - D_4) \tag{4.96}$$

Substituting for F_{cx} from (4.85) and for F_{cy} from (4.86) in (4.93), the equation for the semitrailer yaw is obtained as

$$I_s\dot{\omega}_2 + b_4m_t[(\dot{v}_x - v_y\omega_1)\sin\gamma + (\dot{v}_y + v_x\omega_1)\cos\gamma]$$

$$= d_6(D_5 - D_6) - b_5(F_5 + F_6)$$

$$+ b_4\sin\gamma[-(F_1 + F_2)\sin\delta + (D_1 + D_2)\cos\delta + D_3 + D_4]$$

$$+ b_4\cos\gamma[(F_1 + F_2)\cos\delta + (D_1 + D_2)\sin\delta + F_3 + F_4] \tag{4.97}$$

Hence, the translation of the tractor–semitrailer is represented by (4.94) and (4.95), the yaw of the tractor by (4.96), and the yaw of the semitrailer by (4.97).

To complete the formulation, a mathematical model of the pneumatic tire should be employed to obtain expressions for the driving and side forces. For further details, the reader may consult reference [7].

4.11 SUMMARY

The major objective of this chapter has been the derivation of the equations of motion of a rigid body by direct application of Newton's second law. It is seen that the moment of inertia matrix becomes time invariant when a body coordinate system is employed (i.e., the angular velocity of the coordinate system is the angular velocity of the body). However, when a body has an axis of symmetry, it is possible to choose the angular velocity of the coordinate system that is different from the angular velocity of the body and still have the inertia matrix time invariant.

In the study of dynamics of a rigid body, the best choice for the origin of the coordinate system is either the center of mass of the body or a point in body that is fixed in inertial space, in case such a point does exist. In this manner, the rotational equations of motion are uncoupled from the translational equations and may be studied separately, if so desired. The translational equations of motion, however, remain coupled to the rotational equations through the angular velocities of the body. The rotational equations of motion may be further simplified by selecting the principal axes for the coordinate system as done in the Euler's equations of motion.

The latter part of this chapter has been concerned with the application of work–energy and impluse–momentum principles to the dynamics of rigid bodies. By employing these principles, answers can be obtained to some simple problems without formulating and solving the equations of motion. Finally, gyroscopic motion has been discussed. The Lagrangian method of derivation of equations of motion for rigid bodies by employing Euler angles will be studied in the next chapter.

PROBLEMS

4.1. A disk rolls without slipping on a horizontal surface with variable angular speed ω. The point P is fixed to the disk as shown in Fig. P4.1. Determine the velocity and acceleration of P.

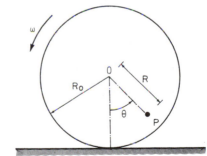

Figure P4.1

4.2. The inertia matrix of an airplane with respect to the xyz coordinate system at its mass center as shown in Fig. P4.2 is given in the following. Locate the principal axes and the principal moments of inertia. Note that the x axis is longitudinal and the y axis is lateral.

$$[I] = \begin{bmatrix} 120{,}000 & 0 & 20{,}000 \\ 0 & 150{,}000 & 0 \\ 20{,}000 & 0 & 250{,}000 \end{bmatrix}$$

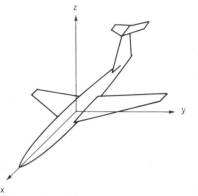

Figure P4.2

4.3. One end of a slender uniform rod of mass m and length L is welded to a shaft rotating at a constant angular speed as shown in Fig. P4.3. Determine the moment exerted by the rod on the shaft in terms of m, L, β, and ω_o. Rod radius is a.

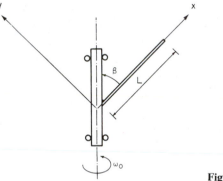

Figure P4.3

4.4. A disk of mass m and radius R is welded to the shaft of a motor which is fixed to a turntable. Initially, the turntable is rotating at the rate $\omega_1 = 40$ rad/s and the motor is rotating at $\omega_2 = 100$ rad/s, as shown in Fig. P4.4. If the turntable is now accelerated at a constant rate of 10 rad/s², what force and moment will the motor shaft exert on the disk?

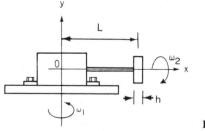

Figure P4.4

4.5. The assembly shown in Fig. P4.5 is rotating about the vertical axis at a constant speed ω_o. The slender bar of mass m is supported by a pin at a point O. Derive the equation relating the constant angle θ to m, L, ω_o, and g. Employ the principal axes coordinate system xyz with origin at center of mass C.

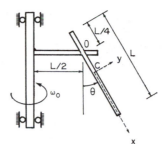

Figure P4.5

4.6. Solve Problem 4.5 but employ the principal axis coordinate system *xyz* with origin at the moving point *O*, as shown in Fig. P4.6.

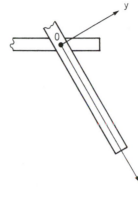

Figure P4.6

4.7. A ($a \times a$) square plate is pinned at one corner and released from the position shown in Fig. P4.7. Use the principle of conservation of energy to obtain the differential equation of motion.

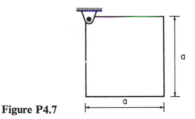

Figure P4.7

4.8. A solid homogeneous cylinder of radius r_o rolls without slipping on a cylindrical surface of radius R (Fig. P4.8). If the cylinder starts from rest at $\theta = 0$, determine the angle θ_m where it will lose contact with the cylindrical surface.

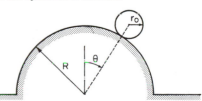

Figure P4.8

4.9. A shaft is rotating about the vertical axis at angular speed ω_o and angular acceleration $\dot{\omega}_o = \alpha_o$ (Fig. P4.9). Two bars of square cross section ($a_1 \times a_1$) and ($a_2 \times a_2$), respectively, are pin-jointed at A and B. Derive the equations of motion of both bars.

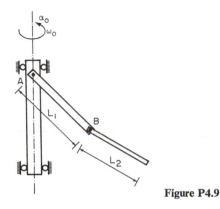

Figure P4.9

REFERENCES

1. Meirovitch, L., *Methods of Analytical Dynamics*, McGraw-Hill Book Company, New York, 1970.

2. Kane, T. R., *Dynamics*, Holt, Rinehart and Winston, New York, 1968.

3. Crandall, S. H., Karnopp, D. C., Kurtz, E. F., and Pridmore-Brown, D. C., *Dynamics of Mechanical and Electromechanical Systems*, McGraw-Hill Book Company, New York, 1968.

4. Halfman, R. L., *Dynamics*, Vol. 1, Addison-Wesley Publishing Company, Inc., Reading, Mass., 1962.

5. McCuskey, S. W., *An Introduction to Advanced Dynamics*, Addison-Wesley Publishing Company, Inc., Reading, Mass., 1962.

6. Beer, F. P., and Johnston, E. R., *Vector Mechanics for Engineers, Dynamics*, 3rd ed., McGraw-Hill Book Company, New York, 1977.

7. D'Souza, A. F., and Eshleman, R. L., "Maneuverability Limits and Handling Criterion of Articulated Vehicles," *Computational Methods in Ground Transportation Vehicles*, ASME, AMD-Vol. 50, Nov. 1982, pp. 117–132.

5

LAGRANGIAN DYNAMICS

5.1 INTRODUCTION

In the previous chapters, the derivation of the equations of motion has been based on direct application of Newton's laws. This chapter deals with the formulation of the equations of motion by employing variational methods. The variational techniques provide an elegant formulation by employing principles containing physical quantities whose definition does not depend on the use of a particular coordinate system; that is, the variational form is invariant under coordinate transformation. The principles of variational dynamics, including Hamilton's principle and Lagrange's equations, are analogous to similar physical principles in other areas of engineering, such as the principle of minimum strain energy and Castigliano's theorems in elasticity.

There are several advantages in employing variational methods in dynamics. These are as follows:

1. The system of particles and rigid bodies is considered as a whole rather than being separated into its individual components.
2. Problems are formulated in terms of kinetic energy and work, both of which are scalar quantities.
3. Forces of constraint that do not perform work are not included.
4. Use of generalized coordinates, instead of physical coordinates, affords ease and makes the formulation versatile.

It should be noted that in some cases, the values of the constraint forces are required for the purpose of stress analysis and design. In such cases, it becomes necessary to employ additional formulation, such as Lagrange multipliers, to include constraint forces or application of Newton's law.

In this chapter, we first discuss generalized coordinates, degrees of freedom, and constraints. Next, the principle of virtual work and Hamilton's principle are considered. Then the derivation of Lagrange's equations of motion and Hamilton's equations is described. In the earlier parts of the chapter, the applications are restricted to a system of particles only. In a later part of the chapter, we define Euler angles. These are then included among the generalized coordinates for the study of dynamics of rigid bodies.

5.2 GENERALIZED COORDINATES, DEGREES OF FREEDOM, AND CONSTRAINTS

The position of a system of particles is called its *configuration*. A set of coordinates is called *complete* if their values corresponding to an arbitrary admissible configuration of the system are sufficient to locate all parts of the system. A set of coordinates is called *independent* when all but one of the coordinates are fixed, there still remains a range of values for that one coordinate which corresponds to a range of admissible configuration. If n number of coordinates form a complete and independent set, the degrees of freedom of the system is said to be n.

In a dynamic system, kinematic constraints often arise due to the relationships among displacements. A single unconstrained particle has three degrees of translational freedom. In a system of N particles, if there are R constraints, the degrees of freedom of the system is given by $n = 3N - R$.

The choice of generalized coordinates is not unique. Generalized coordinates may include physical coordinates but they may also include angles, functions of physical coordinates, and other variables which have no association with physical coordinates. However, the number of generalized coordinates is equal to the degrees of freedom. Hence, when the degrees of freedom is n, generalized coordinates $q_1, \ldots, q_n$ form a complete and independent set.

Example 5.1

To illustrate dynamic system with constraints, we consider a rigid body connected to a fixed point by a massless spring. In three-dimensional space, the configuration of the body would be described by six coordinates: three translations and three rotations. In this case, the degrees of freedom for the system are six. Let us suppose that the system is constrained and it undergoes motion in the xy plane only, as shown in Fig. 5.1. The rigid body in plane motion configuration would require three coordinates. These coordinates may be r, θ, ϕ, or x, y, ϕ. The degrees of freedom for the system are reduced to three. Other choices for coordinates are possible. However, the number of coordinates will always be three. Now if the flexible spring is replaced by a rigid bar of

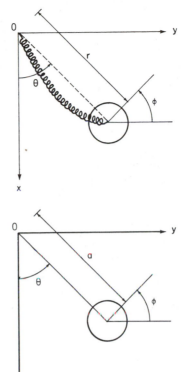

Figure 5.1 Rigid-body motion.

Figure 5.2 Rigid-body motion.

fixed length a as shown in Fig. 5.2, two additional constraints are introduced. These are

$$r = a, \qquad \phi = \theta \tag{5.1}$$

The number of coordinates required to describe the system is reduced to one.

5.2.1 Constraints

Sometimes it is not possible to eliminate the excess coordinates by employing the constraint equations. In that case, the number of coordinates employed is greater than the number of degrees of freedom. Suppose that we choose M coordinates $x_1, x_2, \ldots, x_M$ to represent the configuration of a system. These M coordinates are not independent but are related by R constraints. A general form of constraint is expressed in the form of differentials called Pfaffian. In the form of Pfaffians, let the R constrained be given by

$$a_{j_0} \, dt + \sum_{k=1}^{M} a_{jk} \, dx_k = 0, \qquad j = 1, \ldots, R \tag{5.2}$$

where the coefficients a_{jk} for $k = 0, 1, \ldots, M$ are known and differentiable functions of $x_1, \ldots, x_M, t$. It is assumed that the R constraints are linearly independent; that is, the rank of the matrix $R \times (M + 1)$ is R. Depending on these constraints, the dynamic system is classified as follows:

1. *Catastatic or acatastatic.* If all coefficients a_{j_o} for $j = 1, \ldots, R$ are zero, the system is called catastatic. Otherwise, if at least one of the coefficients a_{j_o} is not zero, it is called acatastatic.

2. *Holonomic or nonholonomic.* If all the Pfaffians of (5.2) are integrable and hence reducible to perfect differentials $df_j(x_1, \ldots, x_M, t) = 0$ for $j = 1, \ldots, R$, the system is called holonomic. Otherwise, if at least one of the Pfaffians is not integrable, the system is called nonholonomic.

3. *Scleronomic or rheonomic.* If the system is holonomic and in addition time t does not appear explicitly in all the integrated forms $f_j(x_1, \ldots, x_M)$, the system is called scleronomic. Otherwise, if the system is holonomic and time t appears explicitly in at least one of the functions $f_j(x_1, \ldots, x_M, t)$, the system is called rheonomic.

The M coordinates $x_1, \ldots, x_M$ chosen here are not independent since they are related by R constraints (5.2). The degree of freedom is $n = M - R$. In a nonholonomic system, the excess coordinates cannot be eliminated by employing the constraints since all the Pfaffians are not integrable. In this case, it becomes necessary to employ the number of coordinates that exceeds the degree of freedom, but the number of excess coordinates must equal the number of constraints that are retained.

Example 5.2

A bead is free to slide along a rod which rotates in the xy plane with a constant angular velocity ω_o about the z axis, as shown in Fig. 5.3.

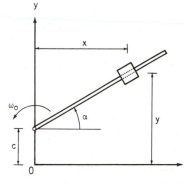

Figure 5.3 Bead sliding on a rotating rod.

It can be seen that if two coordinates x and y are employed to determine the position of the bead, they are related by the angle α that the rod makes with the x axis, so that

$$\tan \alpha = \frac{y - c}{x}$$

Since $\alpha = \omega_o t$, this constraint can be expressed as

$$(\tan \omega_o t)x - y + c = 0 \tag{5.3}$$

This constraint is already in the integrated form $f(x, y, t) = 0$. The dynamic system is then holonomic and rheonomic. The two coordinates x and y are related by one constraint (5.3), and the degree of freedom $n = 1$. Now, we can eliminate y by using (5.3) and use x as the single generalized coordinate, or vice versa. It is of interest to obtain the Pfaffian

$$a_o \, dt + a_1 \, dx + a_2 \, dy = 0 \qquad (5.4)$$

which when integrated out yields (5.3). The Pfaffian is obtained easily by noting that

$$df = \frac{\partial f}{\partial t} \, dt + \frac{\partial f}{\partial x} \, dx + \frac{\partial f}{\partial y} \, dy = 0$$

or

$$[(\sec^2 \omega_o t) \omega_o x] \, dt + (\tan \omega_o t) \, dx - dy = 0 \qquad (5.5)$$

From (5.5) we note that since the coefficient of dt is not zero, the system is acatastatic. In order to be integrable to a perfect differential, the Pfaffian (5.5) has to satisfy the integrability requirements that

$$\frac{\partial}{\partial x}\left(\frac{\partial f}{\partial t}\right) = \frac{\partial}{\partial t}\left(\frac{\partial f}{\partial x}\right)$$

$$\frac{\partial}{\partial y}\left(\frac{\partial f}{\partial t}\right) = \frac{\partial}{\partial t}\left(\frac{\partial f}{\partial y}\right) \qquad (5.6)$$

$$\frac{\partial}{\partial x}\left(\frac{\partial f}{\partial y}\right) = \frac{\partial}{\partial y}\left(\frac{\partial f}{\partial x}\right)$$

It can be easily verified that these requirements are indeed satisfied by (5.5) and hence it can be integrated to the form (5.3).

Example 5.3

We consider the two-dimensional motion of a boat in a plane. The roll, pitch, and heave (up and down) motions of the boat are neglected. As shown in Fig. 5.4, we choose x and y to represent the position of its mass center and the yaw angle ψ to represent its orientation with respect to the x axis. The constraint is that any translation of the center of mass of the boat must be in the direction of its heading. This constraint can be expressed by the equation $\tan \psi = dy/dx$. The Pfaffian is therefore given by

$$(\tan \psi) \, dx - dy = 0 \qquad (5.7)$$

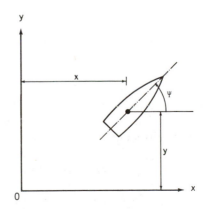

Figure 5.4   Motion of a boat in the plane.

On comparing (5.7) with

$$df = a_1\,dx + a_2\,dy + a_3\,d\psi$$

where $a_1 = \tan\psi$, $a_2 = -1$, and $a_3 = 0$, it can be verified that (5.7) does not satisfy the integrability requirements and hence it cannot be integrated to the form $f(x, y, \psi) = 0$. The system is catastatic and nonholonomic. The boat has only two degrees of freedom since $\psi = \tan^{-1}(\dot{y}/\dot{x})$, but it becomes necessary to employ one excess coordinate along with the constraint (5.7).

Example 5.4

We consider a particle falling from the top of a spherical radome of radius c as shown in Fig. 5.5. The motion is in the plane. Choosing x and y as its position coordinates, they are related by the inequality constraint $x^2 + y^2 - c^2 \geq 0$. This constraint is expressed in the integrated form. However, because of the inequality, the system would be considered as nonholonomic. The system can be made holonomic by describing its motion separately in the two regions. In the first region, where the particle remains on the surface of the sphere, the constraint is an equality and there is one degree of freedom. A single coordinate, which may be x or y or the angle θ, may be chosen as the generalized coordinate to describe the motion in this region. In the second region, where the particle is no longer on the surface of the sphere, it has two degrees of freedom and x and y may be chosen as generalized coordinates.

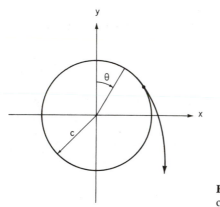

Figure 5.5   Particle falling from the top of a sphere.

5.3 PRINCIPLE OF VIRTUAL WORK

The concept of virtual work is a very useful tool in the field of classical mechanics. The principle of virtual work was first stated by Bernoulli for a system in static equilibrium. This principle has been extended to dynamics by employing d'Alembert's principle. First, we consider the concept of virtual displacements. Virtual displacements, which may not be true displacements, are infinitesimal changes in coordinates, consistent with the constraints without any change in time. We consider the configuration of the system at a certain time t and by freezing the time at that value give infinitesimal changes to the coordinates without violating the constraints. Let M coordinates $x_1, x_2, \ldots, x_M$ be chosen to represent the configuration of a system. Also, let there be R constraints in the form of Pfaffians of (5.2):

$$a_{j_o} dt + \sum_{k=1}^{M} a_{jk} dx_k = 0, \qquad j = 1, \ldots, R \qquad (5.2)$$

True displacements dx_k have to satisfy (5.2). On the other hand, since the time is frozen and $dt = 0$, the virtual displacements δx_k satisfy

$$\sum_{k=1}^{M} a_{jk} \delta x_k = 0, \qquad j = 1, \ldots, R \qquad (5.8)$$

The virtual displacements are denoted by δx_k in order to distinguish them from true displacements dx_k. It should be noted that the virtual displacements cannot violate the constraints (5.8). In case the system is catastatic [i.e., all coefficients a_{j_o} in (5.2) are zero], no distinction need be made between true displacements and virtual displacements.

Example 5.5

We consider the bead that is free to slide along a rotating rod of Example 5.2. At a certain time t, let the configuration be as shown in Fig. 5.3. Freezing the time to this value, we give small virtual displacements δx and δy to the bead along the rod as shown in Fig. 5.6(a). From (5.5), we see that the virtual displacements of Fig. 5.6(a) satisfy the constraint

$$(\tan \omega_o t) \, \delta x - \delta y = 0 \qquad (5.9)$$

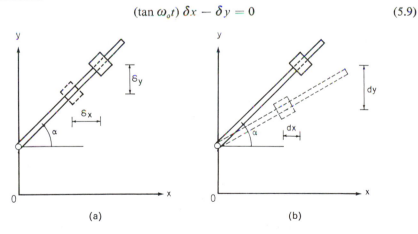

Figure 5.6   (a) Virtual displacements; (b) true displacements.

Since the system is acatastatic, the true displacements satisfy the constraint given by (5.5). Figure 5.6(b) shows the true displacements dx and dy. Here, since the time is not frozen, the angle α of the rod increases by $\omega_o \, \Delta t$ in time increment Δt and the bead has true displacements dx and dy along the rod.

5.3.1 Principle of Virtual Work in Statics

This principle was stated by Bernoulli for a system in static equilibrium. First we consider a single particle whose position is $\vec{r}$ and which is subjected to a resultant force $\vec{F}$. If the particle is given a virtual displacement $\delta\vec{r}$, after noting that $\vec{F} = 0$ for static equilibrium, the virtual work is

$$\delta\bar{W} = \vec{F} \cdot \delta\vec{r} = 0 \tag{5.10}$$

Let the resultant force vector $\vec{F}$ be decomposed into an impressed force vector $\vec{F}*$ and a constraint force $\vec{R}$ so that $\vec{F} = \vec{F}* + \vec{R}$. Substituting for the force in (5.10), we obtain

$$\delta\bar{W} = \vec{F}* \cdot \delta\vec{r} + \vec{R} \cdot \delta\vec{r} = 0 \tag{5.10a}$$

Since virtual displacements do not violate the constraints, the work done by the constraint forces in virtual displacement is zero (i.e., $\vec{R} \cdot \delta\vec{r} = 0$). Hence, from (5.10a), we obtain

$$\delta\bar{W} = \vec{F}* \cdot \delta\vec{r} = 0 \tag{5.11}$$

In Cartesian coordinates, this expression may be written as

$$\delta\bar{W} = F_x^* \, \delta x + F_y^* \, \delta y + F_z^* \, \delta z = 0 \tag{5.12}$$

If the particle is not constrained, then δx, δy, and δz are completely arbitrary. It also follows that $\vec{R} = 0$ and $\vec{F} = \vec{F}*$. We can then choose $\delta y = \delta z = 0$ and $\delta x \neq 0$ but arbitrary. It follows that $F_x = 0$. Employing a similar argument, we get

$$F_x = F_y = F_z = 0 \tag{5.13}$$

When the motion of the particle is constrained, (5.12) still is valid but since the displacements are no longer arbitrary, we cannot conclude that F_x^*, F_y^*, and F_z^* are each zero. For a system of N particles in static equilibrium, the equations corresponding to (5.10) and (5.11), respectively, become

$$\delta\bar{W} = \sum_{i=1}^{N} \vec{F}_i \cdot \delta\vec{r}_i = 0 \tag{5.14}$$

and

$$\delta\bar{W} = \sum_{i=1}^{N} \vec{F}_i^* \cdot \delta\vec{r}_i = 0 \tag{5.15}$$

Let M coordinates $x_1, \ldots, x_M$ subject to R number of constraints be chosen to represent the configuration of a system of N particles in static equilibrium. Let $F_1^*, F_2^*, \ldots, F_M^*$ be the components of the impressed forces along the corresponding coordinates. The expression (5.15) for the virtual work then becomes

$$\delta\bar{W} = \sum_{i=1}^{M} F_i^* \, \delta x_i = 0 \tag{5.16}$$

Since the coordinates are constrained, δx_i are not completely arbitrary and we cannot conclude that each F_i^* is individually equal to zero.

5.3.2 Extension of the Principle of Virtual Work
to Dynamics

The principle of virtual work can be extended to dynamics by employing d'Alembert's principle. Considering the ith particle from a system of N particles and using Newton's second law, we get

$$\vec{F}_i^* + \vec{R}_i - \frac{d}{dt}(m_i \dot{\vec{r}}_i) = 0 \qquad (5.17)$$

The equation for virtual work in dynamics, analogous to (5.15) of statics, then becomes

$$\delta \bar{W} = \sum_{i=1}^{N} \left[\vec{F}_i^* - \frac{d}{dt}(m_i \dot{\vec{r}}_i) \right] \cdot \delta \vec{r}_i = 0 \qquad (5.18)$$

In (5.17) and (5.18), it has been assumed that the position vector $\vec{r}$ is with reference to an inertial system of coordinates; otherwise, it is necessary to express the acceleration by employing (2.65). The quantity $-d/dt(m_i \dot{\vec{r}}_i)$ is referred to as inertia force and $\vec{F}_i^* - d/dt(m_i \dot{\vec{r}}_i)$ as the effective impressed force. In scleronomic systems, we can choose $\delta \vec{r}_i = d\vec{r}_i$ and the principle of virtual work expressed by (5.18) reduces to the work–energy principle, and in conservative systems it leads to the principle of conservation of mechanical energy. The principle of virtual work will be employed in the next section to prove Hamilton's principle. As illustrated by the following example, it can also be employed in its own right to obtain simple answers to simple problems without formulating the equations of motion.

Example 5.6
A bead of mass m is free to slide in the gravity field on a circular hoop of radius c rotating about a vertical axis at a constant angular velocity ω_o as shown in Fig. 5.7. Determine all positions θ at which the bead is in equilibrium.

This question can be answered after formulating the equations of motion as shown later in this chapter. Here, it is resolved by employing the principle of virtual work expressed by (5.18). Axes xyz form noninertial coordinate system with angular velocity $\vec{\omega} = \omega_o \vec{j}$. Its origin O is fixed in space. We choose two coordinates x and y to represent the position of the bead on the rotating hoop. The two coordinates are related by one holonomic constraint

$$x^2 + y^2 = c^2 \qquad (5.19)$$

The position $\vec{r}$ of the bead is denoted by

$$\vec{r} = x\vec{i} + y\vec{j} \qquad (5.20)$$

The first objective is to determine the acceleration of the bead and then the inertia force. From (2.65) the acceleration of the bead is expressed by

$$\vec{a} = \vec{a}_1 + \ddot{\vec{r}} + 2\vec{\omega} \times \dot{\vec{r}} + \dot{\vec{\omega}} \times \vec{r} + \vec{\omega} \times (\vec{\omega} \times \vec{r})$$

where the acceleration of the origin $\vec{a}_1 = 0$. Also, $\ddot{\vec{r}} = \dot{\vec{r}} = \dot{\vec{\omega}} = 0$. Hence, $\vec{a} = \vec{\omega} \times (\vec{\omega} \times \vec{r}) = -\omega_o^2 x\vec{i}$. The inertia force becomes $m\omega_o^2 x\vec{i}$. The only impressed force

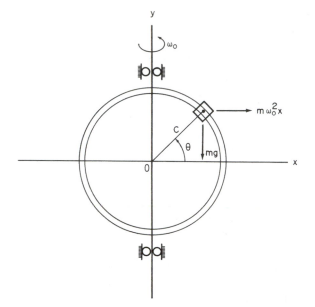

Figure 5.7   Equilibrum position of bead on rotating hoop.

is due to gravity and it is $-mg\vec{j}$, as shown in Fig. 5.7. The constraint force, which does no work in a virtual displacement, is the reaction force between the bead and the hoop. Now consider virtual displacements δx and δy. The virtual work of (5.18) becomes

$$\delta \bar{W} = (m\omega_o^2 x)\, \delta x - mg\, \delta y = 0 \qquad (5.21)$$

Since the two coordinates are related by one constraint given by (5.19), it is not possible to set $\delta x = 0$ and $\delta y \neq 0$ but arbitrary and then conclude that the individual coefficients of δx and δy are each zero. The bead has only one degree of freedom and we choose θ as the generalized coordinate. Now, $x = c\cos\theta$, $y = c\sin\theta$, $\delta x = -c\sin\theta\, \delta\theta$, and $\delta y = c\cos\theta\, \delta\theta$. Substituting this result in (5.21), we obtain

$$\delta \bar{W} = -[m\omega_o^2 c^2 \cos\theta \sin\theta + mgc\cos\theta]\, \delta\theta = 0$$

Now $\delta\theta$ is arbitrary and setting its coefficient to zero in the foregoing equation, it follows that

$$\cos\theta[m\omega_o^2 c^2 \sin\theta + mgc] = 0 \qquad (5.22)$$

The solutions of (5.22) are given by

$$\cos\theta = 0; \quad \text{that is, } \theta = n\frac{\pi}{2}, \quad n = 1, 3, 5, \ldots$$

$$\sin\theta = -\frac{g}{\omega_o^2 c}; \quad \text{that is, } \theta_1 = -\sin^{-1}\frac{g}{\omega_o^2 c} + 2n\pi, \quad n = 0, 1, 2, \ldots$$

or $\theta_2 = \theta_1 - \pi/2$.

It should be realized that some of these equilibriums may be unstable. The investigation of stability is discussed in Chapter 9.

5.4 HAMILTON'S PRINCIPLE

Hamilton's principle is one of the best known variational principles of mechanics. It is an integral principle and considers the configuration of a system between the time interval (t_0, t_1). Using Hamilton's principle, the problems of dynamics are reduced to the evaluation of a scalar definite integral. The formulation has an advantage as it does not depend on the coordinate system used to express the integrand.

We consider a system of N particles. Using d'Alembert's principle and the principle of virtual work, it is seen from (5.18) that

$$\sum_{i=1}^{N} \left[\vec{F}_i^* - \frac{d}{dt}(m_i \vec{r}_i) \right] \cdot \delta \vec{r}_i = 0 \tag{5.23}$$

In the foregoing equation, $\vec{F}_i^*$ is the impressed force acting on the ith particle. The constraint forces are not included since the virtual displacements $\delta \vec{r}_i$ are compatible with the system constraints and the virtual work done by the constraint forces is zero. The second term in (5.23) may be written as

$$\sum_{i=1}^{N} \frac{d}{dt}(m_i \dot{\vec{r}}_i) \cdot \delta \vec{r}_i = \sum_{i=1}^{N} \frac{d}{dt}(m_i \dot{\vec{r}}_i \cdot \delta \vec{r}_i) - \sum_{i=1}^{N} m_i \dot{\vec{r}}_i \cdot \delta \dot{\vec{r}}_i \tag{5.24}$$

In the second term on the right-hand side of the foregoing equation, we have written $d/dt(\delta \vec{r}_i)$ as $\delta \dot{\vec{r}}_i$ by interchanging the operations d/dt and δ. This term can be transformed using the kinetic energy T of the system. We have

$$T = \tfrac{1}{2} \sum_{i=1}^{N} m_i \dot{\vec{r}}_i \cdot \dot{\vec{r}}_i$$

The variation of T can be written as

$$\delta T = \sum_{i=1}^{N} m_i \dot{\vec{r}}_i \cdot \delta \dot{\vec{r}}_i \tag{5.25}$$

Hence, (5.24) becomes

$$\sum_{i=1}^{N} \frac{d}{dt}(m_i \dot{\vec{r}}_i) \cdot \delta \vec{r}_i = \sum_{i=1}^{N} \frac{d}{dt}(m_i \dot{\vec{r}}_i \cdot \delta \vec{r}_i) - \delta T \tag{5.26}$$

Denoting the work done by the impressed forces as

$$\delta \bar{W}^* = \sum_{i=1}^{N} \vec{F}_i^* \cdot \delta \vec{r}_i \tag{5.27}$$

and employing (5.26) and (5.27) in (5.23), we obtain

$$\delta \bar{W}^* + \delta T = \sum_{i=1}^{N} \frac{d}{dt}(m_i \dot{\vec{r}}_i \cdot \delta \vec{r}_i) \tag{5.28}$$

On integrating (5.28) with respect to time over the interval from t_0 to t_1, it follows that

$$\int_{t_0}^{t_1} (\delta \bar{W}^* + \delta T) \, dt = \left[\sum_{i=1}^{N} m_i \dot{\vec{r}}_i \cdot \delta \vec{r}_i \right]_{t_0}^{t_1} \tag{5.29}$$

The system configuration changes with time, tracing a *true path*. A slightly different path known as *varied path* is obtained on giving virtual displacements $\delta \vec{r}_i$ without involving change in time (i.e., $\delta t = 0$). The varied path, however, coincides with the true path at the two end points t_0 and t_1 as shown in Fig. 5.8. Under these conditions, it follows that

$$\delta \vec{r}_i(t_0) = \delta \vec{r}_i(t_1) = \vec{0}$$

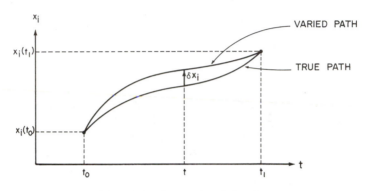

Figure 5.8   True and varied paths.

Now, (5.29) may be stated as

$$\int_{t_0}^{t_1} (\delta \bar{W}^* + \delta T)\, dt = 0 \tag{5.30}$$

Equation (5.30) represents Hamilton's principle in its most general form. It states that the true path followed by the dynamic system to go from $\vec{r}(t_0)$ at time t_0 to $\vec{r}(t_1)$ at time t_1 is such that the time integral of the sum of the virtual kinetic energy change and virtual work vanishes when subjected to virtual displacements from the true path. This general principle is applicable to non-holonomic and nonconservative systems. In case all the impressed forces are conservative, the virtual work of (5.27) is related to the change in potential energy U by $\delta \bar{W} = -\delta U$. Defining a scalar function L called the Lagrangian as

$$L = T - U \tag{5.31}$$

a special case of (5.30) can be expressed as

$$\int_{t_0}^{t_1} \delta L\, dt = 0 \tag{5.32}$$

Furthermore, if the system is holonomic, then (5.32) becomes

$$\delta I = \delta \int_{t_0}^{t_1} L\, dt = 0 \tag{5.33}$$

Equation (5.33) states that the true path followed by a conservative, holonomic system to go from $\vec{r}(t_0)$ at time t_0 to $\vec{r}(t_1)$ at time t_1 is such that the time integral

$$I = \int_{t_0}^{t_1} L\, dt \tag{5.34}$$

is extremized. Of course, it should be noted that the most general form of Hamilton's principle is expressed by (5.30). This principle will be employed in the next section to obtain the Lagrange equations of motion.

5.5 LAGRANGE EQUATIONS OF MOTION

We now derive Lagrange equations of motion using the following two approaches: (1) the application of d'Alembert's principle, and (2) application of Hamilton's principle. First, we treat only holonomic systems and later generalize the results to nonholonomic systems.

5.5.1 Application of d'Alembert's Principle to Holonomic Systems

We consider a dynamic system of N particles. Using d'Alembert's principle and the principle of virtual work, it is seen from (5.23) that

$$\sum_{i=1}^{N} \left[\vec{F}_i^* - \frac{d}{dt}(m_i \vec{r}_i) \right] \cdot \delta \vec{r}_i = 0 \tag{5.35}$$

where $\vec{F}_i^*$ is the impressed force on ith particle of mass m_i. Let the system have n degrees of freedom. Choosing $q_1, \ldots, q_n$ as n generalized coordinates for this holonomic system, we have the transformation equation

$$\vec{r}_i = \vec{r}_i(q_1, \ldots, q_n, t) \tag{5.36}$$

between the vector coordinates of the particles and the n generalized coordinates. Rigid bodies will be considered later by including angular coordinates among the generalized coordinates. The velocities of the particles are then

$$\dot{\vec{r}}_i = \sum_{k=1}^{n} \frac{\partial r_i}{\partial q_k} \frac{dq_k}{dt} + \frac{\partial \vec{r}_i}{\partial t} \tag{5.37}$$

and since the variation in time δt is not considered, the virtual displacements are

$$\delta \vec{r}_i = \sum_{k=1}^{n} \frac{\partial \vec{r}_i}{\partial q_k} \delta q_k$$

Considering the first term in (5.35), the virtual work of the impressed forces $\vec{F}_i^*$, both externally applied and internal, becomes

$$\sum_{i=1}^{N} \vec{F}_i^* \cdot \delta \vec{r}_i = \sum_{i=1}^{N} \sum_{k=1}^{n} \vec{F}_i^* \cdot \frac{\partial \vec{r}_i}{\partial q_k} \delta q_k = \sum_{k=1}^{n} Q_k \, \delta q_k \tag{5.38}$$

where

$$Q_k = \sum_{i=1}^{N} \vec{F}_i^* \cdot \frac{\partial \vec{r}_i}{\partial q_k} \tag{5.39}$$

is called the generalized force in the direction of the kth generalized coordinate. The second term in (5.35) involving the accelerations becomes

$$\sum_{i=1}^{N} m_i \ddot{\vec{r}}_i \cdot \delta \vec{r}_i = \sum_{i=1}^{N} \sum_{k=1}^{n} m_i \ddot{\vec{r}}_i \cdot \frac{\partial \vec{r}_i}{\partial q_k} \delta q_k \tag{5.40}$$

where we can then write

$$\sum_{i=1}^{N} m_i \ddot{\vec{r}}_i \cdot \frac{\partial \vec{r}_i}{\partial q_k} = \sum_{i=1}^{N} \left\{ \frac{d}{dt}\left(m_i \dot{\vec{r}}_i \cdot \frac{\partial \vec{r}_i}{\partial q_k}\right) - m_i \dot{\vec{r}}_i \cdot \frac{d}{dt}\left(\frac{\partial \vec{r}_i}{\partial q_k}\right)\right\} \qquad (5.41)$$

The last term in the foregoing equation becomes

$$\frac{d}{dt}\left(\frac{\partial \vec{r}_i}{\partial q_k}\right) = \sum_{j=1}^{n} \frac{\partial^2 \vec{r}_i}{\partial q_k\, \partial q_j}\dot{q}_j + \frac{\partial^2 \vec{r}_i}{\partial q_k\, \partial t}$$

$$= \frac{\partial}{\partial q_k}\left(\frac{d\vec{r}_i}{dt}\right) = \frac{\partial \dot{\vec{r}}_i}{\partial q_k} \qquad (5.42)$$

It can also be shown from the expression for $\dot{\vec{r}}$ that

$$\frac{\partial \dot{\vec{r}}_i}{\partial \dot{q}_k} = \frac{\partial \vec{r}_i}{\partial q_k} \qquad (5.43)$$

Employing (5.42) and (5.43) in (5.41) and then substituting the result in (5.40), it is seen that the virtual work of the inertia terms can be represented as

$$-\sum_{i=1}^{N} m_i \ddot{\vec{r}}_i \cdot \delta\vec{r}_i = -\sum_{i=1}^{N}\sum_{k=1}^{n}\left\{\frac{d}{dt}\left(m_i\dot{\vec{r}}_i\cdot\frac{\partial\dot{\vec{r}}_i}{\partial\dot{q}_k}\right) - m_i\dot{\vec{r}}_i\cdot\frac{\partial\dot{\vec{r}}_i}{\partial q_k}\right\}\delta q_k$$

$$= -\sum_{k=1}^{n}\left\{\frac{d}{dt}\frac{\partial}{\partial\dot{q}_k}\left(\sum_{i=1}^{N}\frac{1}{2}m_i\dot{\vec{r}}_i\cdot\dot{\vec{r}}_i\right) - \frac{\partial}{\partial q_k}\left(\sum_{i=1}^{N}\frac{1}{2}m_i\dot{\vec{r}}_i\cdot\dot{\vec{r}}_i\right)\right\}\delta q_k$$

$$= -\sum_{k=1}^{n}\left\{\frac{d}{dt}\left(\frac{\partial T}{\partial\dot{q}_k}\right) - \frac{\partial T}{\partial q_k}\right\}\delta q_k \qquad (5.44)$$

Substituting in (5.35) the expressions for the virtual work done by the impressed forces and inertia forces from (5.38) and (5.44), respectively, we obtain

$$\sum_{k=1}^{n}\left[\frac{d}{dt}\left(\frac{\partial T}{\partial\dot{q}_k}\right) - \frac{\partial T}{\partial q_k} - Q_k\right]\delta q_k = 0 \qquad (5.45)$$

For a holonomic system, $q_1, \ldots, q_n$ are independent. Therefore, we can let all δq's except one be zero. Then the coefficient of that nonzero δq must be zero. Employing this argument, we obtain

$$\frac{d}{dt}\left(\frac{\partial T}{\partial\dot{q}_i}\right) - \frac{\partial T}{\partial q_i} = Q_i, \qquad i = 1, 2, \ldots, n \qquad (5.46)$$

These n equations are known as Lagrange equations of motion. They consist of n second-order differential equations which in general are nonlinear but may be linear in some cases.

5.5.2 Application of Hamilton's Principle to Holonomic Systems

We have seen that the general form of Hamilton's principle is expressed by (5.30), namely

$$\int_{t_0}^{t_1} (\delta\bar{W} + \delta T)\, dt = 0$$

where the end points are fixed, that is, $\delta \vec{r}_i(t_0) = \delta \vec{r}_i(t_1) = \vec{0}$. For a holonomic system of N particles with n degrees of freedom, we choose $q_1, q_2, \ldots, q_n$ as the generalized coordinates. The transformation equations between the vector coordinates are given by (5.36). The total kinetic energy of the system is

$$T = \tfrac{1}{2} \sum_{i=1}^{n} m_i \dot{\vec{r}}_i \cdot \dot{\vec{r}}_i \tag{5.47}$$

Substituting for the velocities of the particles from (5.37), we obtain

$$T = \frac{1}{2} \sum_{i=1}^{N} m_i \left(\sum_{j=1}^{n} \frac{\partial \vec{r}_i}{\partial q_j} \dot{q}_j + \frac{\partial \vec{r}_i}{\partial t} \right) \cdot \left(\sum_{k=1}^{n} \frac{\partial \vec{r}_i}{\partial q_k} \dot{q}_k + \frac{\partial \vec{r}_i}{\partial t} \right)$$

$$= \frac{1}{2} \sum_{i=1}^{N} m_i \left(\sum_{j=1}^{n} \sum_{k=1}^{n} \frac{\partial \vec{r}_i}{\partial q_j} \cdot \frac{\partial \vec{r}_i}{\partial q_k} \dot{q}_j \dot{q}_k + 2 \frac{\partial \vec{r}_i}{\partial t} \cdot \sum_{j=1}^{n} \frac{\partial \vec{r}_i}{\partial q_j} \dot{q}_j + \frac{\partial \vec{r}_i}{\partial t} \cdot \frac{\partial \vec{r}_i}{\partial t} \right) \tag{5.48}$$

We now introduce the following coefficients as

$$\alpha_{jk} = \sum_{i=1}^{N} m_i \frac{\partial \vec{r}_i}{\partial q_j} \cdot \frac{\partial \vec{r}_i}{\partial q_k}$$

$$\beta_j = \sum_{i=1}^{N} m_i \frac{\partial \vec{r}_i}{\partial t} \cdot \frac{\partial \vec{r}_i}{\partial q_j} \tag{5.49}$$

$$\gamma = \frac{1}{2} \sum_{i=1}^{N} m_i \frac{\partial \vec{r}_i}{\partial t} \cdot \frac{\partial \vec{r}_i}{\partial t}$$

Then (5.48) can be written in the form

$$T = T_2 + T_1 + T_0 \tag{5.50}$$

where

$$T_2 = \tfrac{1}{2} \sum_{j=1}^{n} \sum_{k=1}^{n} \alpha_{jk} \dot{q}_j \dot{q}_k$$

is a quadratic function in the generalized velocities,

$$T_1 = \sum_{j=1}^{n} \beta_j \dot{q}_j$$

is a linear function in the generalized velocities, and

$$T_0 = \gamma$$

is a nonnegative function of only the generalized coordinates and time but is not a function of generalized velocities. It should be noted that α_{jk}, β_j, and γ are in general functions of generalized coordinates and time. Thus, using generalized coordinates, the expression for the kinetic energy takes the form

$$T = T(q_1, q_2, \ldots, q_n, \dot{q}_1, \dot{q}_2, \ldots, \dot{q}_n, t) \tag{5.51}$$

We have seen earlier that the expression for the virtual work done by the impressed forces is given by (5.38) as

$$\delta \bar{W} = \sum_{i=1}^{n} Q_i \, \delta q_i \tag{5.52}$$

Our concern now is to determine those q_i which satisfy the equation

$$\int_{t_0}^{t_1} \left(\delta T + \sum_{i=1}^{n} Q_i\,\delta q_i \right) dt = 0 \tag{5.53}$$

Taking the variation of T employing (5.51) and noting that $\delta t = 0$, we get

$$\delta T = \sum_{i=1}^{n} \frac{\partial T}{\partial q_i}\,\delta q_i + \sum_{i=1}^{n} \frac{\partial T}{\partial \dot{q}_i}\,\delta \dot{q}_i \tag{5.54}$$

Substituting this result in (5.53), we obtain

$$\int_{t_0}^{t_1} \sum_{i=1}^{n} \left[\left(\frac{\partial T}{\partial q_i} + Q_i \right) \delta q_i + \frac{\partial T}{\partial \dot{q}_i}\,\delta \dot{q}_i \right] dt = 0 \tag{5.55}$$

Integrating the last term in the foregoing equation by parts, we get

$$\int_{t_0}^{t_1} \sum_{i=1}^{n} \frac{\partial T}{\partial \dot{q}_i}\,\delta \dot{q}_i\,dt = \left[\sum_{i=1}^{n} \frac{\partial T}{\partial \dot{q}_i}\,\delta q_i \right]_{t_0}^{t_1} - \int_{t_0}^{t_1} \sum_{i=1}^{n} \frac{d}{dt}\left(\frac{\partial T}{\partial \dot{q}_i} \right) \delta q_i\,dt$$

$$= -\int_{t_0}^{t_1} \sum_{i=1}^{n} \frac{d}{dt}\left(\frac{\partial T}{\partial \dot{q}_i} \right) \delta q_i\,dt \tag{5.56}$$

The foregoing equation follows from the fact that $\delta q(t_0) = \delta q_i(t_1) = 0$ for $i = 1, \ldots, n$. Substituting for the last term in (5.55) from (5.56), we finally obtain

$$\int_{t_0}^{t_1} \sum_{i=1}^{n} \left[-\frac{d}{dt}\left(\frac{\partial T}{\partial \dot{q}_i} \right) + \frac{\partial T}{\partial q_i} + Q_i \right] \delta q_i\,dt = 0 \tag{5.57}$$

Since for an holonomic system the generalized coordinates are independent, the coefficient of each δq_i in (5.57) must be zero. Thus, it follows that

$$\frac{d}{dt}\left(\frac{\partial T}{\partial \dot{q}_i} \right) - \frac{\partial T}{\partial q_i} = Q_i, \qquad i = 1, 2, \ldots, n \tag{5.58}$$

It is seen that these Lagrange equations of motion are identical to those given by (5.46).

Example 5.7

Two masses m_1 and m_2 are connected as shown in Fig. 5.9. Mass m_1 is attached to a rigid massless link OA of length a which is free to rotate at bearing O. Mass m_2 is attached to a rigid massless link AB of length b which is free to rotate at bearing A. The motion is constrained to the vertical plane. The bearings are assumed frictionless and mass m_1 is acted on by a force P in the horizontal direction and a force Q in the vertical direction as shown. Obtain the Lagrange equations of motion.

Choosing a Cartesian coordinate system Oxy to represent the motion, the position coordinates of the mass particles are given by

$$\vec{r}_1 = x_1 \vec{i} + y_1 \vec{j}$$
$$\vec{r}_2 = x_2 \vec{i} + y_2 \vec{j}$$

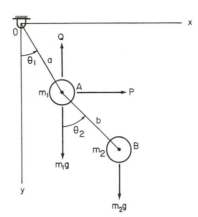

Figure 5.9 Motion of masses m_1 and m_2.

Here, there are four coordinates x_1, x_2, y_1, and y_2 which are related by the holonomic constraints

$$x_1^2 + y_1^2 = a^2$$
$$(x_2 - x_1)^2 + (y_2 - y_1)^2 = b^2$$

Hence, this system has only two degrees of freedom. We choose angles θ_1 and θ_2 as the two generalized coordinates. The transformation equations (5.36) become

$$\vec{r}_1 = a \sin \theta_1 \vec{i} + a \cos \theta_1 \vec{j} \tag{5.59}$$
$$\vec{r}_2 = (a \sin \theta_1 + b \sin \theta_2)\vec{i} + (a \cos \theta_1 + b \cos \theta_2)\vec{j} \tag{5.60}$$

The total kinetic energy is given by

$$T = \tfrac{1}{2} \sum_{i=1}^{2} m_i \vec{r}_i \cdot \vec{r}_i$$
$$= \tfrac{1}{2}m_1\dot{x}_1^2 + \tfrac{1}{2}m_1\dot{y}_1^2 + \tfrac{1}{2}m_2\dot{x}_2^2 + \tfrac{1}{2}m_2\dot{y}_2^2 \tag{5.61}$$

The velocities $\vec{r}_1$ and $\vec{r}_2$ are obtained from (5.37). It is easier here to express the kinetic energy in terms of the generalized coordinates by making the following substitutions directly in (5.61):

$$\dot{x}_1 = a(\cos \theta_1)\dot{\theta}_1$$
$$\dot{y}_1 = -a(\sin \theta_1)\dot{\theta}_1$$
$$\dot{x}_2 = (a \cos \theta_1)\dot{\theta}_1 + b(\cos \theta_2)\dot{\theta}_2$$
$$\dot{y}_2 = -a(\sin \theta_1)\dot{\theta}_1 - b(\sin \theta_2)\dot{\theta}_2$$

Hence, in terms of the generalized coordinates, the expression for the kinetic energy becomes

$$T = \tfrac{1}{2}m_1a^2\dot{\theta}_1^2 \cos^2 \theta_1 + \tfrac{1}{2}m_1a^2\dot{\theta}_1^2 \sin^2 \theta_1$$
$$+ \tfrac{1}{2}m_2(a\dot{\theta}_1 \cos \theta_1 + b\dot{\theta}_2 \cos \theta_2)^2 + \tfrac{1}{2}m_2(a\dot{\theta}_1 \sin \theta_1 + b\dot{\theta}_2 \sin \theta_2)^2 \tag{5.62}$$

It is noted here that for this expression, we have $T = T_2$; that is, T is a quadratic function of the generalized velocities $\dot{\theta}_1$ and $\dot{\theta}_2$ and T_1 and T_0 are both zero. The generalized forces are obtained from expression (5.39), which becomes

$$Q_i = \vec{F}_1 \cdot \frac{\partial \vec{r}_1}{\partial \theta_i} + \vec{F}_2 \cdot \frac{\partial \vec{r}_2}{\partial \theta_i}, \qquad i = 1, 2 \tag{5.63}$$

where $\vec{F}_1$ and $\vec{F}_2$ are the impressed forces on masses m_1 and m_2, respectively. These are given by

$$\vec{F}_1 = P\vec{i} + (m_1 g - Q)\vec{j} \tag{5.64}$$

$$\vec{F}_2 = m_2 g \vec{j} \tag{5.65}$$

There also exist constraint forces which consist of forces in the links and the reactions at the bearings. But since the constraints are not violated, the constraint forces do no work in virtual displacements and are ignored. From (5.59) and (5.60), we obtain

$$\frac{\partial \vec{r}_1}{\partial \theta_1} = \frac{\partial \vec{r}_2}{\partial \theta_1} = a \cos \theta_1 \vec{i} - a \sin \theta_1 \vec{j}$$

$$\frac{\partial \vec{r}_1}{\partial \theta_2} = 0 \quad \text{and} \quad \frac{\partial \vec{r}_2}{\partial \theta_2} = b \cos \theta_2 \vec{i} - b \sin \theta_2 \vec{j}$$

Employing these results and those of (5.64) and (5.65) in (5.63), the generalized forces in the θ_1 and θ_2 directions are given, respectively, by

$$Q_1 = (Q - m_1 g - m_2 g)a \sin \theta_1 + Pa \cos \theta_1 \tag{5.66}$$

and

$$Q_2 = -m_2 gb \sin \theta_2 \tag{5.67}$$

The Lagrange equations of motion are obtained by substituting from (5.62), (5.66), and (5.67) in (5.58). These are given by

$$\frac{d}{dt}\left(\frac{\partial T}{\partial \dot{\theta}_1}\right) - \frac{\partial T}{\partial \theta_1} = Q_1$$

$$\frac{d}{dt}\left(\frac{\partial T}{\partial \dot{\theta}_2}\right) - \frac{\partial T}{\partial \theta_2} = Q_2$$

or

$$\frac{d}{dt}[(m_1 + m_2)a^2\dot{\theta}_1 + m_2 ab\dot{\theta}_2 \cos(\theta_1 - \theta_2)] + m_2 ab\dot{\theta}_1\dot{\theta}_2 \sin(\theta_1 - \theta_2)$$
$$= (Q - m_1 g - m_2 g)a \sin \theta_1 + Pa \cos \theta_1 \tag{5.68}$$

$$\frac{d}{dt}[m_2 ab\dot{\theta}_1 \cos(\theta_1 - \theta_2) + m_2 b^2\dot{\theta}_2] - m_2 ab\dot{\theta}_1\dot{\theta}_2 \sin(\theta_1 - \theta_2)$$
$$= -m_2 gb \sin \theta_2 \tag{5.69}$$

Equations (5.68) and (5.69) are the two equations of motion. It is seen that each equation is nonlinear and of second order.

Example 5.8

We wish to obtain the Lagrange equations of motion for the bead of Example 5.6 shown in Fig. 5.7. We employ the rotating coordinate system $Oxyz$ of Fig. 5.7 with angular velocity $\vec{\omega} = \omega_0 \vec{j}$. The position vector $\vec{r}$ of the bead is denoted by

$$\vec{r} = x\vec{i} + y\vec{j} \tag{5.70}$$

The two coordinates x and y are related by one holonomic constraint,

$$x^2 + y^2 = c^2 \tag{5.71}$$

The velocity of the bead with respect to this coordinate system becomes

$$\vec{v} = \dot{\vec{r}} + \vec{\omega} \times \vec{r}$$
$$= \dot{x}\vec{i} + \dot{y}\vec{j} + \omega_0\vec{j} \times (x\vec{i} + y\vec{j})$$
$$= \dot{x}\vec{i} + \dot{y}\vec{j} - \omega_0 x\vec{k}$$

The kinetic energy is given by

$$T = \tfrac{1}{2}m\vec{v} \cdot \vec{v} = \tfrac{1}{2}m(\dot{x}^2 + \dot{y}^2 + \omega_0^2 x^2) \tag{5.72}$$

The bead has only one degree of freedom since x and y are related by constraint given by (5.71). Choosing θ as the single generalized coordinate, the transformation equations (5.36) become

$$x = c \cos \theta$$
$$y = c \sin \theta$$

and hence $\dot{x} = -c\dot{\theta} \sin \theta$ and $\dot{y} = c\dot{\theta} \cos \theta$. The expression for the kinetic energy in terms of θ is obtained by employing this result in (5.72). We obtain

$$T = \tfrac{1}{2}mc^2(\dot{\theta}^2 + \omega_0^2 \cos^2 \theta) \tag{5.73}$$

It is noted that for this expression we have $T = T_2 + T_0$ and $T_1 = 0$. The generalized force is obtained from the expression

$$Q = \vec{F} \cdot \frac{\partial \vec{r}}{\partial \theta} \tag{5.74}$$

where $\vec{F}$ is the impressed force on the bead. If friction is neglected, the only impressed force is due to gravity and we get

$$\vec{F} = -mg\vec{j}$$

Also,

$$\frac{\partial \vec{r}}{\partial \theta} = -c \sin \theta \vec{i} + c \cos \theta \vec{j} \tag{5.75}$$

Hence, the generalized force becomes $Q = -mgc \cos \theta$. The Lagrange equation of motion is obtained from

$$\frac{d}{dt}\left(\frac{\partial T}{\partial \dot{\theta}}\right) - \frac{\partial T}{\partial \theta} = Q$$

or

$$\frac{d}{dt}(mc^2\dot{\theta}) - mc^2\omega_0^2 \cos \theta(-\sin \theta) = -mgc \cos \theta$$

Hence, the equation of motion becomes

$$\ddot{\theta} + \omega_0^2 \cos \theta \sin \theta + \frac{g}{c} \cos \theta = 0 \tag{5.76}$$

Now if θ is constant, then $\ddot{\theta} = 0$ and the equilibrium is described by

$$\omega_0^2 \cos \theta \sin \theta + \frac{g}{c} \cos \theta = 0 \tag{5.77}$$

It is seen that (5.77) is identical to (5.22), which was obtained by the direct application of the principle of virtual work without formulating the differential equation of motion.

It should be noted that the constraint force which consists of the reaction between the bead and the hoop was ignored in the formulation of the equation of motion. Hence, the value of the reaction is unknown. Sometimes, some impressed forces such as frictional forces depend on the constraint forces. Then it becomes necessary to obtain explicit expressions for the constraint forces by direct application of Newton's law. To illustrate this point, we now include Coulomb friction opposing the sliding of the bead on the hoop. Letting N denote the reaction on the bead, we now have an additional impressed force $-\mu N$ sgn $\dot{\theta}$ acting on the bead in the θ direction. A free-body diagram of the bead is shown in Fig. 5.10.

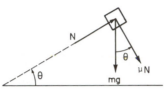

Figure 5.10   Free-body diagram of bead.

The acceleration of the bead is given by

$$\vec{a} = \ddot{\vec{r}} + 2\vec{\omega} \times \dot{\vec{r}} + \dot{\vec{\omega}} \times \vec{r} + \vec{\omega} \times (\vec{\omega} \times \vec{r})$$

where $\ddot{\vec{r}} = \ddot{x}\vec{i} + \ddot{y}\vec{j}$, $2\vec{\omega} \times \dot{\vec{r}} = -2\omega_0\dot{x}\vec{k}$, $\dot{\vec{\omega}} \times \vec{r} = 0$, and $\vec{\omega} \times (\vec{\omega} \times \vec{r}) = -\omega_0^2 x\vec{i}$. Hence, we get

$$\vec{a} = (-c\ddot{\theta}\sin\theta - c\dot{\theta}^2\cos\theta - m\omega_0^2 c\cos\theta)\vec{i}$$
$$+ (c\ddot{\theta}\cos\theta - c\dot{\theta}^2\sin\theta)\vec{j} + 2\omega_0 c\dot{\theta}\sin\theta\vec{k}$$

The component of N in the radial direction is obtained from the first two terms in the preceding equation as

$$N_r = mg\sin\theta - m\omega_0^2 c\cos^2\theta - mc\dot{\theta}^2$$

The component of N in the z direction is given by

$$N_z = 2m\omega_0 c\dot{\theta}\sin\theta$$

Hence, the total normal force on the bead is obtained as

$$N = [(mg\sin\theta - m\omega_0^2 c\cos^2\theta - mc\dot{\theta}^2)^2 + (2m\omega_0 c\dot{\theta}\sin\theta)^2]^{1/2} \qquad (5.78)$$

The additional impressed force in the θ direction due to Coulomb friction becomes

$$F_f = -\mu N \text{ sgn } \dot{\theta}$$

The total impressed force on the bead is due to the gravity force $-mg\vec{j}$ and the friction force F_f. The generalized force in the θ direction is obtained as

$$Q = -mgc\cos\theta - \mu N \text{ sgn } \dot{\theta}$$

Substituting these results in the Lagrange equation, the equation of motion becomes

$$\ddot{\theta} + \frac{\mu N}{mc^2} \text{ sgn } \dot{\theta} + \omega_0^2\cos\theta\sin\theta + \frac{g}{c}\cos\theta = 0 \qquad (5.79)$$

where the normal force N is given by (5.78)

5.5.3 Lagrange Equations of Motion for Nonholonomic Systems

The foregoing development of the Lagrange equations of motion can be easily extended to nonholonomic systems. Let n coordinates $q_1, \ldots, q_n$ be chosen to describe the motion, and these coordinates are related by R nonholonomic constraints of the form

$$a_{j_0}\, dt + \sum_{k=1}^{n} a_{jk}\, dq_k = 0, \qquad j = 1, \ldots, R \tag{5.80}$$

where a_{jk} $(k = 0, 1, \ldots, n)$ are functions of q_k. The degrees of freedom now are given by $n - R$, and the coordinates q_k are not all independent. Hence, the argument employed to set the coefficient of each δq_k in (5.45) or (5.57) to zero becomes invalid. The virtual displacements are related by the equations

$$\sum_{k=1}^{n} a_{jk}\, \delta q_k = 0, \qquad j = 1, \ldots, R \tag{5.81}$$

Here, we employ the method of Lagrange multipliers. Multiplying each of the R equations of (5.81) by an as yet unknown Lagrange multiplier λ_j, we add the sum to the left-hand side of either (5.45) or (5.57). Now, (5.57) is modified to

$$\int_{t_0}^{t_1} \sum_{i=1}^{n} \left[-\frac{d}{dt}\left(\frac{\partial T}{\partial \dot{q}_i}\right) + \frac{\partial T}{\partial q_i} + Q_i + \sum_{j=1}^{R} \lambda_j a_{ji} \right] \delta q_i\, dt = 0$$

This modification is permissible since the right-hand side of each of the R equations of (5.81) is zero. We now choose the Lagrange multipliers λ_j $(j = 1, \ldots, R)$ such that the coefficients of δq_i for $i = 1, \ldots, R$ are equal to zero. The remaining $(n - R)$ δq_i are independent and can be chosen arbitrarily. Hence, we obtain the Lagrange equations of motion

$$\frac{d}{dt}\left(\frac{\partial T}{\partial \dot{q}_i}\right) - \frac{\partial T}{\partial q_i} = Q_i + \sum_{j=1}^{R} \lambda_j a_{ji}, \qquad i = 1, 2, \ldots, n \tag{5.82}$$

The n Lagrange equations of motion (5.82) together with the R constraint equations (5.80) together constitute $(n + R)$ equations in $(n + R)$ unknowns, namely, the n coordinates q_i and the R Lagrange multipliers λ_j. It is noted from (5.82) that the term $\sum_{j=1}^{R} \lambda_j a_{ji}$ is equivalent to an additional generalized force in the direction of the ith coordinate contributed by the constraint force. This procedure permits the solution of not only the coordinates q_i but also the constraint forces associated with each of the R constraints (5.80). The method, however, does not include nonholonomic systems where the constraints are expressed in the form of inequalities as in Example 5.4. As pointed out in that example, such systems can be treated as piecewise holonomic in the different regions.

Example 5.9

Two masses m_1 and m_2 are constrained to move in the xy plane as shown in Fig. 5.11. It is assumed that the pulleys are massless and frictionless and that the rope is inextensible. Let x denote the displacement of mass m_1 from its position where the spring is unstretched, and y the displacement of mass m_2 from its corresponding position.

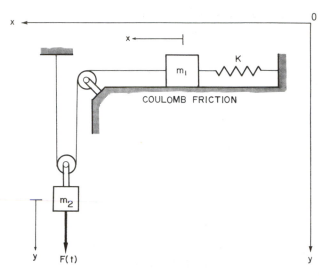

Figure 5.11   Two masses constrained to move in a plane.

Choosing x and y as the two coordinates to describe the motion, we find that they are related by one holonomic constraint $x - 2y = 0$. Hence, it is possible to eliminate one excess coordinate and employ either x or y as the generalized coordinate for this single-degree-of-freedom system. However, we treat the system as nonholonomic for the purpose of determining the constraint force, which in this case is the tension in the rope. The expression for the kinetic energy becomes

$$T = \tfrac{1}{2}m_1\dot{x}^2 + \tfrac{1}{2}m_2\dot{y}^2 \tag{5.83}$$

The spring force and Coulomb friction constitute the impressed forces on mass m_1 and the applied force F and weight $m_2 g$ are the impressed forces on mass m_2. Hence, from (5.39), the generalized forces in the directions of the x and y coordinates are given, respectively, by

$$Q_x = -kx - \mu m_1 g \text{ sgn } \dot{x} \tag{5.84}$$

and

$$Q_y = mg + F \tag{5.85}$$

The Pfaffian corresponding to the constraint is given by $a_1\,dx + a_2\,dy = 0$, where $a_1 = 1$ and $a_2 = -2$. Choosing one Lagrange multiplier λ, the λa_i in the direction of the x and y coordinates becomes

$$\lambda a_1 = \lambda \tag{5.86}$$

$$\lambda a_2 = -2\lambda \tag{5.87}$$

Substitution from (5.83) to (5.87) in (5.82) yields the two Lagrange equations of motion. These equations and the constraint equation become

$$m_1\ddot{x} = -kx - \mu m_1 g \text{ sgn } \dot{x} + \lambda$$

$$m_2\ddot{y} = mg + F - 2\lambda \tag{5.88}$$

$$\dot{x} = 2\dot{y}$$

Equations (5.88) are the three equations in the three unknowns, x, y, and λ. It is seen that λa_1 and λa_2 are the effective constraint force components in the x and y directions, respectively, which would do work if the constraint was relaxed. Obviously, λ here is the value of the tension in the rope.

5.5.4 Alternative Forms of the Lagrange Equations of Motion

Let the impressed force acting on jth particle of a system of N particles be separated into conservative and nonconservative forces as

$$\vec{F}_j^* = \vec{F}_{c,j}^* + \vec{F}_{nc,j}^*$$

Equation (5.38) for the virtual work done by all the impressed forces may now be expressed as

$$\sum_{j=1}^{N} \vec{F}_j^* \cdot \delta \vec{r}_j = \sum_{i=1}^{n} -\frac{\partial U}{\partial q_i} \delta q_i + \sum_{i=1}^{n} Q_{nc,i} \, \delta q_i \qquad (5.89)$$

where U is a scalar potential energy which is a function of position only expressed in terms of the generalized coordinates [i.e., $U = U(q_1, \ldots, q_n)$] and

$$Q_{nc,i} = \sum_{j=1}^{N} \vec{F}_{nc,j}^* \cdot \frac{\partial \vec{r}_j}{\partial q_i}$$

which is the generalized force in the ith coordinate direction contributed by the nonconservative impressed forces only. For a holonomic system, the Lagrange equations of motion may now be written as

$$\frac{d}{dt}\left(\frac{\partial T}{\partial \dot{q}_i}\right) - \frac{\partial T}{\partial q_i} = -\frac{\partial U}{\partial q_i} + Q_{nc,i} \qquad (5.90)$$

A scalar Lagrangian function L is defined as stated in (5.31) by $L = T - U$. Since U is not a function of $\dot{q}_i$, we have

$$\frac{\partial T}{\partial \dot{q}_i} = \frac{\partial L}{\partial \dot{q}_i}$$

Substituting this result in (5.90), the Lagrange equations of motion for a holonomic system when expressed by a Lagrangian become

$$\frac{d}{dt}\left(\frac{\partial L}{\partial \dot{q}_i}\right) - \frac{\partial L}{\partial q_i} = Q_{nc,i} \qquad i = 1, \ldots, n \qquad (5.91)$$

and for a nonholonomic system, we obtain

$$\frac{d}{dt}\left(\frac{\partial L}{\partial \dot{q}_i}\right) - \frac{\partial L}{\partial q_i} = Q_{nc,i} + \sum_{j=1}^{R} \lambda_j a_{ji} \qquad (5.92)$$

For a conservative holonomic system, it follows from (5.91) that

$$\frac{d}{dt}\left(\frac{\partial L}{\partial \dot{q}_i}\right) - \frac{\partial L}{\partial q_i} = 0, \qquad i = 1, \ldots, n \qquad (5.93)$$

Sometimes, the frictional forces acting on a particle are separated into viscous and nonviscous friction forces. Viscous friction forces are proportional to the velocity of a given particle and resist the motion, since they act in a direction opposite to that of the velocity. Nonviscous friction forces are nonlinear functions of the velocity and resist the motion. For the viscous friction forces, we define a scalar function F, known as Rayleigh's dissipation function, which is a quadratic function of the generalized velocities as

$$F = \tfrac{1}{2} \sum_{i=1}^{n} \sum_{j=1}^{n} c_{i,\,j} \dot{q}_i \dot{q}_j \qquad (5.94)$$

Then the virtual work done by viscous friction forces becomes

$$\delta \bar{W}_v = \sum_{i=1}^{n} Q_{v,\,i}\, \delta q_i = -\sum_{i=1}^{n} \frac{\partial F}{\partial \dot{q}_i}\, \delta q_i$$

that is, the viscous friction generalized force in the direction of the ith coordinate becomes

$$Q_{v,\,i} = -\frac{\partial F}{\partial \dot{q}_i} \qquad (5.95)$$

Hence, the Lagrange equations of motion (5.91) for a holonomic system with viscous friction may be expressed as

$$\frac{d}{dt}\left(\frac{\partial L}{\partial \dot{q}_i}\right) - \frac{\partial L}{\partial q_i} + \frac{\partial F}{\partial \dot{q}_i} = Q_{nc,\,i}, \qquad i = 1,\ldots,n \qquad (5.96)$$

where the generalized force $Q_{nc,\,i}$ now does not include the contribution of the viscous friction forces.

Example 5.10

Derive the equations of motion governing the free vibrations of the system shown in Fig. 5.12. Assume the springs and the rigid bar to be massless.

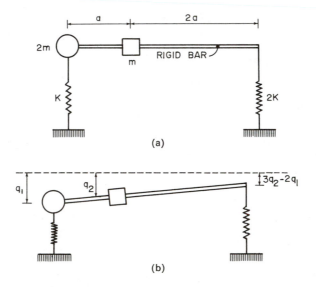

Figure 5.12   Free vibrations of a system of two particles.

We select q_1 and q_2 as the generalized displacements for the masses [Fig. 5.12(b)]. The system has two degrees of freedom. The kinetic energy of the system is expressed by

$$T = \tfrac{1}{2}(2m\dot{q}_1^2 + m\dot{q}_2^2)$$

and the potential energy U of the system is

$$U = \frac{kq_1^2}{2} + \frac{2k}{2}(3q_2 - 2q_1)^2$$

The Lagrangian becomes

$$L = \frac{1}{2}(2m\dot{q}_1^2 + m\dot{q}_2^2) - \frac{kq_1^2}{2} - \frac{2k}{2}(3q_2 - 2q_1)^2$$

and for this holonomic conservative system with two degrees of freedom, the Lagrange equations (5.93) become

$$\frac{d}{dt}\left(\frac{\partial L}{\partial \dot{q}_1}\right) - \frac{\partial L}{\partial q_1} = 0$$

$$\frac{d}{dt}\left(\frac{\partial L}{\partial \dot{q}_2}\right) - \frac{\partial L}{\partial q_2} = 0$$

Now,

$$\frac{d}{dt}\left(\frac{\partial L}{\partial \dot{q}_1}\right) = 2m\ddot{q}_1, \qquad \frac{\partial L}{\partial q_i} = kq_1 - 4k(3q_2 - 2q_1) = 9kq_1 - 12kq_2$$

$$\frac{d}{dt}\left(\frac{\partial L}{\partial \dot{q}_2}\right) = m\ddot{q}_2, \qquad \frac{\partial L}{\partial q_2} = 6k(3q_2 - 2q_1)$$

Substituting these results in the Lagrange equations of motion, we obtain

$$2m\ddot{q}_1 + 9kq_1 - 12kq_2 = 0 \qquad (5.97a)$$

$$m\ddot{q}_2 - 12kq_1 + 18kq_2 = 0 \qquad (5.97b)$$

For this linear system, we define mass and stiffness matrices as

$$[M] = \begin{bmatrix} 2m & 0 \\ 0 & m \end{bmatrix}, \qquad [K] = k\begin{bmatrix} 9 & -12 \\ -12 & 18 \end{bmatrix}$$

and in the matrix notation, (5.97a) and (5.97b) can be expressed as

$$[M]\begin{Bmatrix} \ddot{q}_1 \\ \ddot{q}_2 \end{Bmatrix} + [K]\begin{Bmatrix} q_1 \\ q_2 \end{Bmatrix} = 0 \qquad (5.98)$$

Example 5.11

A spring pendulum as shown in Fig. 5.13 has a mass m suspended by an elastic spring of stiffness k and free length a. Derive the equations of motion of the pendulum. Assume viscous frictional moment at the pivot resisting the motion in this vertical plane.

The system has two degrees of freedom. We select r and θ as the generalized displacements in the polar coordinate system. The generalized velocities of the mass are given by $r\dot{\theta}$ and $\dot{r}$. Hence, the kinetic energy T of the system becomes

$$T = \tfrac{1}{2}m(r\dot{\theta})^2 + \tfrac{1}{2}m\dot{r}^2$$

The potential energy of the spring and mass is given by

$$U = \tfrac{1}{2}k(r - a)^2 + (c_1 - r\cos\theta)mg$$

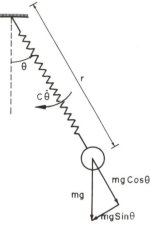

Figure 5.13   Spring pendulum.

where c_1 is a constant and the Lagrangian function becomes

$$L = \tfrac{1}{2}m(r\dot{\theta})^2 + \tfrac{1}{2}m\dot{r}^2 - \tfrac{1}{2}k(r - a)^2 - (c_1 - r\cos\theta)mg \qquad (5.99)$$

The Rayleigh's dissipation function is

$$F = \tfrac{1}{2}c\dot{\theta}^2$$

and the Lagrange equations of motion (5.96) become

$$\frac{d}{dt}\left(\frac{\partial L}{\partial \dot{r}}\right) - \frac{\partial L}{\partial r} + \frac{\partial F}{\partial \dot{r}} = 0 \qquad (5.100)$$

$$\frac{d}{dt}\left(\frac{\partial L}{\partial \dot{\theta}}\right) - \frac{\partial L}{\partial \theta} + \frac{\partial F}{\partial \dot{\theta}} = 0 \qquad (5.101)$$

Now,

$$\frac{d}{dt}\left(\frac{\partial L}{\partial \dot{r}}\right) = m\ddot{r}, \qquad\qquad \frac{\partial L}{\partial r} = mr\dot{\theta}^2 - k(r - a) + mg\cos\theta$$

$$\frac{d}{dt}\left(\frac{\partial L}{\partial \dot{\theta}}\right) = mr^2\ddot{\theta} + 2mr\dot{r}\dot{\theta}, \qquad \frac{\partial L}{\partial \theta} = -mgr\sin\theta$$

$$\frac{\partial F}{\partial \dot{r}} = 0, \qquad \frac{\partial F}{\partial \dot{\theta}} = c\dot{\theta}$$

Substituting this result in (5.100) and (5.101), the resulting equations of motion are given by

$$m\ddot{r} - mr\dot{\theta}^2 + k(r - a) - mg\cos\theta = 0 \qquad (5.102)$$

$$mr^2\ddot{\theta} + 2mr\dot{r}\dot{\theta} + mgr\sin\theta + c\dot{\theta} = 0 \qquad (5.103)$$

Example 5.12

Figure 5.14 represents a mass m which is suspended by an inextensible weightless string of length R. The string constrains the mass to a spherical surface with center O and radius R. The position of the mass m is completely defined by spherical coordinates, θ and ϕ. Determine the equations of motion of the mass.

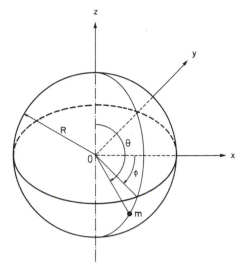

Figure 5.14   Spherical pendulum.

The generalized coordinates of the pendulum are (θ, ϕ). The kinetic and potential energies are given by (Note that $x = R \sin \theta \cos \phi$, $y = R \sin \theta \sin \phi$, and $z = R \cos \theta$)

$$T = \frac{mR^2}{2}(\dot{\theta}^2 + \dot{\phi}^2 \sin^2 \theta) = \frac{1}{2}m(\dot{x}^2 + \dot{y}^2 + \dot{z}^2)$$

$$U = mgR \cos \theta \quad (V|_{\theta=\pi/2} = 0 \text{ is taken as datum})$$

The Lagrangian becomes

$$L = \frac{mR^2}{2}(\dot{\theta}^2 + \dot{\phi}^2 \sin^2 \theta) - mgR \cos \theta$$

and differentiation yields

$$\frac{\partial L}{\partial \dot{\theta}} = mL^2\dot{\theta}, \qquad \frac{\partial L}{\partial \theta} = mR^2\dot{\phi}^2 \sin \theta \cos \theta + mgR \sin \theta$$

$$\frac{\partial L}{\partial \dot{\phi}} = mR^2\dot{\phi} \sin^2 \theta, \qquad \frac{\partial L}{\partial \phi} = 0$$

For this two-degree-of-freedom conservative systems, the equations of motion are obtained from (5.93) as

$$\frac{d}{dt}\left(\frac{\partial L}{\partial \dot{\theta}}\right) - \frac{\partial L}{\partial \theta} = 0$$

$$\frac{d}{dt}\left(\frac{\partial L}{\partial \dot{\phi}}\right) - \frac{\partial L}{\partial \phi} = 0$$

Substitution of these results in the foregoing equations yields the equation of motion as

$$R\ddot{\theta} - R\dot{\phi}^2 \sin \theta \cos \theta - mg \sin \theta = 0 \tag{5.104}$$

$$\ddot{\phi} \sin \theta + 2\dot{\theta}\dot{\phi} \cos \theta = 0 \tag{5.105}$$

5.5.5 State-Space Formulation of the Lagrange Equations of Motion

For a holonomic system with n degrees of freedom, the Lagrange equations of motion consist of a set of n simultaneous second-order differential equations which are generally nonlinear in the generalized coordinates and velocities. We may associate with these n generalized coordinates an n-dimensional Euclidean space which is called the Lagrangian configuration space, where the solution of the dynamic system may be represented. However, for the geometrical representation of the dynamic system and for the application of the analytical tools which are discussed in later chapters, it is convenient to formulate the problem as a set of $2n$ first-order equations in $2n$-dimensional Euclidean space called the state space. For this purpose, we define a $2n$ state-variable vector $\{x\}$ consisting of n generalized coordinates and n generalized velocities as

$$\begin{Bmatrix} x_1 \\ x_2 \\ \cdot \\ \cdot \\ \cdot \\ x_n \\ x_{n+1} \\ \cdot \\ \cdot \\ \cdot \\ x_{2n} \end{Bmatrix} = \begin{Bmatrix} q_1 \\ q_2 \\ \cdot \\ \cdot \\ \cdot \\ q_n \\ \dot{q}_1 \\ \cdot \\ \cdot \\ \cdot \\ \dot{q}_n \end{Bmatrix} = \begin{Bmatrix} \{q\} \\ \cdots \\ \{\dot{q}\} \end{Bmatrix} \tag{5.106}$$

Since the Lagrange equations of motion are of second order, the accelerations appear at most to the first order. Hence, these equations may be written in the form

$$[M]\{\ddot{q}\} = \{g(q_1, \ldots, q_n, \dot{q}_1, \ldots, \dot{q}_n, Q_1, \ldots, Q_n, t)\} \tag{5.107}$$

where $[M]$ is a $n \times n$ matrix whose elements are functions of q_i, $\dot{q}_i$, and time t and $\{g\}$ is a n-dimensional vector function of q_i, $\dot{q}_i$, t, and generalized forces Q_i. Because a general expression for the kinetic energy is as given by (5.50), $[M]$ in (5.107) is a positive-definite matrix and hence has an inverse. Inverting this matrix, we get

$$\{\ddot{q}\} = [M]^{-1}\{g\} = \begin{Bmatrix} f_{n+1}(q_1, \ldots, q_n, \dot{q}_1, \ldots, \dot{q}_n, Q_1, \ldots, Q_n, t) \\ \cdot \\ \cdot \\ \cdot \\ f_{2n}(q_1, \ldots, q_n, \dot{q}_1, \ldots, \dot{q}_n, Q_1, \ldots, Q_n, t) \end{Bmatrix} \tag{5.108}$$

The state equations in the generalized coordinates and generalized velocities of (5.106) are expressed as

$$\{\dot{x}\} = \{f(x_1, \ldots, x_{2n}, Q_1, \ldots, Q_n, t)\} \tag{5.109}$$

where $f_1 = x_{n+1}$, $f_2 = x_{n+2}, \ldots, f_n = x_{2n}$ and the functions $f_{n+1}, \ldots, f_{2n}$ are given by (5.108), where the state variables x_i have been substituted for q_i and $\dot{q}_i$. In (5.109), the first n equations are purely kinematic and obtained from the definition of (5.106) as $\dot{x}_i = x_{n+i}$ for $i = 1, \ldots, n$. The remaining n equations are obtained from the Lagrange equations of motion as seen from (5.108).

Example 5.13

We consider the dynamic system of Example 5.11. The Lagrange equations of motion for this system are given by (5.102) and (5.103). In order to obtain the state variable formulation, we write these equations as

$$\begin{bmatrix} m & 0 \\ 0 & mr^2 \end{bmatrix} \begin{Bmatrix} \ddot{r} \\ \ddot{\theta} \end{Bmatrix} = \begin{Bmatrix} -k(r-a) + mg\cos\theta + mr\dot{\theta}^2 \\ -2mr\dot{r}\dot{\theta} - mgr\sin\theta - c\dot{\theta} \end{Bmatrix}$$

Hence, we obtain

$$\begin{Bmatrix} \ddot{r} \\ \ddot{\theta} \end{Bmatrix} = \begin{bmatrix} m & 0 \\ 0 & mr^2 \end{bmatrix}^{-1} \begin{Bmatrix} -k(r-a) + mg\cos\theta + mr\dot{\theta}^2 \\ -2mr\dot{r}\dot{\theta} - mgr\sin\theta - c\dot{\theta} \end{Bmatrix}$$

$$= \begin{Bmatrix} -\dfrac{k}{m}(r-a) + g\cos\theta + r\dot{\theta}^2 \\ -2\dfrac{\dot{r}\dot{\theta}}{r} - \dfrac{g}{r}\sin\theta - \dfrac{c}{mr^2}\dot{\theta} \end{Bmatrix} \tag{5.110}$$

Choosing the generalized coordinates and generalized velocities as state variables, we define

$$x_1 = r, \qquad x_2 = \theta, \qquad x_3 = \dot{r}, \qquad x_4 = \dot{\theta}$$

Hence, we obtain the state equations as

$$\dot{x}_1 = x_3$$
$$\dot{x}_2 = x_4$$
$$\dot{x}_3 = -\frac{k}{m}x_1 + g\cos x_2 + x_1 x_4^2 + \frac{k}{m}a \tag{5.111}$$
$$\dot{x}_4 = -2\frac{x_3 x_4}{x_1} - g\frac{\sin x_2}{x_1} - \frac{c}{mx_1^2}x_4$$

It is noted that the first two of the foregoing equations are kinematic and are obtained from the definition of state variables. The last two equations are obtained from the Lagrange equations of motion after solving for the acceleration vector as in (5.110). Equations (5.111) are expressed in the form

$$\{\dot{x}\} = \{f(x_1, x_2, x_3, x_4)\} \tag{5.112}$$

5.6 HAMILTON'S CANONIC EQUATIONS OF MOTION

In the preceding section, the Lagrange equations of motion have been expressed as a set of $2n$ first-order equations by choosing generalized coordinates and generalized velocities as the state variables. Of course, this choice of state variables is not unique. Another choice is to select generalized coordinates and generalized momenta as the state variables as shown in the following. For a dynamic system, we define n additional variables called generalized momenta as

$$p_i = \frac{\partial T}{\partial \dot{q}_i} = \frac{\partial L}{\partial \dot{q}_i}, \qquad i = 1, 2, \ldots, n \tag{5.113}$$

A scalar Hamiltonian function H is defined by

$$H = \sum_{i=1}^{n} p_i \dot{q}_i - L \tag{5.114}$$

The variation in the Hamiltonian is then given by

$$\delta H = \sum_{i=1}^{n} \left[\delta p_i \dot{q}_i + p_i \, \delta \dot{q}_i - \frac{\partial L}{\partial q_i} \delta q_i - \frac{\partial L}{\partial \dot{q}_i} \delta \dot{q}_i \right]$$

and after noting that $p_i \, \delta \dot{q}_i - (\partial L / \partial \dot{q}_i) \, \delta \dot{q}_i = 0$ from (5.113), we obtain

$$\delta H = \sum_{i=1}^{n} \left[\delta p_i \dot{q}_i - \frac{\partial L}{\partial q_i} \delta q_i \right] \tag{5.115}$$

We now solve for the generalized velocities $\dot{q}_i$ in terms of the generalized momenta p_i from (5.113) and substitute the result in (5.114) such that the Hamiltonian becomes

$$H = H(q_1, q_2, \ldots, q_n, p_1, p_2, \ldots, p_n, t) \tag{5.116}$$

Employing (5.116), the variation in the Hamiltonian can be expressed also as

$$\delta H = \sum_{i=1}^{n} \left[\frac{\partial H}{\partial p_i} \delta p_i + \frac{\partial H}{\partial q_i} \delta q_i \right] \tag{5.117}$$

On comparing (5.115) and (5.117), we note that

$$\dot{q}_i = \frac{\partial H}{\partial p_i} \tag{5.118a}$$

$$-\frac{\partial L}{\partial q_i} = \frac{\partial H}{\partial q_i} \tag{5.118b}$$

From the defining equation (5.113), we get

$$\dot{p}_i = \frac{d}{dt}\left(\frac{\partial L}{\partial \dot{q}_i}\right) \tag{5.119}$$

and after noting that for a holonomic system from (5.91) we have

$$\frac{d}{dt}\left(\frac{\partial L}{\partial \dot{q}_i}\right) = \frac{\partial L}{\partial q_i} + Q_{nc,i}$$

it follows that

$$\dot{p}_i = \frac{\partial L}{\partial q_i} + Q_{nc,i}$$

After employing (5.118b) in the foregoing equation, the result becomes

$$\dot{p}_i = -\frac{\partial H}{\partial q_i} + Q_{nc,i} \tag{5.120}$$

Equations (5.118a) and (5.120) now constitute a set of $2n$ first-order equations

$$\dot{q}_i = \frac{\partial H}{\partial p_i}$$

$$\dot{p}_i = -\frac{\partial H}{\partial q_i} + Q_{nc,i} \qquad i = 1, \ldots, n \tag{5.121}$$

These equations are known as Hamilton's canonic equations of motion. It is noted that the first half of the foregoing equations is a result of the definition of the Hamiltonian and the second half reflects the Lagrange equations of motion. For a nonholonomic system, it follows from (5.82) that

$$\frac{d}{dt}\left(\frac{\partial L}{\partial \dot{q}_i}\right) = \frac{\partial L}{\partial q_i} + Q_{nc,i} + \sum_{j=1}^{R} \lambda_j a_{ji}$$

Substituting this result in (5.119), the Hamilton's canonic equations of motion for a nonholonomic system are expressed as

$$\dot{q}_i = \frac{\partial H}{\partial p_i}$$

$$\dot{p}_i = -\frac{\partial H}{\partial q_i} + Q_{nc,i} + \sum_{j=1}^{R} \lambda_j a_{ji} \qquad i = 1, 2, \ldots, n \tag{5.122}$$

Equations (5.121) or (5.122) have been expressed as a set of $2n$ first-order equations

$$\{\dot{x}\} = \{f(x_1, \ldots, x_{2n}, Q_1, \ldots, Q_n, t)\}$$

where the state-variable vector is defined by

$$\{x\} = \left\{ \begin{array}{c} \{q\} \\ \hline \{p\} \end{array} \right\}$$

We note that the Lagrangian function has been defined by

$$L = T - U$$

$$= T_2 + T_1 + T_0 - U$$

where we have employed (5.50) for the general expression for the kinetic energy. Hence, it follows that

$$\frac{\partial L}{\partial \dot{q}_i}\dot{q}_i = 2T_2 + T_1$$

Substituting this result in (5.114), we obtain

$$H = 2T_2 + T_1 - L$$

$$= 2T_2 + T_1 - T_2 - T_1 - T_0 + U$$

$$= T_2 - T_0 + U$$

Now, if $T_1 = T_0 = 0$ such that $T = T_2$, we have

$$H = T + U = E$$

that is, under these restricted conditions, the Hamiltonian can be defined as the total mechanical energy.

Example 5.14

We consider again the dynamic system of Example 5.11. In Example 5.13 we have obtained the state equations for this system by employing the generalized coordinates and generalized velocities as state variables. In this example, we obtain the Hamilton's canonic equations, that is, the state equations by employing the generalized coordinates and generalized momenta as the state variables. We note from (5.99) that the Lagrangian has been obtained as

$$L = \tfrac{1}{2}mr^2\dot{\theta}^2 + \tfrac{1}{2}m\dot{r}^2 - \tfrac{1}{2}k(r - a)^2 - (c_1 - r\cos\theta)mg$$

We define two generalized momenta coordinates as

$$p_1 = \frac{\partial L}{\partial \dot{r}} = m\dot{r}$$

$$p_2 = \frac{\partial L}{\partial \dot{\theta}} = mr^2\dot{\theta}$$

The Hamiltonian function becomes

$$H = p_1\dot{r} + p_2\dot{\theta} - L$$
$$= p_1\dot{r} + p_2\dot{\theta} - \tfrac{1}{2}mr^2\dot{\theta}^2 - \tfrac{1}{2}m\dot{r}^2 + \tfrac{1}{2}k(r - a)^2 + (c_1 - r\cos\theta)\,mg \quad (5.123)$$

Solving for $\dot{r}$ and $\dot{\theta}$ in terms of p_1 and p_2 from the foregoing equations, we obtain

$$\dot{r} = \frac{1}{m}p_1$$

$$\dot{\theta} = \frac{1}{mr^2}p_2$$

Substituting this result in (5.123), the Hamiltonian is expressed as a function of r, θ, p_1, and p_2. After simplification, we obtain

$$H = \frac{1}{2}\frac{1}{m}p_1^2 + \frac{1}{2}\frac{1}{mr^2}p_2^2 + \frac{1}{2}k(r - a)^2 + (c_1 - r\cos\theta)mg$$

Since in this example, $T_1 = T_0 = 0$ and $T = T_2$, it follows that here the Hamiltonian is the total mechanical energy. Now, the Hamilton's equations (5.121) become

$$\dot{r} = \frac{\partial H}{\partial p_1} = \frac{1}{m}p_1$$

$$\theta = \frac{\partial H}{\partial p_2} = \frac{1}{mr^2}p_2$$

$$\dot{p}_1 = -\frac{\partial H}{\partial r} + Q_{nc,\,r} = \frac{1}{mr^3}p_2^2 - k(r - a) + mg\cos\theta$$

$$\dot{p}_2 = -\frac{\partial H}{\partial \theta} + Q_{nc,\,\theta} = -mgr\sin\theta - \frac{c}{mr^2}p_2$$

where $Q_{nc,\,r} = 0$ and $Q_{nc,\,\theta} = -c\dot{\theta} = -(c/mr^2)p_2$.

Choosing state variables as $x_1 = r$, $x_2 = \theta$, $x_3 = p_1$, and $x_4 = p_2$, the state equations may also be written as

$$\dot{x}_1 = \frac{1}{m} x_3$$

$$\dot{x}_2 = \frac{1}{mx_1^2} x_4$$

$$\dot{x}_3 = \frac{1}{mx_1^3} x_4^2 - k(x_1 - a) + mg \cos x_2 \tag{5.124}$$

$$\dot{x}_4 = -mgx_1 \sin x_2 - \frac{c}{mx_1^2} x_4$$

On comparing (5.111) and (5.124), it is seen that these two sets of equations are quite different from each other even though they describe the equations of motion of the same dynamic system. Only for linear time-invariant equations, the state equations can be expressed in the form

$$\{\dot{x}\} = [A]\{x\} + [B]\{Q\}$$

where $[A]$ and $[B]$ are constant matrices. Selection of different state variables to describe the motion of the same dynamic system leads to different $[A]$ matrices which are all similar matrices and reduce to the same Jordan normal form.

5.7 EULER ANGLES AND LAGRANGE EQUATIONS FOR RIGID BODIES

We now extend the development of the Lagrange equations of motion to include the general rotation of rigid bodies. In order to describe the orientation of a rigid body, we need in general three independent coordinates. We have seen in Chapter 2 that angular displacements are compounded by the law of matrix multiplication, which is not commutative. Hence, a finite angular displacement is a directed line segment but not a vector. Consequently, the angular velocity components ω_1, ω_2, and ω_3 about the body axes cannot be integrated to obtain the angular displacements about those axes. The direction cosines also cannot be used as generalized coordinates since they are not independent but are related by a constraint. A set of generalized coordinates that may be selected to describe the orientation of a rigid body consists of Euler angles.

The choice of Euler angles is not unique but they involve three successive angular displacements for the transformation from a set of Cartesian coordinates to another. The rotations, however, are not about three orthogonal axes. Then, the three components of the angular velocity of a rigid body are expressed in terms of Euler angles and their time derivatives. In the following, we describe a commonly employed method for the selection of Euler angles.

5.7.1 Euler Angles

We select the origin of all the coordinate systems at a point in the rigid body. If the origin is fixed in the *inertial space*, the xyz coordinate system is referred to the inertial reference frame, whereas if the origin is moving, then xyz coordinate system is a moving frame. We assume that the moving frame always remains parallel to some inertial reference frame.

The rectangular frame 1–2–3 of Fig. 5.15 is assumed to be a *body coordinate* system (i.e., rigid body is rigidly connected to this frame). We also assume that the body coordinates represent the principal directions of the rigid body. We are interested in describing the location of the 1–2–3 frame with respect to the xyz frame. We assume an auxiliary frame $x'y'z'$. A sequence of rotations is used for the xyz frame in order that it coincide with the 1–2–3 frame. First a rotation ϕ about z axis is given to bring axis x into coincidence with x' axis. Next, a rotation θ about the x' axis is used to bring the moving frame into coincidence with the $x'y'z'$ frame. Finally, a rotation ψ about the z' axis is used to bring the moving frame into coincidence with the body frame 1–2–3. The three individual rotation angles, ϕ, θ, and ψ are called *Euler's angles*.

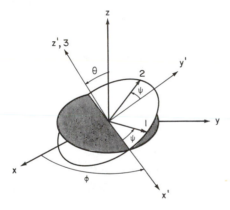

Figure 5.15 Euler angles.

Conversely, if the body frame is in a given orientation, the corresponding Euler's angles are determined as follows. The angle θ is measured directly between the z and 3 axes. The x' axis is the perpendicular to the plane formed by the z and 3 axes. The angles ϕ and ψ are measured from the x' axis to the x axis and to the 1 axis, respectively. When the body changes its orientation, the Euler's angles change. Their time rate of change (i.e., $\dot{\phi}$, $\dot{\theta}$, and $\dot{\psi}$) are the angular velocity components directed along z, x' and 3 axes, respectively. The resultant angular velocity of the body with respect to xyz reference frame is

$$\vec{\Omega} = \dot{\phi}\vec{u}_z + \dot{\theta}\vec{u}_{x'} + \dot{\psi}\vec{u}_3 \qquad (5.125)$$

where $\vec{u}_z, \vec{u}_{x'}$, and $\vec{u}_3$ are the unit vectors along the respective coordinate directions.

The vector $\vec{\Omega}$ can be decomposed into components with respect to any convenient frame of reference. We can note the following relationships between the unit vectors directed along the axes in Fig. 5.15:

$$\vec{u}_z = \vec{u}_1 \sin\theta \sin\psi + \vec{u}_2 \sin\theta \cos\psi + \vec{u}_3 \cos\theta$$
$$\vec{u}_{x'} = \vec{u}_1 \cos\psi - \vec{u}_2 \sin\psi \qquad (5.126)$$
$$\vec{u}_{y'} = \vec{u}_1 \sin\psi + \vec{u}_2 \cos\psi$$

In fact we can transform from one coordinate system to another by employing the rotational transformation matrices discussed in Sections 2.6 and 2.7. We have

$$[C_1(\phi)] = \begin{bmatrix} \cos\phi & \sin\phi & 0 \\ -\sin\phi & \cos\phi & 0 \\ 0 & 0 & 1 \end{bmatrix}$$

$$[C_2(\theta)] = \begin{bmatrix} 1 & 0 & 0 \\ 0 & \cos\theta & \sin\theta \\ 0 & -\sin\theta & \cos\theta \end{bmatrix}$$

$$[C_3(\psi)] = \begin{bmatrix} \cos\psi & \sin\psi & 0 \\ -\sin\psi & \cos\psi & 0 \\ 0 & 0 & 1 \end{bmatrix}$$

The transformation between the $x'y'z'$ axes and the 1–2–3 axes is given by

$$\begin{Bmatrix} u_1 \\ u_2 \\ u_3 \end{Bmatrix} = [C_3(\psi)] \begin{Bmatrix} u_{x'} \\ u_{y'} \\ u_{z'} \end{Bmatrix}$$

or

$$\begin{Bmatrix} u_{x'} \\ u_{y'} \\ u_{z'} \end{Bmatrix} = [C_3(\psi)]^T \begin{Bmatrix} u_1 \\ u_2 \\ u_3 \end{Bmatrix}$$

because $[C_3(\psi)]$ represents an orthonormal transformation between two Cartesian systems. The transformation between the xyz and the $x'y'z'$ axes is given by

$$\begin{Bmatrix} u_{x'} \\ u_{y'} \\ u_{z'} \end{Bmatrix} = [C_2(\theta)][C_1(\phi)] \begin{Bmatrix} u_x \\ u_y \\ u_z \end{Bmatrix}$$

where $\qquad [C_2(\theta)][C_1(\phi)] = \begin{bmatrix} \cos\phi & \sin\phi & 0 \\ -\sin\phi\cos\theta & \cos\phi\cos\theta & \sin\theta \\ \sin\phi\sin\theta & -\cos\phi\sin\theta & \cos\theta \end{bmatrix}$

The transformation between the xyz axes and the 1–2–3 axes is obtained from

$$\begin{Bmatrix} u_1 \\ u_2 \\ u_3 \end{Bmatrix} = [C_3(\psi)][C_2(\theta)][C_1(\phi)] \begin{Bmatrix} u_x \\ u_y \\ u_z \end{Bmatrix}$$

$$= [C] \begin{Bmatrix} u_x \\ u_y \\ u_z \end{Bmatrix}$$

or

$$\begin{Bmatrix} u_x \\ u_y \\ u_z \end{Bmatrix} = [C]^T \begin{Bmatrix} u_1 \\ u_2 \\ u_3 \end{Bmatrix}$$

where $[C] = \begin{bmatrix} \cos\phi\cos\psi - \sin\phi\cos\theta\sin\psi & \sin\phi\cos\psi + \cos\phi\cos\theta\sin\psi & \sin\theta\sin\psi \\ -\cos\phi\sin\psi - \sin\phi\cos\theta\cos\psi & -\sin\phi\sin\psi + \cos\phi\cos\theta\cos\psi & \sin\theta\cos\psi \\ \sin\phi\sin\theta & -\cos\phi\sin\theta & \cos\theta \end{bmatrix}$

Substituting the result from (5.126) in (5.125), we can obtain the angular velocity in terms of components along the body axes 1, 2, and 3 as

$$\vec{\Omega} = \vec{\omega} = (\dot{\theta}\cos\psi + \dot{\phi}\sin\psi\sin\theta)\vec{u}_1 + (\dot{\phi}\sin\theta\cos\psi - \dot{\theta}\sin\psi)\vec{u}_2$$
$$+ (\dot{\phi}\cos\theta + \dot{\psi})\vec{u}_3 \qquad (5.127)$$

where $\vec{\omega}$ is the angular velocity of the body axes, or along the inertial axes x, y, and z as

$$\vec{\Omega} = (\dot{\theta}\cos\psi + \dot{\psi}\sin\theta\sin\psi)\vec{u}_x + (\dot{\theta}\sin\phi - \dot{\psi}\sin\theta\cos\phi)\vec{u}_y$$
$$+ (\dot{\psi}\cos\theta + \dot{\phi})\vec{u}_z \qquad (5.128)$$

or along the auxiliary axes x', y', and z' in the form

$$\vec{\Omega} = \dot{\theta}\vec{u}_{x'} + \dot{\phi}\sin\theta\vec{u}_{y'} + (\dot{\phi}\cos\theta + \dot{\psi})\vec{u}_{z'} \qquad (5.129)$$

5.7.2 Euler's Equation for a Rigid Body

We consider the rotation of the rigid body about O fixed in the body. The body is subjected to an external torque $\vec{M}_o$ with components M_1, M_2, and M_3 along the three principal directions. Let the principal moments of inertia about O be I_1, I_2, and I_3. The kinetic energy T of the body is

$$T = \tfrac{1}{2}I_1\omega_1^2 + \tfrac{1}{2}I_2\omega_2^2 + \tfrac{1}{2}I_3\omega_3^2$$
$$= \tfrac{1}{2}I_1(\dot{\theta}\cos\psi + \dot{\phi}\sin\psi\sin\theta)^2 + \tfrac{1}{2}I_2(\dot{\phi}\sin\theta\cos\psi - \dot{\theta}\sin\psi)^2$$
$$+ \tfrac{1}{2}I_3(\dot{\phi}\cos\theta + \dot{\psi})^2 \qquad (5.130)$$

Substituting for T from (5.130) in the Lagrange's equations (5.58), the equations of motion in the three generalized coordinates ϕ, θ, and ψ are obtained as

$$I_1\dot{\omega}_1 + (I_3 - I_2)\omega_3\omega_2 = M_1$$
$$I_2\dot{\omega}_2 + (I_1 - I_3)\omega_1\omega_3 = M_2 \qquad (5.131)$$
$$I_3\dot{\omega}_3 + (I_2 - I_1)\omega_2\omega_1 = M_3$$

These equations are the same Euler equations that were obtained in Chapter 4.

Example 5.15

Using Hamilton's equations, derive an equation of motion for a top on a horizontal surface (Fig. 5.16). Assume the tip of the top to remain at a fixed point O.

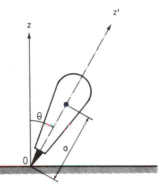

Figure 5.16   Spinning top.

Since the tip of the top is considered to remain at a fixed point, this requires the existence of a horizontal reactive force by the surface on which the top spins. The axes (x', y', z') are fixed in the top with the origin at the tip O; the z' axis is the axis of symmetry. Let the moments of inertia of the top about (x', y', z') axes be I_1, I_2, and I_3, respectively. But because of symmetry $I_1 = I_2$. The components of the angular velocity $\overline{\omega}$ in the (x', y', z') axes are ω_1, ω_2, and ω_3. The kinetic energy T of the top is given as

$$T = \tfrac{1}{2}(I_1\omega_1^2 + I_2\omega_2^2 + I_3\omega_3^2) \qquad (5.132)$$

Using the Euler angles θ, ϕ, and ψ as the generalized coordinates, the angle of the axis of the top with the vertical is θ. The angle of plane zOz' with a fixed vertical plane is ϕ. The angle of rotation of the top about z' axis is ψ.

Substituting the values of ω_1, ω_2, and ω_3 from (5.127) in (5.132), we get

$$T = \tfrac{1}{2}I_1(\dot{\theta}_2 + \dot{\phi}^2\sin^2\theta) + \tfrac{1}{2}I_3(\dot{\phi}\cos\theta + \dot{\psi})^2$$

The potential energy of the top is

$$U = mga\cos\theta$$

where m is the mass of the top and a is the distance from the origin O to the center of mass.

The Lagrangian function $L = T - U$. As U does not depend on the time derivatives $(\dot{\theta}, \dot{\phi}, \dot{\psi})$, $\partial L/\partial\dot{\theta} = \partial T/\partial\dot{\theta}$, and so on. The components of the generalized momenta are

$$p_1 = \frac{\partial T}{\partial \dot{\theta}} = I_1 \dot{\theta}$$

$$p_2 = \frac{\partial T}{\partial \dot{\phi}} = (I_1 \sin^2 \theta + I_3 \cos^2 \theta)\dot{\phi} + I_3 \dot{\psi} \cos \theta$$

$$p_3 = \frac{\partial T}{\partial \dot{\psi}} = I_3(\dot{\psi} + \dot{\phi} \cos \theta)$$

Solving for $\dot{\theta}$, $\dot{\phi}$, and $\dot{\psi}$ in terms of p_1, p_2, and p_3, we get

$$\dot{\theta} = \frac{p_1}{I_1}$$

$$\dot{\phi} = \frac{p_2 - p_3 \cos \theta}{I_1 \sin^2 \theta}$$

$$\dot{\psi} = \frac{p_3}{I_3} - \left(\frac{p_2 - p_3 \cos \theta}{I_1 \sin^2 \theta}\right) \cos \theta$$

The Hamilton function H is given as

$$H = p_1 \dot{\theta} + p_2 \dot{\phi} + p_3 \dot{\psi} - \tfrac{1}{2}I_1(\dot{\theta}^2 + \dot{\phi}^2 \sin^2 \theta) - \tfrac{1}{2}I_3(\dot{\phi} \cos \theta + \dot{\psi})^2 + mga \cos \theta$$

Eliminating $\dot{\theta}$, $\dot{\phi}$, and $\dot{\psi}$, we get

$$H = \frac{p_1^2}{2I_1} + \frac{(p_2 - p_3 \cos \theta)^2}{2I_1 \sin^2 \theta} + \frac{p_3^2}{2I_3} + mga \cos \theta$$

Hamilton's equations (5.121) become

$$\dot{\theta} = \frac{\partial H}{\partial p_1}, \qquad \dot{\phi} = \frac{\partial H}{\partial p_2}, \qquad \dot{\psi} = \frac{\partial H}{\partial p_3}$$

$$\dot{p}_1 = -\frac{\partial H}{\partial \theta}, \qquad \dot{p}_2 = -\frac{\partial H}{\partial \phi}, \qquad \dot{p}_3 = -\frac{\partial H}{\partial \psi}$$

Thus, the last three equations are written as

$$\dot{p}_1 = -\frac{(p_2 - p_3 \cos \theta)(p_3 - p_2 \cos \theta)}{I_1 \sin^3 \theta} + mga \sin \theta$$

$$\dot{p}_2 = 0$$

$$\dot{p}_3 = 0$$

Therefore, $p_2 = I_1 C = $ constant and $p_3 = I_3 D = $ constant. Hence,

$$\dot{\psi} + \dot{\phi} \cos \theta = D = \text{constant}$$

$$\dot{\phi} \sin^2 \theta + bD \cos \theta = C = \text{constant}; \qquad b = \frac{I_3}{I_1}$$

but

$$\dot{p}_1 = I_1 \ddot{\theta}$$

Hence, it follows that

$$\ddot{\theta} = -\frac{(I_1 C - I_3 D \cos \theta)(I_3 D - I_1 C \cos \theta)}{I_1^2 \sin^3 \theta} + \frac{mga \sin \theta}{I_1}$$

We obtain by integration

$$\dot{\theta}^2 = -\frac{(C^2 + b^2 D^2 - 2bCD \cos \theta)}{\sin^2 \theta} - \gamma \cos \theta + \text{constant}$$

where

$$\gamma = \frac{2mga}{I_1}$$

Assuming the additive constant in the preceding equation as $N + b^2 D^2$ and using the substitution $u = \cos \theta$, we obtain

$$\dot{u}^2 = (1 - u^2)(N - \gamma u) - (C - bDu)^2$$

The other equations are

$$\dot{\psi} + \dot{\phi}u = D$$
$$\dot{\phi}(1 - u^2) + bDu = C$$

These are the three first-order differential equations which determine θ, ϕ, and ψ as functions of time.

5.8 SUMMARY

In this chapter the equations of motion have been derived by methods based on variational principles. As discussed earlier, this formulation has several advantages. The use of generalized coordinates instead of the physical coordinates makes the formulation quite versatile. The Lagrange method offers a powerful alternative approach to the method of direct application of Newton's laws to obtain the equations of motion. In the method of direct application of Newton's laws, the constraint forces appear in the equations and have to be eliminated to obtain the equations of motion. In the Lagrangian formulation, constraint forces are ignored since they do not perform work in virtual displacement. However, some of the impressed forces, such as frictional force, may depend on the constraint forces, and also in some applications it may be necessary to evaluate the constraint forces for the purpose of stress analysis and design. Simultaneous use of both methods could be employed in such cases. Both methods could also be used as a check of the results.

The Lagrange equations of motion consist of a set of simultaneous second-order differential equations which are generally nonlinear in the generalized coordinates and velocities. By choosing generalized coordinates and generalized velocities as the state variables, the equations can be expressed as a set of coupled first-order equations. An alternative approach is offered by the Hamiltonian formulation, where generalized coordinates and generalized momenta are chosen as the state variables to represent Lagrange equations as a set of first-order equations. General rotation of rigid bodies can be studied by including Euler's angles among the generalized coordinates.

PROBLEMS

5.1. Consider the bead of Problem 3.1. Determine all equilibrium positions x_e of the bead by the principle of virtual work:
   (a) Neglecting friction between the bead and wire.
   (b) Including friction between the bead and wire.

5.2. A uniform rigid bar of mass m and length b is supported as shown in Fig. P5.2. Neglecting friction at the supports, determine the equilibrium position θ_e by the principle of virtual work.

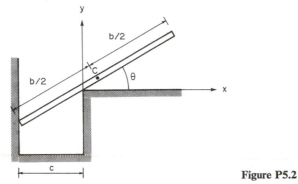

Figure P5.2

5.3. Obtain the Lagrange equation of motion for the bead of Problem 3.1, including friction between the bead and wire.

5.4. Obtain the Lagrange equations of motion for the system of Example 3.4.

5.5. Mass m_2 is pivoted at the center of mass m_1 by a rigid massless link of length R (Fig. P5.5). Neglect friction at the pivot. The motion is in the vertical plane. Choosing y and θ as the generalized coordinates, obtain the Lagrange equations of motion.

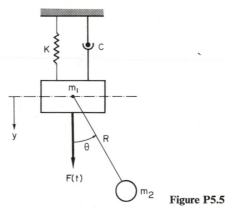

Figure P5.5

5.6. A uniform rod of mass m and length b moves on the horizontal xy plane without friction (P5.6). At one end A, it has a knife-edge constraint which prevents a velocity component perpendicular to the rod at that point.
   (a) Write the nonholonomic constraint relating x, y, and θ in the form of a Pfaffian.
   (b) Using x, y, and θ as coordinates, obtain the Lagrange equations of motion.
   (c) Show that the Lagrange multiplier λ represents the transverse force of constraint at end A.

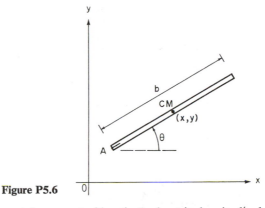

Figure P5.6

5.7. A cylinder of mass m_2 and moment of inertia I_2 about its longitudinal axis rolls without slipping on a wedge (Fig. P5.7). The wedge slides on the floor under the action of an applied force $F(t)$. There is friction between the cylinder and wedge and the coefficient of sliding friction between the wedge and the floor is μ. Choosing x_1 and x_2 as generalized coordinates, obtain the Lagrange equations of motion.

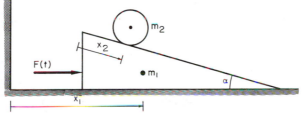

Figure P5.7

5.8. A uniform rod of mass m and length b is released from rest and slides in the vertical plane (Fig. P5.8). The coefficient of sliding friction between the rod and the ground is μ. Obtain the Lagrange equations of motion for the rod.

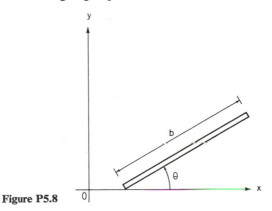

Figure P5.8

5.9. A particle of mass m is connected by a massless spring of stiffness k and unstressed length r_o to a point P which is moving along a circular path of radius a at a uniform angular velocity ω_0 (Fig. P5.9). The coefficient of friction between the particle and the horizontal plane on which it moves is μ. Obtain the Lagrange equations of motion.

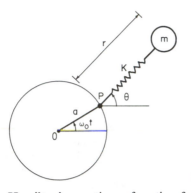

Figure P5.9

5.10. Obtain Hamilton's equations of motion for the system of Problem 5.4.

5.11. Obtain Hamilton's equations of motion for the system of Problem 5.7.

REFERENCES

1. Goldstein, H., *Classical Mechanics*, Addison-Wesley Publishing Company, Inc., Reading, Mass., 1950.

2. Meirovitch, L., *Methods of Analytical Dynamics*, McGraw-Hill Book Company, New York, 1970.

3. Rosenberg, R. M., *Analytical Dynamics of Discrete Systems*, Plenum Press, New York, 1977.

4. Greenwood, D. T., *Classical Dynamics*, Prentice-Hall, Inc., Englewood Cliffs, N. J., 1977.

5. Synge, J. L., and Griffith, B. A., *Principles of Mechanics*, 3rd ed., McGraw-Hill Book Company, New York, 1959.

6

RESPONSE OF DYNAMIC SYSTEMS

6.1 INTRODUCTION

The main objectives of the previous chapters have been the mathematical modeling and formulation of the equations of motion. This chapter is concerned with the solution of the equations of motion that were formulated. We first discuss the state-space formulation of the equations of motion. This formulation permits the application of many mathematical techniques, such as Lyapunov stability theory, which have been developed for a system of first-order ordinary differential equations. The existence and uniqueness of the solution to the equations of motion are studied next.

It should be noted that in most nonlinear problems, it is not possible to obtain a closed-form analytic solution to the equations of motion. Hence in such cases, a computer simulation is generally used for the response analysis and these techniques are disucssed in the next chapter. In this chapter, we concentrate on linear time-invariant equations of motion. Such equations usually represent perturbations from an equilibrium state or from a stationary motion.

For this restricted case of linear time-invariant equations of motions we employ the state transition matrix in order to obtain the response to time-varying forces and moments. Finally, in order to gain insight into the dynamic behavior of such systems, we consider coordinate transformation so that the coefficient matrix of the state equations is represented in the Jordan canonical or normal form, thereby exhibiting its eigenvalues along the main diagonal.

Such transformation will also be useful in Chapter 8 for the normal-mode analysis of linear vibrations.

6.2 STATE-SPACE REPRESENTATION

When time-domain analysis of the equations of motion is desired, it is preferable to express the equations as a set of first-order differential equations. This formulation permits direct application of many mathematical methods that have been developed for a set of first-order ordinary differential equations. Consider a system with k degrees of freedom and let $q_1, \ldots, q_k$ be the generalized displacement coordinates. We recall that the equations of motion that were derived in Chapters 3 and 4 by the direct application of Newton's law and the Lagrange equations derived in Chapter 5 consist of a set of k second-order coupled equations in the generalized coordinates. Defining generalized velocity coordinates $\dot{q}_1, \ldots, \dot{q}_k$ and an n-dimensional vector $\{x\}$, where $n = 2k$, as

$$\{x\} = \left\{ \frac{\{q\}}{\{\dot{q}\}} \right\} \tag{6.1}$$

the equations of motion can be expressed as a set of first-order coupled equations in the form

$$\{\dot{x}\} = \{f(x_1, \ldots, x_n, Q_1, \ldots, Q_m, t)\} \tag{6.2}$$

In the foregoing equation, $Q_1, \ldots, Q_m$ are the input forces and moments and f_i, being an explicit function of time t, indicates that the parameters such as mass may be time varying. The n-dimensional column vector is called the state vector. In the Hamiltonian formulation of Chapter 5, the state vector $\{x\}$ consists of k generalized coordinates $q_1, \ldots, q_k$, and k generalized momenta coordinates $p_1, \ldots, p_k$ in the form

$$\{x\} = \left\{ \frac{\{q\}}{\{p\}} \right\} \tag{6.3}$$

and Hamilton's equations have been already expressed in the state-variable form of (6.2). In some cases considered in the previous chapters, some of the generalized displacement coordinates are ignorable and need not appear in the equations of motion, which, however, include the generalized velocities. In such cases, the state-variable vector $\{x\}$ need not include the ignorable displacement coordinates and its dimension n will be less than $2k$, where k are the degrees of freedom. For example, the Euler equations of motion of a rigid body were derived in Chapter 4 and are described by

$$\dot{\omega}_1 = -\frac{I_3 - I_2}{I_1}\omega_2\omega_3 + \frac{M_1}{I_1}$$

$$\dot{\omega}_2 = -\frac{I_1 - I_3}{I_2}\omega_3\omega_1 + \frac{M_2}{I_2} \tag{6.4}$$

$$\dot{\omega}_3 = -\frac{I_2 - I_1}{I_3}\omega_1\omega_2 + \frac{M_3}{I_3}$$

where ω_i is the angular velocity, M_i the applied external moment, and I_i is the principal moment of inertia for $i = 1, 2, 3$. Here, the angular displacements do not appear in the equations of motion and are ignorable coordinates. Even though we consider three degrees of freedom, we define only a three-dimensional state-variable vector $\{x\}$ as $\{x\} = \{\omega\}$ and then (6.4) are already in the standard form of state equations given by (6.2) with $Q_i = M_i$ ($i = 1, 2, 3$) and f_i is not an explicit function of time since the parameter I_i is a constant.

Each state $\{x\}$ of a system may be represented as a point in an n-dimensional Euclidean space whose coordinates are $x_1, \ldots, x_n$ as shown in Fig. 6.1 and may be viewed as an n-dimensional vector $\mathbf{x}$. The Euclidean space E^n is a linear vector space which is complete, normed, and where an inner product has been defined. The norm and inner product are defined, respectively, by

$$\| \mathbf{x} \| = (\sum_{i=1}^{n} x_i^2)^{1/2} \tag{6.5}$$

and

$$\langle \mathbf{x}^T, \mathbf{x} \rangle = \sum_{i=1}^{n} x_i^2 = \| \mathbf{x} \|^2 \tag{6.6}$$

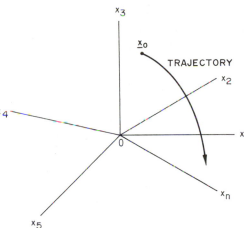

Figure 6.1 State space.

The norm of a state is the distance in the n-dimensional space of the state from the origin. This space, which is called the state space of the system, permits us to extend the concepts of the geometry of motion of a single particle in a physical space which is at most three-dimensional to the motion of a dynamic system in an n-dimensional space. When all the state variables consist of generalized displacements and generalized velocities as in (6.1) or of generalized displacements and generalized momenta as in (6.3), the state space is also referred to as phase space, after Gibbs.

Given an initial state $\mathbf{x}_0$ at time $t = t_0$ and specified input forces and moments $\mathbf{Q}(t)$, the system state will change from $\mathbf{x}_0$ with time. The set of values that the state takes at times $t > t_0$ is denoted by $\mathbf{x}(t)$ or, more specifically, by $\mathbf{x}(\mathbf{x}_0, t_0; \mathbf{Q}, t)$. The set of points traced out by $\mathbf{x}(\mathbf{x}_0, t_0; \mathbf{Q}, t)$ is called the state trajectory of the system. Hence, a state trajectory of the dynamic system of (6.2) is a particular solution when the initial conditions $\mathbf{x}_0$ at time t_0 and the inputs $\mathbf{Q}(t)$ are specified. It should be noted that time t plays the role of a parameter along system trajectory in state space. It is possible to introduce an additional time coordinate t and to define an $(n + 1)$-dimensional space $(\mathbf{x}, t)$ called the motion space. We shall employ the state space, not the motion space, in our analysis.

When all the n initial conditions $\mathbf{x}_0$ for the system of (6.2) are specified at the initial time t_0, the problem is called the initial value problem of ordinary differential equations. There are some applications where some of the conditions are specified at the initial time t_0 and some at the terminal time t_f. For example, in certain control problems the inputs $\mathbf{Q}(t)$ must be synthesized such that the state of the system of (6.2) is changed from a certain initial state to a terminal state which may be partly specified. This class of problems is called the boundary value problem of ordinary differential equations. In our analysis, we shall be concerned only with the initial value problem of ordinary differential equations.

6.3 EXISTENCE AND UNIQUENESS OF SOLUTIONS

In this section we are concerned with the existence and uniqueness of solutions of initial value problems of ordinary differential equations. The existence of a solution to the equations of motion cannot always be implied from the fact that a dynamic system will respond to external forces and moments. Many assumptions are made in modeling dynamic systems, and when the equations of motion have no solution, this may be an indication that the equations do not adequately represent the dynamic system. It is more important to consider the uniqueness of a solution since a dynamic system can have nonunique modes of behavior. As pointed out earlier, a computer simulation is generally employed to obtain the response of nonlinear equations of motion and when a solution has been obtained, its uniqueness should not be taken for granted.

In the following, we consider the conditions that are sufficient to guarantee the existence and uniqueness of the initial value problem of ordinary differential equations. Again, it should be noted that these conditions are not necessary and sufficient but only sufficient conditions and when they are not satisfied, it does not imply that there is no solution or that it is not unique. After completely specifying the input forces and moments $Q_1, \ldots, Q_m$ for time $t > t_0$, the initial value problem of (6.2) is expressed as

$$\dot{\mathbf{x}} = \mathbf{f}(\mathbf{x}, t), \qquad \mathbf{x} = \mathbf{x}_0 \quad \text{at } t = t_0 \qquad (6.7)$$

Theorem 6.1: Local Existence and Uniqueness.   For the system of (6.7), let $\mathbf{f}(\mathbf{x}, t)$ be continuous with respect to $\mathbf{x}$ and t in a region R of the state space defined by the ball $\|\mathbf{x} - \mathbf{x}_0\| \leq a$ and in the time interval $|t - t_0| \leq b$, where $a, b > 0$. If there exist finite positive constants k and h, where $0 < k, h < \infty$ such that for any two vectors $\mathbf{x}_\alpha$ and $\mathbf{x}_\beta$ in region R, the conditions

$$\|\mathbf{f}(\mathbf{x}_\alpha, t) - \mathbf{f}(\mathbf{x}_\beta, t)\| \leq k \|\mathbf{x}_\alpha - \mathbf{x}_\beta\|, \qquad \mathbf{x}_\alpha, \mathbf{x}_\beta \in R, \quad |t - t_0| \leq b \qquad (6.8)$$

$$\max \|\mathbf{f}(\mathbf{x}, t)\| = h, \qquad \mathbf{x} \in R, \quad |t - t_0| \leq b \qquad (6.9)$$

are satisfied, then there exists a unique solution to (6.7) in R for $|t - t_0| \leq c$ with c obeying

$$c \leq \min \left\{ b, \frac{a}{h} \right\} \qquad (6.10)$$

Remarks

1. The condition (6.8) is known as a Lipschitz condition and the constant k is known as a Lipschitz constant. If k is a Lipschitz constant for the function $\mathbf{f}(\mathbf{x}, t)$ so is any constant larger than k. To satisfy the condition (6.8), every component $f_i(\mathbf{x}, t)$ of the vector function $\mathbf{f}(\mathbf{x}, t)$ must satisfy a Lipschitz condition, where the Lipschitz constant may be different for each component.

2. We note that condition (6.8) is a local Lipschitz condition because it holds for all $\mathbf{x}_\alpha$ and $\mathbf{x}_\beta$ in some ball around $\mathbf{x}_0$ (i.e., in the region R and for time such that $|t - t_0| \leq b$). Accordingly, Theorem 6.1 is a local existence and uniqueness theorem because it guarantees the existence and uniqueness of solution only in that interval around $\mathbf{x}_0$ and t_0. Stronger conditions for global existence and uniqueness are given by the following theorem.

Theorem 6.2: Global Existence and Uniqueness.   If conditions (6.8) and (6.9) of Theorem 6.1 are satisfied throughout the entire state space E^n and for time $t_0 \leq t < \infty$ (i.e., the constants a and b of Theorem 6.1 are both infinite), then system (6.7) has a unique solution throughout the entire state space for all time $t_0 \leq t < \infty$.

Example 6.1

To illustrate the theorem, we consider a very simple example of a scalar, linear, unforced differential equation described by

$$\dot{x} = -3x$$

Here, $f(x) = -3x$ and hence we obtain

$$\|f(x_\alpha) - f(x_\beta)\| = |f(x_\alpha) - f(x_\beta)|$$
$$= 3|-x_\alpha + x_\beta|$$

Any number k such that $k \gtrless 3$ can be found as the Lipschitz constant for $t_0 \leq t < \infty$ and for any x_α and x_β in the entire state space. Hence, the conditions (6.8) and (6.9) are satisfied with a and b both infinite. We then conclude that this example has a unique solution throughout the entire state space (one-dimensional in this case) for all time $0 \leq t < \infty$. In fact, it is shown later by considering equation (6.70) that a linear time-invariant system always satisfies the conditions for global existence and uniqueness of solution.

Example 6.2

A vehicle is moving down a plane inclined at angle α to the horizontal as shown in Fig. 6.2, with a resistive force proportional to the square of the velocity. Consider the vehicle as a point mass with a single degree of freedom. The equation of motion may be written as

$$m\ddot{q} + c\dot{q}^2 = mg \sin \alpha \tag{6.11}$$

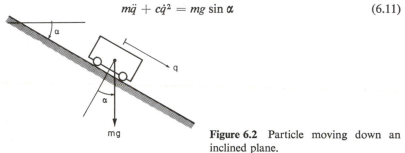

Figure 6.2   Particle moving down an inclined plane.

Letting the state variable $x = \dot{q}$ and $F_0 = mg \sin \alpha$, the equation becomes

$$\dot{x} = -\frac{c}{m}x^2 + \frac{F_0}{m} \tag{6.11a}$$

with initial condition $x_0 = 0$ at time $t_0 = 0$. Here, we have a scalar first-order equation and $f(x) = -(c/m)x^2 + (F_0/m)$ and a graph of $f(x)$ versus x is shown in Fig. 6.3.

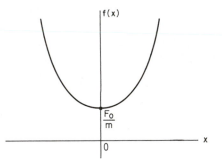

Figure 6.3   Plot of **f(x)** versus **x**.

For a first-order time-invariant system for any two values x_α and x_β of x, the Lipschitz condition can be written as

$$\frac{|f(x_\alpha) - f(x_\beta)|}{|x_\alpha - x_\beta|} \le k \tag{6.12}$$

Condition (6.12) implies that on a plot of $f(x)$ versus x, a straight line joining any two points of $f(x)$ cannot have a slope whose absolute value is greater than k. It is not required that $f(x)$ be continuously differentiable. However, if $f(x)$ is differentiable and the maximum value of $|df/dx|$ in the region R is k, then k is a Lipschitz constant. From this discussion it is clear that a local Lipschitz constant can be found for any interval of finite length about x_0 for this example. However, a finite positive $k < \infty$ cannot be found to satisfy a global Lipschitz condition. Hence, this example satisfies the conditions of Theorem 6.1 but not those of Theorem 6.2. Since these conditions are not necessary, this fact does not imply that this example has a unique solution only in a finite region around the initial condition x_0 and not throughout the state space. An exact closed-form solution for this example can be obtained by separating the variables. Noting that $x_0 = 0$ and $t_0 = 0$, we get

$$\int_0^x \frac{dx'}{1 - (c/F_0)(x')^2} = \frac{F_0}{m} \int_0^t dt' \tag{6.13}$$

Integrating both sides of the foregoing equation, it follows that

$$\frac{1 + \sqrt{c/F_0}\,x}{1 - \sqrt{c/F_0}\,x} = \exp\left(\frac{2}{m}\sqrt{cF_0}\,t\right) \tag{6.14}$$

Solving the equation above for x, we obtain

$$x(t) = \sqrt{\frac{F_0}{c}} \frac{\exp\left[(2/m)\sqrt{cF_0}\,t\right] - 1}{\exp\left[(2/m)\sqrt{cF_0}\,t\right] + 1}$$

$$= \sqrt{\frac{F_0}{c}} \tanh \frac{\sqrt{cF_0}}{m} t \tag{6.15}$$

Example 6.3

Consider a first-order differential equation

$$\dot{x} = \frac{1}{x - 3} \tag{6.16}$$

Here, $f(x) = 1/(x - 3)$ and if a region R around the initial condition x_0 contains the point $x = 3$, it is clear that no finite h can be found at the point $x = 3$ in this region to satisfy condition (6.9) of Theorem 6.1. For $x \neq 3$, we have $df/dx = -1/(x - 3)^2$ and $|df/dx| = 1/(x - 3)^2$, which is finite for $x \neq 3$. Hence, this example does not satisfy the conditions of Theorem 6.2 but only those of Theorem 6.1 in a region that does not contain the point $x = 3$.

Example 6.4

A function $f(x)$ that is discontinuous does not satisfy a Lipschitz condition at the point of discontinuity. Consider a function $f(x)$ which has a unit jump at the point $x = 3$ as shown in Fig. 6.4. Let x_α and x_β be any two values of x on either side of the point $x = 3$ such that $|x_\alpha - x_\beta| = 0.001$. Then from (6.12), the Lipschitz constant is given by

$$k \geq 10^3 |f(x_\alpha) - f(x_\beta)| \tag{6.17}$$

Now let both x_α and x_β tend to the point $x = 3$ from the left and right, respectively. Then, we get $\lim |x_\alpha - x_\beta| \longrightarrow 0$, whereas $\lim |f(x_\alpha) - f(x_\beta)| \longrightarrow 1$. Hence, no finite k can be found to satisfy the Lipschitz condition at the point $x = 3$.

In many applications, dynamic systems are subjected to impulsive forces and moments and $\mathbf{f}(\mathbf{x}, t)$ may be discontinuous with respect to t at a finite or countably infinite number of points. In such cases Theorem 6.2 would not apply, but according to Theorem 6.1, a unique solution may be guaranteed over those time intervals where the discontinuities do not occur.

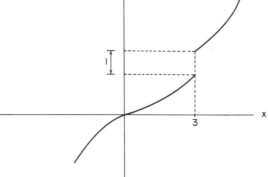

Figure 6.4 Discontinous function.

Proof of Theorem 6.1. A proof of this theorem may be obtained by employing the contraction-mapping fixed-point theorem. However, this method requires some techniques from functional analysis whose knowledge is not assumed here. Hence, we give a proof employing Picard's method of solution of the differential equations in the form of the Liouville–Neumann series. If $\mathbf{x}(t)$ is a solution of (6.7) in the region R defined by $\|\mathbf{x} - \mathbf{x}_0\| \le a$ and in the time interval $|t - t_0| \le b$, then in that region $\mathbf{x}(t)$ also satisfies

$$\mathbf{x}(t) = \mathbf{x}_0 + \int_{t_0}^{t} \mathbf{f}(\mathbf{x}(t'), t') \, dt' \tag{6.18}$$

On the other hand, continuous functions that satisfy (6.18) are differentiable and satisfy (6.7). Equation (6.18) is a nonlinear Volterra integral equation. According to Picard's method, a first approximation to the solution is a function $\mathbf{x}_1$ defined by

$$\mathbf{x}_1 = \mathbf{x}_0 + \int_{t_0}^{t} \mathbf{f}(\mathbf{x}_0, t') \, dt' \tag{6.19}$$

Similarly, we get a sequence of successive approximations as

$$\mathbf{x}_2 = \mathbf{x}_0 + \int_{t_0}^{t} \mathbf{f}(\mathbf{x}_1(t'), t') \, dt'$$

$$\vdots$$

$$\mathbf{x}_m = \mathbf{x}_0 + \int_{t_0}^{t} \mathbf{f}(\mathbf{x}_{m-1}(t'), t') \, dt'$$

We now consider the series

$$\mathbf{x}(t) = \mathbf{x}_0 + (\mathbf{x}_1 - \mathbf{x}_0) + (\mathbf{x}_2 - \mathbf{x}_1) + \cdots + (\mathbf{x}_m - \mathbf{x}_{m-1}) + \cdots \tag{6.20}$$

Equation (6.20) is called the Liouville–Neumann series solution of the nonlinear Volterra integral equation. We now show that when the conditions of Theorem 6.1 are satisfied, the series converses and represents the unique solution of the original equation. First, we prove that any term of the foregoing sequence also belongs to the region R. We have

$$\mathbf{x}_1 - \mathbf{x}_0 = \int_{t_0}^{t} \mathbf{f}(\mathbf{x}_0, t') \, dt'$$

Thus,

$$\|\mathbf{x}_1 - \mathbf{x}_0\| = \left\| \int_{t_0}^{t} \mathbf{f}(\mathbf{x}_0, t') \, dt' \right\|$$

$$\leq \int_{t_0}^{t} \|\mathbf{f}(\mathbf{x}_0, t')\| \, dt'$$

$$\leq h|t - t_0| \qquad \text{from (6.9)} \qquad (6.21)$$

$$\leq hc$$

$$\leq a \qquad \text{from (6.10)} \qquad (6.22)$$

Hence, $\mathbf{x}_1$ belongs to the region R. Similarly, we can show that each $\mathbf{x}_m(t)$ belongs to the region R. Second, we now show that the series (6.20) converges. We have

$$\|\mathbf{x}_2 - \mathbf{x}_1\| \leq \int_{t_0}^{t} \|\mathbf{f}(\mathbf{x}_1(t'), t') - \mathbf{f}(\mathbf{x}_0, t')\| \, dt'$$

$$\leq k \int_{t_0}^{t} \|\mathbf{x}_1 - \mathbf{x}_0\| \, dt' \qquad \text{from (6.8)}$$

$$\leq kh \int_{t_0}^{t} |t' - t_0| \, dt' \qquad \text{from (6.21)}$$

$$\leq kh \frac{(t - t_0)^2}{2}$$

$$\leq kh \frac{c^2}{2}$$

By mathematical induction, we can show that

$$\|\mathbf{x}_m - \mathbf{x}_{m-1}\| \leq \frac{k^{m-1} h c^m}{m!} \qquad (6.23)$$

Hence, by ratio test we conclude that the series (6.20) converges absolutely and uniformly when $\mathbf{x}$ is in region R. Therefore, the limit function $\lim_{m \to \infty} \mathbf{x}_m(t)$ exists and is continuous in R. That $\mathbf{x}(t)$ as defined by (6.20) is indeed a solution of (6.7) can be established by noting that

$$\mathbf{x}(t) = \lim_{m \to \infty} \mathbf{x}_m(t)$$

$$= \mathbf{x}_0 + \lim_{m \to \infty} \int_{t_0}^{t} \mathbf{f}(\mathbf{x}_{m-1}, t') \, dt'$$

$$= \mathbf{x}_0 + \int_{t_0}^{t} \lim_{m \to \infty} \mathbf{f}(\mathbf{x}_{m-1}, t') \, dt'$$

$$= \mathbf{x}_0 + \int_{t_0}^{t} \mathbf{f}(\mathbf{x}, t') \, dt' \qquad (6.24)$$

The interchange of the order of integration and limit in the foregoing can be justified from the fact that

$$\int_{t_0}^{t} \| \mathbf{f}(\mathbf{x}, t') - \mathbf{f}(\mathbf{x}_{m-1}, t') \| \, dt' \le k \int_{t_0}^{t} \| \mathbf{x} - \mathbf{x}_{m-1} \| \, dt'$$

$$\le \frac{k^m h c^m}{m!} \left[1 + \frac{kc}{m+1} + \frac{k^2 c^2}{(m+1)(m+2)} + \cdots \right] |t - t_0| \qquad (6.25)$$

which approaches zero as $m \to \infty$.

Now, it remains to be shown that the solution thus obtained is the unique solution of (6.7). For this purpose, let $\mathbf{y}(t)$ which belongs to region R be another solution of (6.7). Subject to the restriction that $\|\mathbf{y} - \mathbf{x}\| \le d$. Then, we have

$$\mathbf{y} = \mathbf{x}_0 + \int_{t_0}^{t} \mathbf{f}(\mathbf{y}, t') \, dt'$$

$$\mathbf{x}_m = \mathbf{x}_0 + \int_{t_0}^{t} \mathbf{f}(\mathbf{x}_{m-1}, t') \, dt'$$

from which it follows that

$$\| \mathbf{y} - \mathbf{x}_m \| \le \int_{t_0}^{t} \| \mathbf{f}(\mathbf{y}, t') - \mathbf{f}(\mathbf{x}_{m-1}, t') \| \, dt'$$

$$\le k \int_{t_0}^{t} \| \mathbf{y} - \mathbf{x}_{m-1} \| \, dt'$$

$$\le k^m d \frac{(t - t_0)^m}{m!} \qquad (6.26)$$

Since the right-hand member of inequality (6.26) approaches zero as $m \to \infty$, we see that

$$\mathbf{y} = \lim_{m \to \infty} \mathbf{x}_m = \mathbf{x} \qquad (6.27)$$

Hence, when the conditions for the existence and uniqueness of solution are satisfied in a region R and for all times in the interval $|t - t_0| \le b$, for all initial conditions $\mathbf{x}_0$ in the region R there is a trajectory of the system with these initial conditions. This trajectory is unique. Furthermore, the trajectories are continuous with respect to the initial condition $\mathbf{x}_0$ and the initial time t_0.

6.4 LINEARIZED TIME-VARYING SYSTEMS

It can be seen from the previous chapters that the equations of motion are in general nonlinear. We consider perturbations from an equilibrium state or from a nominal motion. When all the nonlinearities are analytic functions of their arguments and the perturbations are sufficiently small, it is possible to linearize the equations that represent the perturbed motion. Let $\mathbf{x}^*(\mathbf{Q}^*, t; \mathbf{x}_0, t_0)$ be a particular trajectory starting at time t_0 with initial conditions $\mathbf{x}_0$ and specified input generalized forces $\mathbf{Q}^*$. In general, this particular motion $\mathbf{x}^*$ can be obtained by computer simulation of the equations of motion of (6.2).

Consider the effect of perturbations $\Delta \mathbf{x}_0$ on the initial state and $\Delta \mathbf{Q}$ on the input forces. Let $\mathbf{x}(\mathbf{Q}^* + \Delta \mathbf{Q}, t; \mathbf{x}_0 + \Delta \mathbf{x}_0, t_0)$ be the resultant perturbed motion. The vector $\Delta \mathbf{x}(t)$ of perturbed state variables is defined by

$$\Delta \mathbf{x}(t) = \mathbf{x}(t) - \mathbf{x}^*(t); \qquad \text{that is, } \mathbf{x}(t) = \mathbf{x}^*(t) + \Delta \mathbf{x}(t) \qquad (6.28)$$

Now, $\mathbf{x}(t)$ from (6.28) is substituted into (6.2) and, assuming that $\mathbf{f}(\mathbf{x}, \mathbf{Q}, t)$ is continuously differentiable with respect to $\mathbf{x}$ and $\mathbf{Q}$, the resulting equation is expanded in Taylor series about the particular motion $\mathbf{x}^*(t)$. Let $\mathbf{A}(t)$ and $\mathbf{B}(t)$ denote the $(n \times n)$ and $(n \times m)$ Jacobian matrices defined, respectively, by

$$\mathbf{A}(t) = \frac{\partial \mathbf{f}}{\partial \mathbf{x}} = \begin{bmatrix} \dfrac{\partial f_1}{\partial x_1} & \cdots & \dfrac{\partial f_1}{\partial x_n} \\ & \vdots & \\ \dfrac{\partial f_n}{\partial x_1} & \cdots & \dfrac{\partial f_n}{\partial x_n} \end{bmatrix}_{\substack{\text{evaluated at} \\ \mathbf{x}(t)=\mathbf{x}^*(t) \text{ and } \mathbf{Q}(t)=\mathbf{Q}^*(t)}} \qquad (6.29)$$

and

$$\mathbf{B}(t) = \frac{\partial \mathbf{f}}{\partial \mathbf{Q}} = \begin{bmatrix} \dfrac{\partial f_1}{\partial Q_1} & \cdots & \dfrac{\partial f_1}{\partial Q_m} \\ & \vdots & \\ \dfrac{\partial f_n}{\partial Q_1} & \cdots & \dfrac{\partial f_n}{\partial Q_m} \end{bmatrix}_{\substack{\text{evaluated at} \\ \mathbf{x}(t)=\mathbf{x}^*(t) \text{ and } \mathbf{Q}(t)=\mathbf{Q}^*(t)}} \qquad (6.30)$$

Taylor series expansion then yields

$$\dot{\mathbf{x}}^* + \Delta \dot{\mathbf{x}} = \mathbf{f}(\mathbf{x}^*, \mathbf{Q}^*, t) + \mathbf{A}(t) \, \Delta \mathbf{x} + \mathbf{B}(t) \, \Delta \mathbf{Q} + \mathbf{h}(\Delta \mathbf{x}, \Delta \mathbf{Q}, t) \qquad (6.31)$$

Assuming that $\mathbf{h}(\Delta \mathbf{x}, \Delta \mathbf{Q}, t)$ contains only terms that are higher than the first degree in $\Delta \mathbf{x}$ and $\Delta \mathbf{Q}$ resulting from the Taylor series expansion about $\mathbf{x}^*(t)$ and noting that the particular trajectory satisfies the equation

$$\dot{\mathbf{x}}^* = \mathbf{f}(\mathbf{x}^*, \mathbf{Q}^*, t)$$

equation (6.31) for small perturbations becomes

$$\Delta \dot{\mathbf{x}} = \mathbf{A}(t) \, \Delta \mathbf{x} + \mathbf{B}(t) \, \Delta \mathbf{Q} \qquad (6.32)$$

This system represents a set of first-order linear equations in the perturbations, where the matrices $\mathbf{A}(t)$ and $\mathbf{B}(t)$ are functions of time. These equations play a very important role in stability and sensitivity analyses and in control system synthesis. The matrix $\mathbf{A}(t)$ is useful for the stability analysis of the particular trajectory for small perturbations and will be employed in Chapter 9. The matrix $\mathbf{B}(t)$ is useful for the sensitivity analysis to investigate the changes in $\Delta \mathbf{x}$ due to changes in the input $\Delta \mathbf{Q}$. In aerospace applications, it is common practice first to determine a nominal trajectory for a spacecraft. The linearized perturbation equations (6.32) about this nominal trajectory are then employed for further analysis. Later in this chapter, we shall employ the linearized perturbation equations about an equilibrium or stationary motion, where the matrices

A and **B** are constant and time invariant, for the analysis of linear vibrations. For the simplicity of notation, we replace Δx by x and ΔQ by Q and represent (6.32) in the form

$$\dot{x} = A(t)x + B(t)Q \tag{6.33}$$

with the understanding that here x and Q represent deviations from their nominal values.

Example 6.5

For the sake of illustration, we consider a simple example of the Euler equations of motion (6.4), where a particular trajectory can be obtained analytically in a closed form. In (6.4), let $I_1 = I_2 = I$, $M_1 = M_2 = M_3 = 0$, and the initial conditions be $\omega_1(0)$, $\omega_2(0)$, and $\omega_3(0)$ at initial time $t_0 = 0$. From the last equation of (6.4), we see that $\omega_3 = \text{const.} = \omega_3(0)$ and this particular motion is described by

$$
\begin{Bmatrix} \omega_1 \\ \omega_2 \\ \omega_3 \end{Bmatrix} = \begin{Bmatrix} \sqrt{\omega_1^2(0) + \omega_2^2(0)} \, \sin(\alpha t + \beta) \\ \sqrt{\omega_1^2(0) + \omega_2^2(0)} \, \cos(\alpha t + \beta) \\ \omega_3(0) \end{Bmatrix} \tag{6.34}
$$

where

$$\alpha = \frac{I - I_3}{I}\omega_3(0) \qquad \text{and} \qquad \beta = \tan^{-1}\frac{\omega_1(0)}{\omega_2(0)}$$

After obtaining the Jacobian matrices $A(t)$ and $B(t)$ defined by (6.29) and (6.30), respectively, for the system of (6.4), the equations representing the linearized perturbations about this particular trajectory are described by

$$
\begin{Bmatrix} \Delta\dot{\omega}_1 \\ \Delta\dot{\omega}_2 \\ \Delta\dot{\omega}_3 \end{Bmatrix} = \begin{bmatrix} 0 & \alpha & \gamma\cos(\alpha t + \beta) \\ \alpha & 0 & -\gamma\sin(\alpha t + \beta) \\ 0 & 0 & 0 \end{bmatrix} \begin{Bmatrix} \Delta\omega_1 \\ \Delta\omega_2 \\ \Delta\omega_3 \end{Bmatrix}
$$

$$
+ \begin{bmatrix} \dfrac{1}{I} & 0 & 0 \\ 0 & \dfrac{1}{I} & 0 \\ 0 & 0 & \dfrac{1}{I_3} \end{bmatrix} \begin{Bmatrix} \Delta M_1 \\ \Delta M_2 \\ \Delta M_3 \end{Bmatrix} \tag{6.35}
$$

where

$$\gamma = [\omega_1^2(0) + \omega_2^2(0)]^{1/2}\frac{I - I_3}{I}$$

6.4.1 Solution of Linear Time-Varying Equations

The linear unforced system corresponding to (6.33) is given by

$$\dot{x} = A(t)x, \qquad x = x_0 \quad \text{at } t = t_0 \tag{6.36}$$

If every element of the matrix $A(t)$ is piecewise continuous* for some

---

*Actually, it is sufficient that $A(t)$ be Riemann integrable. A function that differs from a piecewise-continuous function on a set of zero measure is Riemann integrable and both integrals have the same value.

interval of time, there exists a unique solution of (6.36) in that interval. Since $\mathbf{A}(t)$ is piece wise continuous, we can find a Lipschitz constant k to satisfy condition (6.8) of Theorem 6.1 such that

$$\| \mathbf{A}(t)\mathbf{x}_\alpha - \mathbf{A}(t)\mathbf{x}_\beta \| \leq k \| \mathbf{x}_\alpha - \mathbf{x}_\beta \| \tag{6.37}$$

for any $\mathbf{x}_\alpha$ and $\mathbf{x}_\beta$ in the state space during that interval of time. At the points of discontinuity of $\mathbf{A}(t)$, $\dot{\mathbf{x}}$ and hence (6.36) are not defined. In the case of the forced system (6.33), it is sufficient for the existence and uniqueness of solution that both $\mathbf{A}(t)$ and $\mathbf{B}(t)\mathbf{Q}(t)$ be piecewise continuous.

Assuming that the sufficient conditions for the existence and uniqueness of solution have been satisfied, the actual solution can be obtained as follows. The system

$$\dot{\mathbf{x}} = \mathbf{A}(t)\mathbf{x} + \mathbf{B}(t)\mathbf{Q}, \qquad \mathbf{x} = \mathbf{x}_0 \text{ at } t = t_0 \tag{6.38}$$

may be alternatively expressed as

$$\dot{\mathbf{x}} = \mathbf{A}(t)\mathbf{x} + \mathbf{x}_0 \delta(t - t_0) + \mathbf{B}(t)\mathbf{Q} \tag{6.39}$$

where $\delta(t - t_0)$ is a Dirac delta function. Hence, we have

$$\left[\mathbf{I}\frac{d}{dt} - \mathbf{A}(t) \right]\mathbf{x} = \mathbf{x}_0\, \delta(t - t_0) + \mathbf{B}(t)\mathbf{Q} \tag{6.40}$$

where $\mathbf{I}$ is a $(n \times n)$ identity matrix. Then, it follows that

$$\mathbf{x}(t) = \left[\mathbf{I}\frac{d}{dt} - \mathbf{A}(t) \right]^{-1}\mathbf{x}_0\, \delta(t - t_0) + \left[\mathbf{I}\frac{d}{dt} - \mathbf{A}(t) \right]^{-1}\mathbf{B}(t)\mathbf{Q}(t) \tag{6.41}$$

The inverse of a differential operator is an integral operator and the kernel is called the Green's function matrix and, in this case, is represented by $\mathbf{G}(t, t')$. Equation (6.41) may then be expressed as

$$\mathbf{x}(t) = \int_{t_0}^{\infty} \mathbf{G}(t, t')\mathbf{x}_0\, \delta(t' - t_0)\, dt' + \int_{t_0}^{\infty} \mathbf{G}(t, t')\mathbf{B}(t')\mathbf{Q}(t')\, dt' \tag{6.42}$$

Since the independent variable is time t and the system is causal (i.e., it does not respond in anticipation before an input is applied), we have $\mathbf{G}(t, t') = \mathbf{O}$ for $t' > t$. This Green's function matrix for the initial value problem of ordinary differential equations is called the state transition matrix and is represented by $\mathbf{\Phi}(t, t')$. Hence, (6.42) is expressed as

$$\mathbf{x}(t) = \int_{t_0}^{t} \mathbf{\Phi}(t, t')\mathbf{x}_0\, \delta(t' - t_0)\, dt' + \int_{t_0}^{t} \mathbf{\Phi}(t, t')\mathbf{B}(t')\mathbf{Q}(t')\, dt' \tag{6.43}$$

$$= \mathbf{\Phi}(t, t_0)\mathbf{x}_0 + \int_{t_0}^{t} \mathbf{\Phi}(t, t')\mathbf{B}(t')\mathbf{Q}(t')\, dt' \tag{6.44}$$

The problem now is the determination of the state transition matrix $\mathbf{\Phi}(t, t')$. For this purpose, multiplying both sides of (6.43) by $\left[\mathbf{I}\frac{d}{dt} - \mathbf{A}(t) \right]$, we obtain

$$\left[\mathbf{I}\frac{d}{dt} - \mathbf{A}(t)\right]\mathbf{x}(t) = \left[\mathbf{I}\frac{d}{dt} - \mathbf{A}(t)\right]\int_{t_0}^{t} \mathbf{\Phi}(t, t')\mathbf{x}_0\,\delta(t' - t_0)\,dt'$$

$$+ \left[\mathbf{I}\frac{d}{dt} - \mathbf{A}(t)\right]\int_{t_0}^{t} \mathbf{\Phi}(t, t')\mathbf{B}(t')\mathbf{Q}(t')\,dt'$$

Employing Leibnitz's rule, the foregoing equation becomes

$$\left[\mathbf{I}\frac{d}{dt} - \mathbf{A}(t)\right]\mathbf{x}(t) = \mathbf{\Phi}(t, t')\mathbf{x}_0\delta(t - t_0)$$

$$+ \int_{t_0}^{t} \left[\mathbf{I}\frac{d}{dt} - \mathbf{A}(t)\right]\mathbf{\Phi}(t, t')\mathbf{x}_0\,\delta(t' - t_0)\,dt'$$

$$+ \mathbf{\Phi}(t, t)\mathbf{B}(t)\mathbf{Q}(t)$$

$$+ \int_{t_0}^{t} \left[\mathbf{I}\frac{d}{dt} - \mathbf{A}(t)\right]\mathbf{\Phi}(t, t')\mathbf{B}(t')\mathbf{Q}(t')\,dt' \qquad (6.45)$$

Equations (6.45) and (6.40) must be the same. Hence, comparing these two equations, we obtain

$$\left[\mathbf{I}\frac{d}{dt} - \mathbf{A}(t)\right]\mathbf{\Phi}(t, t') = 0 \qquad (6.46)$$

with conditions

$$\mathbf{\Phi}(t, t) = \mathbf{I} \qquad (6.47)$$

Hence, the state transition matrix $\mathbf{\Phi}(t, t')$ is obtained from the solution of (6.46) with conditions (6.47). A closed-form analytical solution of this linear time-varying parameter equation (6.46) is not possible in most cases and a computer simulation must be employed.

Example 6.6

In this example, given by Hsu and Meyer [2], we consider a second-order Euler linear differential equation

$$t^2\frac{d^2x}{dt^2} + 6t\frac{dx}{dt} + 6x = 0 \qquad (6.48)$$

with initial conditions $x(t_0)$ and $\frac{dx}{dt}(t_0)$. This example may have no practical application in dynamics but is considered here for the purpose of illustration. By substituting $t = e^\tau$ (i.e., $\tau = \ln t$), (6.48) can be reduced to an equation with constant coefficients. We get

$$\frac{dx}{dt} = \frac{dx}{d\tau}\frac{d\tau}{dt} = \frac{1}{t}\frac{dx}{d\tau} \qquad (6.49)$$

$$\frac{d^2x}{dt^2} = \frac{d}{dt}\left(\frac{1}{t}\frac{dx}{d\tau}\right)$$

$$= -\frac{1}{t^2}\frac{dx}{d\tau} + \frac{1}{t^2}\frac{d^2x}{d\tau^2} \qquad (6.50)$$

Employing (6.49) and (6.50) in (6.48), we obtain

$$\frac{d^2x}{d\tau^2} + 5\frac{dx}{d\tau} + 6x = 0 \qquad (6.51)$$

The foregoing equation is a linear time-invariant (i.e., constant-parameter) equation. Choosing the state variables as $x_1 = x$ and $x_2 = dx/d\tau$, the state transition matrix can be obtained by employing the techniques discussed in the next section and is given by

$$\Phi(\tau, \tau_0) = \begin{bmatrix} 3e^{-2(\tau-\tau_0)} - 2e^{-3(\tau-\tau_0)} & e^{-2(\tau-\tau_0)} - e^{-3(\tau-\tau_0)} \\ -6e^{-2(\tau-\tau_0)} + 6e^{-3(\tau-\tau_0)} & -2e^{-2(\tau-\tau_0)} + 3e^{-3(\tau-\tau_0)} \end{bmatrix} \quad (6.52)$$

For equation (6.51), as is the case for all linear time-invariant equations, we note that $\Phi(\tau, \tau_0) = \Phi(\tau - \tau_0)$ or $\Phi(\tau, \tau') = \Phi(\tau - \tau')$. From equation (6.44), since in this case $\mathbf{Q} = 0$, the solution can be written as

$$\begin{Bmatrix} x \\ t\dfrac{dx}{dt} \end{Bmatrix} = \begin{bmatrix} 3\left(\dfrac{t_0}{t}\right)^2 - 2\left(\dfrac{t_0}{t}\right)^3 & \left(\dfrac{t_0}{t}\right)^2 - \left(\dfrac{t_0}{t}\right)^3 \\ -6\left(\dfrac{t_0}{t}\right)^2 + 6\left(\dfrac{t_0}{t}\right)^3 & -2\left(\dfrac{t_0}{t}\right)^2 + 3\left(\dfrac{t_0}{t}\right)^3 \end{bmatrix} \begin{Bmatrix} x(t_0) \\ t_0\dfrac{dx}{dt}(t_0) \end{Bmatrix}$$

or

$$\begin{Bmatrix} x \\ \dfrac{dx}{dt} \end{Bmatrix} = \begin{bmatrix} 3\left(\dfrac{t_0}{t}\right)^2 - 2\left(\dfrac{t_0}{t}\right)^3 & \dfrac{t_0^3}{t^2} - \dfrac{t_0^4}{t^3} \\ -6\dfrac{t_0^2}{t^3} + 6\dfrac{t_0^3}{t^4} & -2\left(\dfrac{t_0}{t}\right)^3 + 3\left(\dfrac{t_0}{t}\right)^4 \end{bmatrix} \begin{Bmatrix} x(t_0) \\ \dfrac{dx}{dt}(t_0) \end{Bmatrix} \quad (6.53)$$

The state transition matrix $\Phi(t, t_0)$ is now obvious by inspection of (6.53) and $\Phi(t, t')$ is obtained by replacing t_0 by t' in $\Phi(t, t_0)$. Here, as is the case for all linear time-varying equations, we note that $\Phi(t, t') \neq \Phi(t - t')$, unlike linear time-invariant systems. As discussed earlier, unless the time-varying linear equation is of a standard form, such as Bessel's equation, it is not possible to obtain a closed-form analytic solution for $\Phi(t, t')$ and computer simulation is generally required.

6.5 LINEARIZED TIME-INVARIANT SYSTEMS

We now consider a special case of the equation of motion

$$\dot{\mathbf{x}} = \mathbf{f}(\mathbf{x}, t, \mathbf{Q}) \quad (6.54)$$

where the input forces and moments are constants or zero and in addition the parameters are time-invariant such that the functions f_i are not explicit functions of time. In that case, (6.54) reduces to the form

$$\dot{\mathbf{x}} = \mathbf{f}(\mathbf{x}) \quad (6.55)$$

The foregoing system of equations is called an autonomous system. A state $\mathbf{x}_e$ of (6.55) is called an equilibrium state if, starting at that state, the system will remain in that state in the absence of forcing functions or disturbances. Since for equilibrium $\dot{\mathbf{x}} = \mathbf{0}$, the equilibrium states are found from the solution of the nonlinear algebraic equations

$$\mathbf{f}(\mathbf{x}) = \mathbf{0} \quad (6.56)$$

A set of nonlinear algebraic equations may have no solution. On the other hand, it may have one or infinite number of solutions. When the equilibrium states are countable, they are called isolated equilibrium states. When every state of a connected region satisfies (6.56), this set of states is called an equilibrium zone. There is a lack of a theory analogous to Theorem 6.1 concerning the solution of a set of nonlinear algebraic equations.

We consider an isolated equilibrium state $\mathbf{x}_e$ and let $\Delta\mathbf{x}(t)$ be perturbed state variables defined by

$$\Delta\mathbf{x}(t) = \mathbf{x}(t) - \mathbf{x}_e; \qquad \text{that is} \qquad \mathbf{x}(t) = \mathbf{x}_e + \Delta\mathbf{x}(t) \qquad (6.57)$$

Assuming that $\mathbf{f}(\mathbf{x}, \mathbf{Q})$ is continuously differentiable with respect to $\mathbf{x}$ and $\mathbf{Q}$, we employ Taylor series expansion about the equilibrium point $\mathbf{x}_e$. Now, the Jacobian matrices of equations (6.29) and (6.30) are evaluated at the constant values $\mathbf{x}_e$ and $\mathbf{Q}_e$ and hence the matrices $\mathbf{A}$ and $\mathbf{B}$ are constant matrices. For small perturbations, the linearized equations analogous to (6.32) become

$$\Delta\dot{\mathbf{x}} = \mathbf{A}\,\Delta\mathbf{x} + \mathbf{B}\,\Delta\mathbf{Q} \qquad (6.58)$$

It should be noted again that the matrices $\mathbf{A}$ and $\mathbf{B}$ are constant when the equations of motion (6.55) are autonomous and the Taylor series expansion is about constant values. For the simplicity of notation, we replace $\Delta\mathbf{x}$ by $\mathbf{x}$ and $\Delta\mathbf{Q}$ by $\mathbf{Q}$ and represent (6.58) in the form

$$\dot{\mathbf{x}} = \mathbf{A}\mathbf{x} + \mathbf{B}\mathbf{Q} \qquad (6.59)$$

with the understanding that here $\mathbf{x}$ and $\mathbf{Q}$ represent deviations from their equilibrium values.

Example 6.7

We consider a mass, linear damping, and nonlinear soft spring described by the equation

$$m\ddot{x} + c\dot{x} + k\left(x - \frac{x^3}{6}\right) = F \qquad (6.60)$$

Choosing the state variables as $x_1 = x$ and $x_2 = \dot{x}$, the state-equation representation becomes

$$\dot{x}_1 = x_2$$

$$\dot{x}_2 = -\frac{k}{m}\left(x_1 - \frac{x_1^3}{6}\right) - \frac{c}{m}x_2 + \frac{1}{m}F$$

Let $F = 0$. The equilibrium states are determined from the solution of the algebraic equations

$$0 = x_2$$

$$0 = -\frac{k}{m}\left(x_1 - \frac{x_1^3}{6}\right) - \frac{c}{m}x_2$$

The solution yields three isolated equilibrium states given by

$$\{x_e\} = \begin{Bmatrix} 0 \\ 0 \end{Bmatrix}, \quad \begin{Bmatrix} \sqrt{6} \\ 0 \end{Bmatrix}, \quad \text{and} \quad \begin{Bmatrix} -\sqrt{6} \\ 0 \end{Bmatrix}$$

We first consider the equilibrium $\begin{Bmatrix} 0 \\ 0 \end{Bmatrix}$ and let Δx_1 and Δx_2 be deviations about this equilibrium in the state variables and ΔF be the deviation in the input force. For small deviations, the linearized equations are described by

$$\begin{Bmatrix} \Delta \dot{x}_1 \\ \Delta \dot{x}_2 \end{Bmatrix} = \begin{bmatrix} \dfrac{\partial f_1}{\partial x_1} & \dfrac{\partial f_1}{\partial x_2} \\ \dfrac{\partial f_2}{\partial x_1} & \dfrac{\partial f_2}{\partial x_2} \end{bmatrix} \begin{Bmatrix} \Delta x_1 \\ \Delta x_2 \end{Bmatrix} + \begin{Bmatrix} \dfrac{\partial f_1}{\partial F} \\ \dfrac{\partial f_2}{\partial F} \end{Bmatrix} \Delta F \tag{6.61}$$

Evaluated at
$x_1 = 0.\ x_2 = 0.$
$F = 0$

Hence,

$$\begin{Bmatrix} \Delta \dot{x}_1 \\ \Delta \dot{x}_2 \end{Bmatrix} = \begin{bmatrix} 0 & 1 \\ -\dfrac{k}{m} & -\dfrac{c}{m} \end{bmatrix} \begin{Bmatrix} \Delta x_1 \\ \Delta x_2 \end{Bmatrix} + \begin{Bmatrix} 0 \\ \dfrac{1}{m} \end{Bmatrix} \Delta F \tag{6.62}$$

Here, the $\mathbf{A}$ and $\mathbf{B}$ matrices are given by (6.62). Since, in this case we have only a single force input, the $\mathbf{B}$ matrix becomes a column matrix.

We now consider the equilibrium states $\begin{Bmatrix} \sqrt{6} \\ 0 \end{Bmatrix}$ and $\begin{Bmatrix} -\sqrt{6} \\ 0 \end{Bmatrix}$. Evaluating the Jacobian matrices of (6.61) at the equilibrium states ($x_1 = \pm\sqrt{6}$, $x_2 = 0$, $F = 0$), for both equilibrium states the linearized equations are given by

$$\begin{Bmatrix} \Delta \dot{x}_1 \\ \Delta \dot{x}_2 \end{Bmatrix} = \begin{bmatrix} 0 & 1 \\ 2\dfrac{k}{m} & -\dfrac{c}{m} \end{bmatrix} \begin{Bmatrix} \Delta x_1 \\ \Delta x_2 \end{Bmatrix} + \begin{Bmatrix} 0 \\ \dfrac{1}{m} \end{Bmatrix} \Delta F \tag{6.63}$$

Example 6.8

In this example, we consider a mass, linear spring, and nonlinear Coulomb friction as shown in Fig. 6.5. This system is unforced and the equation of motion is given by

$$m\ddot{x} + \mu mg \operatorname{sgn} \dot{x} + kx = 0 \tag{6.64}$$

with initial conditions $x(0)$ and $\dot{x}(0)$. The function $\operatorname{sgn} \dot{x} = +1$ for $\dot{x} > 0$ and -1 for $\dot{x} < 0$ as shown in Fig. 6.6. For $\dot{x} = 0$, $-1 \le \operatorname{sgn} \dot{x} \le 1$ but is otherwise undefined. Choosing the state variables as $x_1 = x$ and $x_2 = \dot{x}$, the state equations are described by

$$\dot{x}_1 = x_2$$
$$\dot{x}_2 = -\frac{k}{m} x_1 - \mu g \operatorname{sgn} x_2 \tag{6.65}$$

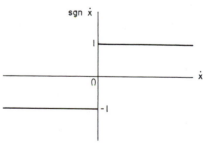

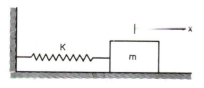

Figure 6.5 Mass, spring, and Coulomb friction.

Figure 6.6 Sgn function.

The equilibrium states are obtained from the solution of the nonlinear algebraic equations

$$0 = x_2$$

$$0 = -\frac{k}{m}x_1 - \mu g \text{ sgn } x_2$$

(6.66)

From the first equation of (6.66), we get $x_{2e} = 0$ and from the second

$$-\frac{\mu mg}{k} \leq x_{1e} \leq \frac{\mu mg}{k}$$

Hence, in this example we get an equilibrium zone as shown in Fig. 6.7. It should be noted that dry sliding friction between two surfaces in contact may be more complicated than Coulomb friction, as static friction can be higher than kinetic friction and the magnitude of the kinetic friction may be a function of the sliding velocity. Also, in this example, linearization of the form (6.58) is not possible even for small changes from the equilibrium zone, as the function

$$f_2(x_1, x_2) = -\frac{k}{m}x_1 - \mu g \text{ sgn } x_2$$

is not an analytic function of its arguments.

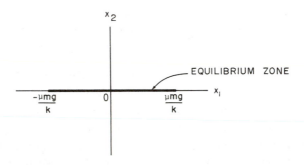

Figure 6.7 Example of equilibrium zone.

Example 6.9:   Stationary Motion

In some cases, the forces and moments acting on a dynamic system have constant values and as a result the generalized velocities may also be constants. Such a motion is called stationary motion and, if the generalized displacements do not enter the equations of motion, only the generalized velocities can be chosen as the state variables. As an example, consider the Euler equations of motion (6.4) when the applied moments M_1, M_2, and M_3 are constants. The constant values of ω_1, ω_2, and ω_3 can be obtained from the solution of the nonlinear algebraic equations

$$0 = -\frac{I_3 - I_2}{I_1}\omega_2\omega_3 + \frac{M_1}{I_1}$$

$$0 = -\frac{I_1 - I_3}{I_2}\omega_3\omega_1 + \frac{M_2}{I_2}$$

$$0 = -\frac{I_2 - I_1}{I_3}\omega_1\omega_2 + \frac{M_3}{I_3}$$

(6.67)

The linearized equations for small perturbations about a stationary motion obtained from the solution of (6.67) can be expressed as

$$\begin{Bmatrix} \Delta\dot{\omega}_1 \\ \Delta\dot{\omega}_2 \\ \Delta\dot{\omega}_3 \end{Bmatrix} = [A] \begin{Bmatrix} \Delta\omega_1 \\ \Delta\omega_2 \\ \Delta\omega_3 \end{Bmatrix} + [B] \begin{Bmatrix} \Delta M_1 \\ \Delta M_2 \\ \Delta M_3 \end{Bmatrix} \tag{6.68}$$

Hence, the linearized equations of perturbations about an equilibrium state or stationary motion can be represented in the form of (6.59).

6.5.1 Solution of Linear Time-Invariant Equations

The linear unforced system corresponding to (6.59) is given by

$$\{\dot{x}\} = [A]\{x\}, \qquad \{x\} = \{x_0\} \quad \text{at } t = t_0 \tag{6.69}$$

Here, $\{f(x)\} = [A]\{x\}$ and condition (6.8) of Theorem 6.1 becomes

$$\| \mathbf{f}(\mathbf{x}_\alpha) - \mathbf{f}(\mathbf{x}_\beta) \| = \| \mathbf{A}\mathbf{x}_\alpha - \mathbf{A}\mathbf{x}_\beta \|$$
$$= \| \mathbf{A}(\mathbf{x}_\alpha - \mathbf{x}_\beta) \| \tag{6.70}$$
$$\leq \| \mathbf{A} \| \| \mathbf{x}_\alpha - \mathbf{x}_\beta \|$$

Now, it is known that $\| \mathbf{A} \| \leq |\lambda_{\max}|$, where $|\lambda_{\max}|$ is the absolute value of the maximum eigenvalue of matrix $[A]$. Hence, a global Lipschitz constant can always be found such that $k \geq |\lambda_{\max}|$. As a result, a linear time-invariant system (6.69) always satisfies the conditions for global existence and uniqueness of solution. The solution of the linear system of equations

$$\dot{\mathbf{x}} = \mathbf{A}\mathbf{x} + \mathbf{B}\mathbf{Q}, \qquad \mathbf{x} = \mathbf{x}_0 \quad \text{at } t = t_0 \tag{6.71}$$

can be expressed, as shown in the preceding section, in the form

$$\mathbf{x}(t) = \mathbf{\Phi}(t - t_0)\mathbf{x}_0 + \int_{t_0}^{t} \mathbf{\Phi}(t - t')\mathbf{B}\mathbf{Q}(t')\, dt' \tag{6.72}$$

where for time-invariant systems, the state transition matrix $\mathbf{\Phi}(t, t_0) = \mathbf{\Phi}(t - t_0)$ and $\mathbf{\Phi}(t, t') = \mathbf{\Phi}(t - t')$. To obtain the state transition matrix, we can set $t' = 0$ in (6.46) and solve the equation

$$\left[\mathbf{I}\frac{d}{dt} - \mathbf{A} \right]\mathbf{\Phi}(t) = 0 \qquad \text{with} \qquad \mathbf{\Phi}(0) = \mathbf{I} \tag{6.73}$$

Many methods are available for obtaining a closed-form solution of (6.73). Here, we employ the method of Laplace transformation. Letting $\hat{\mathbf{\Phi}}(s)$ be the Laplace transformation of $\mathbf{\Phi}(t)$, from (6.73) with initial condition $\mathbf{\Phi}(0) = \mathbf{I}$, we get

$$\mathbf{I}s\hat{\mathbf{\Phi}}(s) - \mathbf{I} - \mathbf{A}\hat{\mathbf{\Phi}}(s) = 0$$

or

$$(s\mathbf{I} - \mathbf{A})\hat{\mathbf{\Phi}}(s) = \mathbf{I} \tag{6.74}$$

Hence,

$$\hat{\mathbf{\Phi}}(s) = [s\mathbf{I} - \mathbf{A}]^{-1}$$
$$\mathbf{\Phi}(t) = L^{-1}([s\mathbf{I} - \mathbf{A}]^{-1}) \tag{6.75}$$

where L^{-1} denotes the operation of inverse Laplace transformation. Then $\mathbf{\Phi}(t - t_0)$ and $\mathbf{\Phi}(t - t')$ are obtained by replacing t by $t - t_0$ and $t - t'$, respectively, in (6.75). The state transition matrix may also be expressed as matrix exponential for time-invariant systems in the form

$$\mathbf{\Phi}(t) = e^{\mathbf{A}t} \tag{6.76}$$

Then $\mathbf{\Phi}(t - t_0) = e^{\mathbf{A}(t-t_0)}$ and $\mathbf{\Phi}(t - t') = e^{\mathbf{A}(t-t')}$. It can be seen easily that the matrix exponential form (6.76) for $\mathbf{\Phi}(t)$ does indeed satisfy (6.73).

Example 6.10

We consider equations (6.62) of Example 6.6. These equations represent deviations about the equilibrium $\begin{Bmatrix} 0 \\ 0 \end{Bmatrix}$ and, replacing $\Delta\mathbf{x}$ by $\mathbf{x}$ and ΔF by F for simplicity of notation, are given by

$$\begin{Bmatrix} \dot{x}_1 \\ \dot{x}_2 \end{Bmatrix} = \begin{bmatrix} 0 & 1 \\ -\dfrac{k}{m} & -\dfrac{c}{m} \end{bmatrix} \begin{Bmatrix} x_1 \\ x_2 \end{Bmatrix} + \begin{Bmatrix} 0 \\ \dfrac{1}{m} \end{Bmatrix} F$$

We define the natural frequency ω_n and the damping ratio ζ as

$$\omega_n = \sqrt{\frac{k}{m}} \quad \text{and} \quad \zeta = \frac{1}{2} \frac{c}{(mk)^{1/2}}$$

Now, the foregoing equations may be written as

$$\begin{Bmatrix} \dot{x}_1 \\ \dot{x}_2 \end{Bmatrix} = \begin{bmatrix} 0 & 1 \\ -\omega_n^2 & -2\zeta\omega_n \end{bmatrix} \begin{Bmatrix} x_1 \\ x_2 \end{Bmatrix} + \begin{Bmatrix} 0 \\ \dfrac{1}{m} \end{Bmatrix} F \tag{6.77}$$

Hence, the $\mathbf{A}$ matrix is given by

$$\mathbf{A} = \begin{bmatrix} 0 & 1 \\ -\omega_n^2 & -2\zeta\omega_n \end{bmatrix} \tag{6.78}$$

It follows that

$$(s\mathbf{I} - \mathbf{A}) = \begin{bmatrix} s & -1 \\ \omega_n^2 & s + 2\zeta\omega_n \end{bmatrix}$$

$$(s\mathbf{I} - \mathbf{A})^{-1} = \begin{bmatrix} \dfrac{s + 2\zeta\omega_n}{\Delta} & \dfrac{1}{\Delta} \\ \dfrac{-\omega_n^2}{\Delta} & \dfrac{s}{\Delta} \end{bmatrix} \tag{6.79}$$

where Δ is the determinant of the $(s\mathbf{I} - \mathbf{A})$ matrix (i.e., $\Delta = s^2 + 2\zeta\omega_n s + \omega_n^2$). Here, $\Delta = 0$ is the characteristic equation whose roots are given by

$$\lambda_1, \lambda_2 = -\zeta\omega_n \pm \omega_n\sqrt{\zeta^2 - 1} \tag{6.80}$$

Employing partial-fraction expansion of each element of matrix (6.79) and then the inverse Laplace transformation, we obtain

$$\mathbf{\Phi}(t) = \begin{bmatrix} \dfrac{\lambda_1 + 2\zeta\omega_n}{\lambda_1 - \lambda_2} e^{\lambda_1 t} + \dfrac{\lambda_2 + 2\zeta\omega_n}{-\lambda_1 + \lambda_2} e^{\lambda_2 t} & \dfrac{1}{\lambda_1 - \lambda_2} e^{\lambda_1 t} + \dfrac{1}{-\lambda_1 + \lambda_2} e^{\lambda_2 t} \\ \dfrac{-\omega_n^2}{\lambda_1 - \lambda_2} e^{\lambda_1 t} + \dfrac{-\omega_n^2}{-\lambda_1 + \lambda_2} e^{\lambda_2 t} & \dfrac{\lambda_1}{\lambda_1 - \lambda_2} e^{\lambda_1 t} + \dfrac{\lambda_2}{-\lambda_1 + \lambda_2} e^{\lambda_2 t} \end{bmatrix} \tag{6.81}$$

The state transition matrix (6.81) can now be employed in (6.72) to obtain the response to any arbitrary initial conditions $\mathbf{x}_0$ and forcing function $\mathbf{Q}(t)$. We now consider two cases, one of which is the overdamped case and the other, underdamped.

Case 1:   Overdamped, $\zeta > 1$. In this case, both roots of the characteristic equation are real and negative and are given by (6.80). We now obtain the response to a step input in the force. Let the initial conditions be zero [i.e., $x_1(0) = 0$ and $x_2(0) = 0$] and $F = H$, a constant, for $t > 0$ and $F = 0$ for $t < 0$ [i.e., $F(t)$ is a Heaviside step function]. The response is obtained from

$$\mathbf{x}(t) = \int_0^t \mathbf{\Phi}(t - t') \begin{Bmatrix} 0 \\ 1 \\ \frac{1}{m} \end{Bmatrix} H \, dt' \tag{6.82}$$

where $\mathbf{\Phi}(t)$ is given by (6.81). The solution for the state variables can be obtained from (6.82) and the response of the displacement is

$$x_1(t) = \frac{H}{m\omega_n^2} + \frac{H}{2m\omega_n^2\sqrt{\zeta^2 - 1}}[-(\zeta + \sqrt{\zeta^2 - 1})e^{-\omega_n(\zeta - \sqrt{\zeta^2 - 1})t}$$
$$+ (\zeta - \sqrt{\zeta^2 - 1})e^{-\omega_n(\zeta + \sqrt{\zeta^2 - 1})t}] \tag{6.83}$$

which is shown in Fig. 6.8.

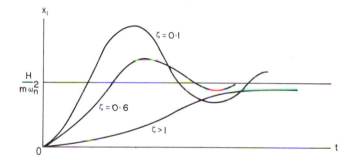

Figure 6.8 Response of a second-order linear system to step input with zero initial conditions.

Case 2:   Underdamped, $0 < \zeta < 1$. Here, the roots of the characteristic equation are complex conjugate and are

$$\lambda_1, \lambda_2 = -\zeta\omega_n \pm j\omega_n\sqrt{1 - \zeta^2} \tag{6.84}$$

Employing these values of λ_1 and λ_2 in (6.81) for the state transition matrix and equation (6.82), we can obtain the response to a step change in the force with zero initial conditions. The response of the displacement is

$$x_1(t) = \frac{H}{m\omega_n^2} + \frac{H}{m\omega_n^2\sqrt{1 - \zeta^2}}e^{-\zeta\omega_n t} \sin(\omega_n\sqrt{1 - \zeta^2}\, t - \psi) \tag{6.85}$$

where $\psi = \tan^{-1}(\sqrt{1 - \zeta^2}/-\zeta)$. This response is shown in Fig. 6.8 for various values of the damping ratio ζ. For low values of the damping ratio ζ, the response is fast, has large overshoot, and is highly oscillatory before it reaches its equilibrium value. For $\zeta \geq 1$, there is no overshoot and no oscillations occur. When $\zeta = 1$, the damping is called critical. The frequency $\omega_d = \omega_n\sqrt{1 - \zeta^2}$ is called the damped natural frequency.

We consider a typical underdamped step response as shown in Fig. 6.9. The period T of the damped oscillations can be measured directly from the crossing points of the steady-state value. Relating the damped natural frequency to the period, we obtain

$$\omega_n\sqrt{1 - \zeta^2} = \frac{2\pi}{T} \quad \text{rad/s} \tag{6.86}$$

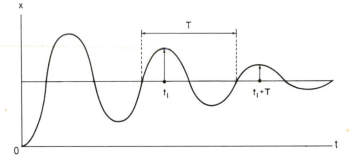

Figure 6.9 Logarithmic decrement.

At time t_1 the amplitude is $y(t_1)$ and at time $t_1 + T$, the amplitude is $y(t_1 + T)$. From (6.85), it can be seen that

$$\frac{y(t_1)}{y(t_1 + T)} = \frac{e^{-\zeta\omega_n t_1}}{e^{-\zeta\omega_n(t_1+T)}}$$

$$= e^{\zeta\omega_n T} \tag{6.87}$$

Employing (6.86) in (6.87), we obtain

$$\ln\frac{y(t_1)}{y(t_1 + T)} = \frac{\zeta(2\pi)}{\sqrt{1 - \zeta^2}} \tag{6.88}$$

The logarithm of the ratio of amplitudes separated by a period is given the name of logarithmic decrement. Knowing the left-hand side of (6.88) from experimental data, we can determine the damping ratio ζ, and then the natural frequency ω_n can be determined from (6.86).

Example 6.11

We consider the stationary motion of Example 6.9, where the linearized equations for small perturbations about a stationary motion are given by (6.68). Let the matrix A be given by

$$[A] = \begin{bmatrix} 0 & 1 & 0 \\ 0 & 0 & 1 \\ -6 & -11 & -6 \end{bmatrix} \tag{6.89}$$

It follows that

$$(s\mathbf{I} - \mathbf{A}) = \begin{bmatrix} s & -1 & 0 \\ 0 & s & -1 \\ 6 & 11 & s+6 \end{bmatrix} \tag{6.90}$$

$$(s\mathbf{I} - \mathbf{A})^{-1} = \frac{1}{\Delta} \begin{bmatrix} s(s+6)+11 & s+6 & 1 \\ -6 & s(s+6) & s \\ -6 & -11s-6 & s^2 \end{bmatrix} \tag{6.91}$$

where Δ is the determinant of the $(s\mathbf{I} - \mathbf{A})$ matrix and is given by

$$\Delta = (s+1)(s+2)(s+3) \tag{6.92}$$

The characteristic equation is $\Delta = 0$ and its roots -1, -2, and -3 are the eigenvalues of matrix $\mathbf{A}$. The state transition matrix $\boldsymbol{\Phi}(t)$ can be obtained by taking the inverse Laplace transformation of (6.91) as shown by (6.74) and becomes

$$\boldsymbol{\Phi}(t) = \begin{bmatrix} 3e^{-t} - 3e^{-2t} + e^{-3t} & \frac{5}{2}e^{-t} - 4e^{-2t} + \frac{3}{2}e^{-3t} & \frac{1}{2}e^{-t} - e^{-2t} + \frac{1}{2}e^{-3t} \\ -3e^{-t} + 6e^{-2t} - 3e^{-3t} & -\frac{5}{2}e^{-t} + 8e^{-2t} - \frac{9}{2}e^{-3t} & -\frac{1}{2}e^{-t} + 2e^{-2t} - \frac{3}{2}e^{-3t} \\ 3e^{-t} - 12e^{-2t} + 9e^{-3t} & \frac{5}{2}e^{-t} - 16e^{-2t} + \frac{27}{2}e^{-3t} & \frac{1}{2}e^{-t} - 4e^{-2t} + \frac{9}{2}e^{-3t} \end{bmatrix} \tag{6.93}$$

This state transition matrix can now be employed in (6.72) in order to obtain the response of the perturbations to arbitrary initial conditions and forcing functions.

6.6 COORDINATE TRANSFORMATION FOR LINEAR TIME-INVARIANT SYSTEMS

In the preceding section, the state transition matrix for linear time-invariant systems has been obtained by employing the Laplace transformation. It should be noted that other methods are available for the direct solution of (6.73) in order to obtain the state transition matrix. One method as given by (6.76) is to express the state transition matrix in the form of a matrix exponential as

$$\boldsymbol{\Phi}(t) = e^{[A]t}$$

$$= [I] + [A]t + [A^2]\frac{t^2}{2!} + [A]^3\frac{t^3}{3!} + \cdots$$

$$= \sum_{n=0}^{\infty} [A]^n \frac{t^n}{n!} \tag{6.94}$$

However, a disadvantage of obtaining the state transition matrix in this manner is that a closed form for its elements may not be apparent from (6.94). But the method may be suitable for machine computation. In fact, for systems with more than two degrees of freedom, hand calculation of the state transition matrix becomes very cumbersome and machine computation becomes a necessity. For this reason, in this section, we consider a method that is quite suitable for the machine computation of the state transition matrix. The method is based on coordinate transformation in order to diagonalize the system $\mathbf{A}$ matrix. Another reason for the study of this method is that it will be employed in Chapter 8 for the normal-mode solution of linear vibration problems.

Given a system described by linear time-invariant state equation

$$\{\dot{x}\} = \mathbf{A}\{x\} + \mathbf{B}\{Q\} \tag{6.95}$$

with initial conditions $\{x(0)\}$, we define new variables $\{y\}$ by the linear transformation

$$\{x\} = \mathbf{P}\{y\} \tag{6.96}$$

where $\mathbf{P}$ is a $(n \times n)$ constant matrix and $\{y\}$ are the new transformed state variables. Substitution of (6.96) in (6.95) yields

$$\mathbf{P}\{\dot{y}\} = \mathbf{A}\mathbf{P}\{y\} + \mathbf{B}\{Q\} \tag{6.97}$$

If $\mathbf{P}$ is a nonsingular matrix, we obtain

$$\{\dot{y}\} = \mathbf{P}^{-1}\mathbf{A}\mathbf{P}\{y\} + \mathbf{P}^{-1}\mathbf{B}\{Q\} \tag{6.98}$$

with initial conditions $\{y(0)\} = \mathbf{P}^{-1}\{x(0)\}$. Letting $\mathbf{P}^{-1}\mathbf{A}\mathbf{P} = \mathbf{\Lambda}$, matrix $\mathbf{A}$ is said to be similar to matrix $\mathbf{\Lambda}$ and the transformation $\{x\} = \mathbf{P}\{y\}$ is called a similarity transformation. Here, we seek a nonsingular matrix $\mathbf{P}$ such that $\mathbf{\Lambda}$ is a diagonal matrix of the form

$$\mathbf{\Lambda} = \begin{bmatrix} \lambda_1 & 0 & \cdots & 0 \\ 0 & \lambda_2 & \cdots & 0 \\ \vdots & & & \vdots \\ 0 & 0 & \cdots & \lambda_n \end{bmatrix} \tag{6.99}$$

where $\lambda_1, \ldots, \lambda_n$ are the n eigenvalues of matrix $\mathbf{A}$.

If there exists such a nonsingular matrix $\mathbf{P}$, the coupled equations (6.95) can be transformed into a set of n uncoupled first-order differential equations (6.98) in the transformed state variables $\{y\}$. The uncoupled state variables $\{y\}$ are said to be in the normal or canonic form. The unforced system corresponding to (6.98) now becomes

$$\begin{Bmatrix} \dot{y}_1 \\ \dot{y}_2 \\ \vdots \\ \dot{y}_n \end{Bmatrix} = \begin{bmatrix} \lambda_1 & 0 & \cdots & 0 \\ 0 & \lambda_2 & \cdots & 0 \\ \vdots & & & \vdots \\ 0 & 0 & \cdots & \lambda_n \end{bmatrix} \begin{Bmatrix} y_1 \\ y_2 \\ \vdots \\ y_n \end{Bmatrix}$$

or

$$\dot{y}_i = \lambda_i y_i$$

The state transition matrix for this sytem of equations can then be obtained by inspection as

$$\mathbf{\Phi}(t) = \begin{bmatrix} e^{\lambda_1 t} & 0 & \cdots & 0 \\ 0 & e^{\lambda_2 t} & \cdots & 0 \\ \cdot & & & \\ \cdot & & & \\ \cdot & & & \\ 0 & 0 & \cdots & e^{\lambda_n t} \end{bmatrix} \qquad (6.100)$$

In case the solution $\{x(t)\}$ of (6.95) is required, it can be obtained from the transformation (6.96). The problem now is to determine when such a non-singular transformation matrix $\mathbf{P}$ exits and how it can be obtained. But first, we shall consider some preliminaries. Given the unforced system $\{\dot{x}\} = \mathbf{A}\{x\}$ corresponding to (6.95), we assume a solution of the form

$$\{x(t)\} = e^{\lambda(t-t_0)}\{x(t_0)\}$$

Then, we have

$$\{\dot{x}(t)\} = \lambda e^{\lambda(t-t_0)}\{x(t_0)\} = \lambda\{x(t)\} \qquad (6.101)$$

Substituting this result in the differential equation, we get

$$\lambda\{x\} = \mathbf{A}\{x\}$$

or

$$[\lambda\mathbf{I} - \mathbf{A}]\{x\} = 0 \qquad (6.102)$$

where $\mathbf{I}$ is the identity matrix. Nontrivial solution $\{x\}$ exists only if $\det[\lambda\mathbf{I} - \mathbf{A}] = 0$, which is called the characteristic equation of matrix $\mathbf{A}$. The characteristic polynomial of $\mathbf{A}$ is $\det[\lambda\mathbf{I} - \mathbf{A}]$, which is an nth-order polynomial. The n values $\lambda_1, \lambda_2, \ldots, \lambda_n$ which are the roots of the characteristic equation, are called the eigenvalues of matrix $\mathbf{A}$. It is noted that there may be some repeated roots. An eigenvalue of $\mathbf{A}$ is said to be distinct if it is not a repeated root. The eigenvector corresponding to the eigenvalue λ_i is obtained from the identity

$$\mathbf{A}\{v_i\} = \lambda_i\{v_i\} \qquad (6.103)$$

where at least one element of $\{v_i\}$ is nonzero.

If $\{v_i\}$ is a solution of (6.103), then $\alpha\{v_i\}$ is also a solution for any scalar α. Hence, only the direction of the eigenvectors can be determined from (6.103) and their length is arbitrary. An eigenvector may be normalized such that its length is unity (i.e., $\|\mathbf{v}_i\| = 1$).

Example 6.12

For the stationary motion of Example 6.9, let the unforced linearized equations be given by

$$\begin{Bmatrix} \dot{x}_1 \\ \dot{x}_2 \\ \dot{x}_3 \end{Bmatrix} = \begin{bmatrix} 8 & -8 & -2 \\ 4 & -3 & -2 \\ 3 & -4 & 1 \end{bmatrix} \begin{Bmatrix} x_1 \\ x_2 \\ x_3 \end{Bmatrix} \qquad (6.104)$$

The characteristic equation is obtained from

$$\det[\lambda\mathbf{I} - \mathbf{A}] = 0$$

or

$$\det\begin{bmatrix} \lambda - 8 & 8 & 2 \\ -4 & \lambda + 3 & 2 \\ -3 & 4 & \lambda - 1 \end{bmatrix} = 0 \tag{6.105}$$

Calculation of the determinant (6.105) and factorization yields

$$(\lambda - 1)(\lambda - 2)(\lambda - 3) = 0$$

The three distinct eigenvalues of $\mathbf{A}$ are therefore given by

$$\lambda_1 = 1, \qquad \lambda_2 = 2, \qquad \lambda_3 = 3$$

The three eigenvectors are obtained from the solution of (6.103). For $\lambda_1 = 1$, we have

$$\begin{bmatrix} 8 & -8 & -2 \\ 4 & -3 & -2 \\ 3 & -4 & 1 \end{bmatrix}\begin{Bmatrix} v_{11} \\ v_{21} \\ v_{31} \end{Bmatrix} = 1\begin{Bmatrix} v_{11} \\ v_{21} \\ v_{31} \end{Bmatrix}$$

or

$$7v_{11} - 8v_{21} - 2v_{31} = 0$$

$$4v_{11} - 4v_{21} - 2v_{31} = 0$$

$$3v_{11} - 4v_{21} \qquad = 0$$

The three equations are not linearly independent, as it can be seen that the first equation can be obtained by adding the second and third equations. Hence, the length of the eigenvector is arbitrary. If we choose $v_{31} = 2$ arbitrarily, we obtain $v_{11} = 4$ and $v_{21} = 3$. Hence, we get

$$\{v_1\} = \begin{Bmatrix} 4 \\ 3 \\ 2 \end{Bmatrix}$$

The eigenvectors corresponding to the eigenvalues $\lambda_2 = 2$ and $\lambda_3 = 3$ are obtained in a similar manner and are

$$\{v_2\} = \begin{Bmatrix} 3 \\ 2 \\ 1 \end{Bmatrix}, \qquad \{v_3\} = \begin{Bmatrix} 2 \\ 1 \\ 1 \end{Bmatrix}$$

where v_{32} and v_{33} have been chosen arbitrarily as 1. The three eigenvectors of $\mathbf{A}$ can be normalized such that the length of each is unity. The normalized eigenvectors are given by

$$\{v_1\} = \frac{1}{\sqrt{29}}\begin{Bmatrix} 4 \\ 3 \\ 2 \end{Bmatrix}, \qquad \{v_2\} = \frac{1}{\sqrt{14}}\begin{Bmatrix} 3 \\ 2 \\ 1 \end{Bmatrix}, \qquad \{v_3\} = \frac{1}{\sqrt{6}}\begin{Bmatrix} 2 \\ 1 \\ 1 \end{Bmatrix}$$

6.6.1 Matrix Diagonalization

We now consider the determination of the nonsingular transformation matrix $\mathbf{P}$, if it exists, in order to diagonalize the $\mathbf{A}$ matrix such that $\mathbf{P}^{-1}\mathbf{A}\mathbf{P} = \mathbf{\Lambda}$. We first state the following two properties of similar matrices:

1. All similar matrices have the same eigenvalues.

2. All similar matrices have the same determinant.

These properties can be proved by noting that

$$\det [\lambda I - P^{-1}AP] = \det [\lambda P^{-1}IP - P^{-1}AP]$$
$$= \det [P^{-1}(\lambda I - A)P]$$
$$= \det P^{-1} \det (\lambda I - A) \det P$$
$$= \det (\lambda I - A) \qquad (6.106)$$

since $\det P^{-1} = (\det P)^{-1}$.

Now, we show that a matrix A can be reduced to a diagonal matrix Λ by a similarity transformation if and only if it has a set of n linearly independent eigenvectors. This can be proved by considering the fact that matrix Λ must have the eigenvalues of A appearing along the diagonal. If $AP = P\Lambda$, then by partitioned matrix multiplication, it follows that $A\{P_i\} = \lambda_i\{P_i\}$, where $\{P_i\}$ are the columns of P. Hence, the transformation matrix P has the eigenvectors of A as its columns:

$$P = [\{v_1\}\{v_2\} \cdots \{v_n\}] \qquad (6.107)$$

and P^{-1} exists if and only if its columns are linearly independent. In the following, we consider three cases that may arise.

Case 1: Matrix *A* has distinct eigenvalues.   This case occurs in most dynamic systems. It can be shown that in this case, the eigenvectors of A are linearly independent; that is, if

$$\alpha_1\{v_1\} + \alpha_2\{v_2\} + \cdots + \alpha_n\{v_n\} = \{0\}$$

for some constant α_i, then this is possible only if $\alpha_1 = \alpha_2 = \cdots = \alpha_n = 0$. The proof is by contradiction [6, 7].

Example 6.13

Consider the A matrix of Example 6.12. This matrix has distinct eigenvalues and the three eigenvectors have been already determined. It can be verified that these eigenvectors are linearly independent. The transformation matrix that we seek is given by

$$P = \begin{bmatrix} \dfrac{4}{\sqrt{29}} & \dfrac{4}{\sqrt{14}} & \dfrac{2}{\sqrt{6}} \\[2mm] \dfrac{3}{\sqrt{29}} & \dfrac{2}{\sqrt{14}} & \dfrac{1}{\sqrt{6}} \\[2mm] \dfrac{2}{\sqrt{29}} & \dfrac{1}{\sqrt{14}} & \dfrac{1}{\sqrt{6}} \end{bmatrix} \qquad (6.108)$$

It can be checked by matrix multiplication that

$$
\mathbf{P^{-1}AP} =
\begin{bmatrix}
\dfrac{4}{\sqrt{29}} & \dfrac{3}{\sqrt{14}} & \dfrac{2}{\sqrt{6}} \\[2mm]
\dfrac{3}{\sqrt{29}} & \dfrac{2}{\sqrt{14}} & \dfrac{1}{\sqrt{6}} \\[2mm]
\dfrac{2}{\sqrt{29}} & \dfrac{1}{\sqrt{14}} & \dfrac{1}{\sqrt{6}}
\end{bmatrix}^{-1}
\begin{bmatrix}
8 & -8 & 2 \\
4 & -3 & -2 \\
3 & -4 & 1
\end{bmatrix}
\begin{bmatrix}
\dfrac{4}{\sqrt{29}} & \dfrac{3}{\sqrt{14}} & \dfrac{2}{\sqrt{6}} \\[2mm]
\dfrac{3}{\sqrt{29}} & \dfrac{2}{\sqrt{14}} & \dfrac{1}{\sqrt{6}} \\[2mm]
\dfrac{2}{\sqrt{29}} & \dfrac{1}{\sqrt{14}} & \dfrac{1}{\sqrt{6}}
\end{bmatrix}
$$

$$
=
\begin{bmatrix}
1 & 0 & 0 \\
0 & 2 & 0 \\
0 & 0 & 3
\end{bmatrix}
$$

In the transformed state variables $\{y\}$, the differential equations are uncoupled and the unforced system is described by

$$
\dot{y}_i = \lambda_i y_i, \qquad i = 1, 2, 3
$$

where λ_i are the three distinct eigenvalues of $\mathbf{A}$ that appear along the diagonal of the $\boldsymbol{\Lambda}$ matrix. The state transition matrix for the normal variables is obtained by inspection as

$$
\boldsymbol{\Phi}(t) =
\begin{bmatrix}
e^t & 0 & 0 \\
0 & e^{2t} & 0 \\
0 & 0 & e^{3t}
\end{bmatrix}
\tag{6.109}
$$

After the solution of the normal state variables $\{y(t)\}$ has been obtained, the solution of the original state variables is obtained as $\{x(t)\} = \mathbf{P}\{y(t)\}$, where $\mathbf{P}$ is given by (6.108).

Case 2: Matrix *A* does not have distinct eigenvalues.   As stated earlier, any $(n \times n)$ matrix can be diagonalized if and only if it has a set of n linearly independent eigenvectors. Hence, in this case the requirement for diagonalization is that for each multiple eigenvalue λ_i of multiplicity m_i there must exist m_i linearly independent eigenvectors corresponding to λ_i. If this condition is not satisfied, a matrix that does not have distinct eigenvalues is called degenerate and cannot be diagonalized.

Example 6.14

Consider the matrix **A** given by

$$
\mathbf{A} =
\begin{bmatrix}
1 & 0 & 0 \\
1 & 1 & 1 \\
-1 & 0 & 0
\end{bmatrix}
\tag{6.110}
$$

The characteristic equation is obtained as

$$
\det [\lambda \mathbf{I} - \mathbf{A}] = \lambda(\lambda - 1)^2 = 0
$$

Hence, the three eigenvalues are 0, 1, and 1. The eigenvector corresponding to zero eigenvalue is obtained from

$$
\begin{bmatrix}
1 & 0 & 0 \\
1 & 1 & 1 \\
-1 & 0 & 0
\end{bmatrix}
\begin{Bmatrix}
v_{11} \\
v_{21} \\
v_{31}
\end{Bmatrix}
= (0)
\begin{Bmatrix}
v_{11} \\
v_{21} \\
v_{31}
\end{Bmatrix}
$$

$$\{v_1\} = \begin{Bmatrix} 0 \\ 1 \\ -1 \end{Bmatrix}$$

where v_{21} has been chosen arbitrarily as 1. For $\lambda = 1$, the eigenvector satisfies

$$\begin{bmatrix} 1 & 0 & 0 \\ 1 & 1 & 1 \\ -1 & 0 & 0 \end{bmatrix} \begin{Bmatrix} v_{12} \\ v_{22} \\ v_{32} \end{Bmatrix} = (1) \begin{Bmatrix} v_{12} \\ v_{22} \\ v_{32} \end{Bmatrix}$$

or

$$0 = 0$$

$$v_{12} + v_{32} = 0$$

$$-v_{12} - v_{32} = 0$$

Therefore, all eigenvectors belonging to unity eigenvalue can be expressed as

$$\begin{Bmatrix} v_{12} \\ v_{22} \\ v_{32} \end{Bmatrix} = \begin{Bmatrix} c_1 \\ 0 \\ -c_1 \end{Bmatrix} + \begin{Bmatrix} 0 \\ c_2 \\ 0 \end{Bmatrix}$$

where c_1 and c_2 are any nonzero constants. Hence, two linearly independent vectors can be found for the eigenvalue 1, which has a multiplicity of 2, as

$$\{v_2\} = \begin{Bmatrix} 1 \\ 0 \\ -1 \end{Bmatrix} \quad \text{and} \quad \{v_3\} = \begin{Bmatrix} 0 \\ 1 \\ 0 \end{Bmatrix}$$

The nonsingular transformation matrix is thus obtained as

$$\mathbf{P} = \begin{bmatrix} 0 & 1 & 0 \\ 1 & 0 & 1 \\ -1 & -1 & 0 \end{bmatrix} \tag{6.111}$$

It can be easily verified that

$$\mathbf{P^{-1}AP} = \begin{bmatrix} -1 & 0 & -1 \\ 1 & 0 & 0 \\ 1 & 1 & 1 \end{bmatrix} \begin{bmatrix} 1 & 0 & 0 \\ 1 & 1 & 1 \\ -1 & 0 & 0 \end{bmatrix} \begin{bmatrix} 0 & 1 & 0 \\ 1 & 0 & 1 \\ -1 & -1 & 0 \end{bmatrix}$$

$$= \begin{bmatrix} 0 & 0 & 0 \\ 0 & 1 & 0 \\ 0 & 0 & 1 \end{bmatrix} \tag{6.112}$$

It has been stated previously that if for each multiple eigenvalue λ_i of multiplicity m_i, there do not exist m_i linearly independent eigenvectors corresponding to λ_i, then the matrix cannot be diagonalized since in this case the transformation matrix $\mathbf{P}$ defined by (6.107) would be singular. Here, the simplest form to which matrix $\mathbf{A}$ can be reduced is called the Jordan normal form. A Jordan normal matrix $\mathbf{J}$ has the following properties:

1. All elements below the principal diagonal are zero.
2. The diagonal elements of matrix $\mathbf{J}$ are the eigenvalues of $\mathbf{A}$.

3. All elements above the principal diagonal are zero except possibly those elements which are adjacent to two equal diagonal elements, depending on the degeneracy of matrix A.

In fact if $n \times n$ matrix A has only r linearly independent eigenvectors, the Jordan normal matrix has $n - r$ ones above the principal diagonal.

Example 6.15

Consider a matrix A given by

$$\mathbf{A} = \begin{bmatrix} 5 & 4 & 0 \\ 0 & 1 & 0 \\ -4 & 4 & 1 \end{bmatrix} \tag{6.113}$$

The characteristic equation is obtained as

$$\det [\lambda \mathbf{I} - \mathbf{A}] = (\lambda - 1)(\lambda - 1)(\lambda - 5) = 0 \tag{6.114}$$

The three eigenvalues are $\lambda_1 = 1$, $\lambda_2 = 1$, and $\lambda_3 = 5$. For the multiple eigenvalue $\lambda_1 = \lambda_2 = 1$, the corresponding eigenvector is obtained from the solution of

$$\begin{bmatrix} 5 & 4 & 0 \\ 0 & 1 & 0 \\ -4 & 4 & 1 \end{bmatrix} \begin{Bmatrix} v_{11} \\ v_{21} \\ v_{31} \end{Bmatrix} = (1) \begin{Bmatrix} v_{11} \\ v_{21} \\ v_{31} \end{Bmatrix}$$

or

$$4v_{11} + 4v_{21} = 0$$
$$0 = 0$$
$$-4v_{11} + 4v_{21} = 0$$

From the foregoing equations, we obtain only one linearly independent eigenvector as

$$\begin{Bmatrix} v_{11} \\ v_{21} \\ v_{31} \end{Bmatrix} = \begin{Bmatrix} 0 \\ 0 \\ c \end{Bmatrix}$$

where c is any nonzero constant. For the eigenvalue $\lambda_3 = 5$, the corresponding eigenvector is obtained from

$$[A]\{v_3\} = 5\{v_3\}$$

or

$$4v_{23} = 0$$
$$-4v_{23} = 0$$
$$-4v_{13} + v_{23} - 4v_{33} = 0$$

This eigenvector is thus obtained as

$$\begin{Bmatrix} v_{13} \\ v_{23} \\ v_{33} \end{Bmatrix} = \begin{Bmatrix} -b \\ 0 \\ b \end{Bmatrix}$$

where b is any nonzero constant. This matrix **A** has a degeneracy of 1 and it follows that the Jordan normal form is given by

$$[J] = \begin{bmatrix} 1 & 1 & 0 \\ 0 & 1 & 0 \\ 0 & 0 & 5 \end{bmatrix} \tag{6.115}$$

Example 6.16

We now consider another matrix **A**, where

$$\mathbf{A} = \begin{bmatrix} -1 & 2 & -1 \\ 0 & -1 & 0 \\ 0 & 0 & -1 \end{bmatrix} \tag{6.116}$$

The characteristic equation becomes

$$\det[\lambda\mathbf{I} - \mathbf{A}] = (\lambda + 1)(\lambda + 1)(\lambda + 1) = 0$$

Here, the eigenvalue -1 is repeated thrice (i.e., $\lambda_1 = \lambda_2 = \lambda_3 = -1$). The eigenvector corresponding to this multiple eigenvalue can be obtained from

$$\begin{bmatrix} -1 & 2 & -1 \\ 0 & -1 & 0 \\ 0 & 0 & -1 \end{bmatrix} \begin{Bmatrix} v_{11} \\ v_{21} \\ v_{31} \end{Bmatrix} = (-1)\begin{Bmatrix} v_{11} \\ v_{21} \\ v_{31} \end{Bmatrix} \tag{6.117}$$

or

$$2v_{21} - v_{31} = 0$$
$$0 = 0$$
$$0 = 0$$

Hence, all eigenvalues belonging to eigenvalue -1 can be expressed as

$$\begin{Bmatrix} v_{11} \\ v_{21} \\ v_{31} \end{Bmatrix} = \begin{Bmatrix} c \\ 0 \\ 0 \end{Bmatrix} + \begin{Bmatrix} 0 \\ b \\ 2b \end{Bmatrix}$$

Two linearly independent eigenvectors can be found for this eigenvalue as

$$\begin{Bmatrix} 1 \\ 0 \\ 0 \end{Bmatrix} \quad \text{and} \quad \begin{Bmatrix} 0 \\ 1 \\ 2 \end{Bmatrix}$$

The degeneracy of matrix **A** is 1 and its Jordan normal form is given by

$$[J] = \begin{bmatrix} -1 & 1 & 0 \\ 0 & -1 & 0 \\ 0 & 0 & -1 \end{bmatrix} \tag{6.118}$$

We have not discussed here the techniques of determining the transformation matrix that will convert a matrix to its Jordan normal form and the reader may consult several references [6, 7].

Case 3: Matrix A is real and symmetric.   Such a matrix can always be diagonalized, even if it has multiple eigenvalues, by using an orthogonal transformation matrix. The eigenvalues of a square, real, and symmetric matrix are always real. It should be noted that this case was encountered in Chapter 4 in connection with the (3×3) moments of inertia matrix. Hence, the principal directions and principal moments of inertia can always be found. In the more general case, we seek an orthogonal transformation matrix $\mathbf{T}$ that will diagonalize an $(n \times n)$ matrix $\mathbf{A}$ such that $\mathbf{T}^{-1}\mathbf{AT} = \mathbf{\Lambda}$. It is recalled that if $\mathbf{T}$ is an orthogonal matrix, then $\mathbf{T}^{-1} = \mathbf{T}'$, where $\mathbf{T}'$ denotes the transpose of $\mathbf{T}$.

We seek an orthogonal transformation matrix $\mathbf{T}$ such that $\mathbf{T}^{-1}\mathbf{AT} = \mathbf{T}'\mathbf{AT} = \mathbf{\Lambda}$. It is noted again that in order for such a transformation matrix $\mathbf{T}$ to exist, matrix $\mathbf{A}$ must be symmetric since if $\mathbf{T}^{-1}\mathbf{AT} = \mathbf{\Lambda}$ with $\mathbf{T}^{-1} = \mathbf{T}'$, we get

$$\mathbf{A} = \mathbf{T\Lambda T}' = \mathbf{A}' \tag{6.119}$$

Any two vectors $\{v_i\}$ and $\{v_j\}$ are said to be orthogonal if

$$\langle\{v_i\}, \{v_j\}\rangle = \{v_i\}'\{v_j\} = 0, \qquad i \neq j$$

It has been mentioned earlier that the eigenvalues of a real symmetric matrix are real and for such a matrix, it can be shown that eigenvectors corresponding to different eigenvalues are mutually orthogonal as follows. Consider

$$[A]\{v_i\} = \lambda_i\{v_i\}$$
$$[A]\{v_j\} = \lambda_j\{v_j\}$$

Multiplying the first equation by $\{v_j\}'$ and the second by $\{v_i\}'$, it follows that

$$\{v_j\}'[A]\{v_i\} - (\{v_i\}'[A]\{v_j\})' = (\lambda_i - \lambda_j)\{v_j\}'\{v_i\}$$
$$= 0 \tag{6.120}$$

The last equality follows from the fact that $\mathbf{A} = \mathbf{A}'$. Now, since $\lambda_i \neq \lambda_j$, we get $\{v_j\}'\{v_i\} = 0$. Even if the eigenvalues are not distinct, a set of n orthogonal eigenvectors can be found for $(n \times n)$ real, symmetric matrix. The proof of this statement is by induction and is given in several references [5–7].

Example 6.17

Consider the following real, symmetric matrix $\mathbf{A}$ given by

$$\mathbf{A} = \begin{bmatrix} 2 & 1 & -1 \\ 1 & 2 & -1 \\ -1 & -1 & 2 \end{bmatrix} \tag{6.121}$$

The characteristic equation yields $\det[\lambda\mathbf{I} - \mathbf{A}] = (\lambda - 1)^2(\lambda - 4) = 0$ with eigenvalues $\lambda_1 = 1$, $\lambda_2 = 1$, and $\lambda_3 = 4$. The eigenvector corresponding to the eigenvalue 1 is obtained from the solution of

$$[A]\{v_1\} = (1)\{v_1\}$$

or

$$v_{11} + v_{21} - v_{31} = 0$$
$$v_{11} + v_{21} - v_{31} = 0$$
$$-v_{11} - v_{21} + v_{31} = 0$$

Hence, only one equation is available in three unknowns. Arbitrarily choosing $v_{11} = 0$ and $v_{21} = 1$, we obtain $v_{31} = 1$. Hence,

$$\{v_1\} = \left\{ \begin{array}{c} 0 \\ 1 \\ 1 \end{array} \right\}$$

The second eigenvector $\{v_2\}$ corresponding to the repeated eigenvalue 1 is obtained such that it is orthogonal to $\{v_1\}$:

$$\langle \{v_1\}, \{v_2\} \rangle = 0 \tag{6.122}$$

and satisfies the equation

$$[A]\{v_2\} = (1)\{v_2\} \tag{6.123}$$

or

$$v_{12} + v_{22} - v_{32} = 0$$

The orthogonality condition (6.122) yields

$$v_{22} + v_{32} = 0 \tag{6.124}$$

Now, (6.123) and (6.124) are two equations in three unknowns. Arbitrarily choosing $v_{12} = 2$, we obtain $v_{22} = -1$ and $v_{32} = 1$. Hence,

$$\{v_2\} = \left\{ \begin{array}{c} 2 \\ -1 \\ 1 \end{array} \right\}$$

The third eigenvector corresponding to the eigenvalue $\lambda = 4$ is obtained from the solution of the equation

$$[A]\{v_3\} = 4\{v_3\}$$

These equations are described by

$$-2v_{13} + v_{23} - v_{33} = 0$$
$$v_{13} - 2v_{23} - v_{33} = 0$$
$$-v_{13} - v_{23} - 2v_{33} = 0$$

There are only two independent equations in three unknowns, as the first equation can be obtained by multiplying the second equation by -1 and adding it to the third. Arbitrarily choosing $v_{13} = 1$, we obtain

$$\{v_3\} = \left\{ \begin{array}{c} 1 \\ 1 \\ -1 \end{array} \right\}$$

Normalizing the three eigenvectors such that the length of each is unity, we obtain the orthonormal transformation matrix as

$$
\mathbf{T} = \begin{bmatrix} 0 & \dfrac{2}{\sqrt{6}} & \dfrac{1}{\sqrt{3}} \\[2mm] \dfrac{1}{\sqrt{2}} & -\dfrac{1}{\sqrt{6}} & \dfrac{1}{\sqrt{3}} \\[2mm] \dfrac{1}{\sqrt{2}} & \dfrac{1}{\sqrt{6}} & -\dfrac{1}{\sqrt{3}} \end{bmatrix} \tag{6.125}
$$

It can be verified that $\mathbf{T}^{-1} = \mathbf{T}'$ and that

$$
\mathbf{T}^{-1}\mathbf{A}\mathbf{T} = \begin{bmatrix} 1 & 0 & 0 \\ 0 & 1 & 0 \\ 0 & 0 & 4 \end{bmatrix} \tag{6.126}
$$

As mentioned earlier, for systems with large degrees of freedom, a computer solution becomes a necessity even for linear time-invariant equations of motion. Different digital computer methods for direct integration of the equations of motion are discussed in the following chapter. The techniques discussed in this section may also be used to obtain a computer solution for a system of linear time-invariant equations of motion. This method involves first determining the eigenvalues of matrix $\mathbf{A}$ by using one of the standard computer programs [4]. If matrix $\mathbf{A}$ has distinct eigenvalues, as would be the case in practice in most applications, the state transition matrix for the normal state variables is then easily obtained as given by (6.100). After determining the transformation matrix $\mathbf{P}$, and referring to the solution (6.72), it can be seen that the solution of the normal state equation (6.98) is given by

$$
\{y(t)\} = \boldsymbol{\Phi}(t)\{y(0)\} + \int_0^t \boldsymbol{\Phi}(t - t')\mathbf{P}^{-1}\mathbf{B}\{\mathbf{Q}(t')\}\, dt' \tag{6.127}
$$

The convolution integral on the right-hand side of (6.127) can be approximated by convolution summation or it can be evaluated by using a method of numerical integration. The solution of the original state equation is obtained as

$$
\begin{aligned}
\{x(t)\} &= \mathbf{P}\{y(t)\} \\
&= \mathbf{P}\boldsymbol{\Phi}(t)\mathbf{P}^{-1}\{x(0)\} + \int_0^t \mathbf{P}\boldsymbol{\Phi}(t - t')\mathbf{P}^{-1}\mathbf{B}\{\mathbf{Q}(t')\}\, dt' \tag{6.128}
\end{aligned}
$$

6.7 SUMMARY

In this chapter the equations of motion have been expressed as a set of first-order differential equations in the state-variable form by choosing the generalized coordinates and generalized velocities or, alternatively, the generalized coordinates and generalized momenta as the state variables. A theorem is stated and proved concerning the existence and uniqueness of the solution. A similar proof of this theorem is given by Davis [1]. An alternative proof using the contraction-mapping fixed-point theorem is given by Hsu and Meyer [2] and by Vidyasagar [3]. Since the solutions of most nonlinear differential equations are obtained by computer simulation as discussed in the next chapter, the existence and uniqueness theorem is useful to verify whether the solution that has been obtained is unique or possibly there are other solutions to be considered. It is noted that a nonunique mode of behavior is not uncommon for nonlinear dynamic systems.

Since the equations of motion are in general nonlinear, linearized equations are considered as perturbations from an equilibrium state or from a nominal motion when the nonlinearities are analytic functions of their arguments. It is seen that linear time-invariant systems always satisfy the sufficient conditions of the existence and uniqueness theorem. Hence, in such cases, it is not required to verify the uniqueness of the solution. The dynamic response of linear equations to initial conditions and forcing functions has been obtained by using the state transition matrix and convolution integral.

In the case of linear time-invariant systems, the state transition matrix has been obtained by using the Laplace transformation. The last section covers a method that is suitable for computer solution of linear time-invariant systems when the degrees of freedom are large. In this method, the state transition matrix is obtained by matrix diagonalization and similarity transformation. The eigenvalues and eigenvectors of the system matrix are required and several computer programs [4] are available for this purpose. A good discussion on matrix analysis and linear algebra is given by Bellman [5]. Further results concerning the state transition matrix and matrix diagonalization may be found in references [6] and [7].

PROBLEMS

6.1. Investigate the existence and uniqueness of the solution of the following systems. State whether local or global conditions are satisfied and mention singular points or regions, if any.

(a) $mb^2\ddot{\theta} + c\dot{\theta} + mgb \sin \theta = a \sin \omega t$

(b) $m\ddot{x} + c \, \text{sgn} \, \dot{x} + k\left(x - \dfrac{x^3}{6}\right) = a \sin \omega t$

(c) The system defined by (6.35)

6.2. The equation $\dot{x} = x^2$ does not obey a global Lipschitz condition. Show by direct integration that for this system it is possible for $x(t)$ to go to infinity as t approaches some finite time t_1 (i.e., it has a finite escape time).

6.3. Let the linearized equation be described by the Euler differential equation

$$t^2\ddot{x} + t\dot{x} + x = 0$$

with initial conditions $x(t_0)$ and $\dot{x}(t_0)$ at initial time t_0. By choosing x and $\dot{x}$ as state variables, obtain the state transition matrix $\mathbf{\Phi}(t, t_0)$ and show that $\mathbf{\Phi}(t, t') \neq \mathbf{\Phi}(t - t')$.

6.4. Let the matrix $[A(t)]$ be given by $[A(t)] = g(t)[C]$, where $[C]$ is a constant matrix and $g(t)$ is a scalar function of time. Show that $[A(t_1)][A(t_2)] = [A(t_2)]A(t_1)]$ and that for this special case, the state transition matrix is given by

$$\mathbf{\Phi}(t, t_0) = \exp\left[\int_{t_0}^{t} \mathbf{A}(t') \, dt'\right]$$

6.5. A projectile of unit mass is fired with initial speed v_0 at an elevation angle α. A gravity acceleration g acts on the projectile and air resistance is neglected. The equations of motion are given by

$$\ddot{x} = 0$$
$$\ddot{y} = -g$$

with initial conditions $x(0) = x_0$, $y(0) = y_0$, $\dot{x}(0) = v_0 \cos \alpha$, and $\dot{y}(0) = v_0 \sin \alpha$. Obtain the state transition matrix by employing (6.75) and the range of the projectile.

6.6. The equations of motion for the system shown in Fig. P6.6 are given by

$$m_2 \ddot{x}_1 \cos \alpha + \tfrac{3}{2} m_2 \ddot{x}_2 = m_2 g \sin \alpha$$
$$(m_1 + m_2) \ddot{x}_1 + m_2 \ddot{x}_2 \cos \alpha + c \dot{x}_1 = F(t)$$

By choosing $x_1, x_2, \dot{x}_1$, and $\dot{x}_2$ as state variables and employing (6.75), obtain the state transition matrix $\mathbf{\Phi}(t)$.

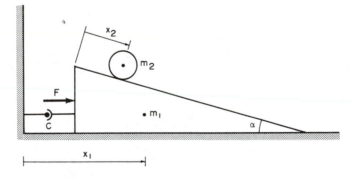

Figure P6.6

6.7. The linearized perturbations in the Euler equations of motion about a stationary motion are described by

$$\begin{Bmatrix} \dot{x}_1 \\ \dot{x}_2 \\ \dot{x}_3 \end{Bmatrix} = \begin{bmatrix} 0 & 0 & 0 \\ 1 & -1 & 0 \\ 0 & -2 & -2 \end{bmatrix} \begin{Bmatrix} x_1 \\ x_2 \\ x_3 \end{Bmatrix}$$

Obtain the state transition matrix $\mathbf{\Phi}(t)$ by employing equation (6.75).

6.8. Reduce the state variables of Problem 6.7 to the Jordan normal form by employing the similarity transformation of (6.96). Obtain the state transition matrix for the state variables $\{x\}$ from the equation

$$\mathbf{\Phi}(t) = \mathbf{P}\mathbf{\Phi}_1(t)\mathbf{P}^{-1}$$

where $\mathbf{\Phi}_1(t)$ is the state transition matrix for the Jordan normal state variables.

REFERENCES

1. Davis, H. T., *Introduction to Nonlinear Differential and Integral Equations*, Dover Publications, Inc., New York, 1962.

2. Hsu, J. C., and Meyer, A. U., *Modern Control Principles and Applications*, McGraw-Hill Book Company, New York, 1968.

3. Vidyasagar, M., *Nonlinear Systems Analysis*, Prentice-Hall, Inc., Englewood Cliffs, N. J., 1978.

4. Smith, B. T., Boyle, J. M., Garbow, B. S., Ikebe, Y., Klema, V. C., and Moler, C. B., *Matrix Eigensystem Routines—EISPACK Guide*, Springer-Verlag, New York, 1970.

5. Bellman, R., *Introduction to Matrix Analysis*, McGraw-Hill Book Company, New York, 1960.

6. Ogata, K., *State Space Analysis of Control Systems*, Prentice-Hall Inc., Englewood Cliffs, N.J., 1967.

7. Wiberg, D. M., *State Space and Linear Systems*, Schaum's Outline Series, McGraw-Hill Book Company, New York, 1971.

7

DYNAMIC RESPONSE BY NUMERICAL METHODS

7.1 INTRODUCTION

Many numerical integration methods are used for the approximate solution of equations of motion or sets of such equations. A complete coverage of numerical integration methods is beyond the scope of this book, and the reader is referred to many available textbooks on the subject. In this chapter we discuss several widely used step-by-step numerical integration schemes for linear and nonlinear dynamic analysis. A brief description of these methods is presented and their application is illustrated. In addition, accuracy, stability, and efficiency of the methods are examined by comparing the results for a sample example.

7.2 FORMULATION OF PROBLEM

It has been shown in the previous chapters that a standard form, in which the equations of motion for a general nonlinear time-varying parameter dynamic system can be expressed, is the state-variable form:

$$\{\dot{x}\} = \{f(x_1, \ldots, x_k, Q_1, Q_2, \ldots, Q_m, t)\} \tag{7.1}$$

where $x_1, \ldots, x_k$ are the components of state-variable vector $\{x\}$ and $\{Q\}$ is a vector of generalized forces. However, in order to decrease the computation time of the digital computer simulation, in many cases it is advantageous to express the equations of motion in the form

$$[m]\{\ddot{q}\} + [c]\{\dot{q}\} + [k]\{q\} = \{Q(t)\} \tag{7.2}$$

198

Equation (7.2) represents a set of n coupled second-order differential equations of motion for the system. Here $[m]$, $[c]$, and $[k]$ are the mass, damping, and stiffness matrices of the system, and $\{Q(t)\}$ is the external force vector. The variables $\{\ddot{q}\}$, $\{\dot{q}\}$, and $\{q\}$ correspond to acceleration, velocity, and displacement vectors of the system. In the case of a linear dynamic system with time-invariant parameters, matrices $[m]$, $[c]$, and $[k]$ are constants and remain unchanged during integration procedure.

For the solution of linear time-invariant equations of motion, one can employ either the normal-mode superposition method as discussed in Chapter 6 or the direct numerical integration methods in case the dimension n is very large. However, for the solution of nonlinear equations of motion, the latter procedure is generally mandatory as the matrices $[m]$, $[c]$, and $[k]$ vary with time* and are required to be modified at each integration step. When certain type of nonlinearities such as hysteresis and Coulomb friction are encountered in the equations of motion, some care is required in expressing the equations of motion in the form of (7.2). This formulation is clarified in Example 7.1.

Example 7.1

A block of mass m_1 is suspended by a linear spring of stiffness k and is constrained to move on a straight bar AB with Coulomb friction. A mass m_2 is suspended from mass m_1 by a rigid rod of length a and is free to move about the pivot O_1 with viscous friction as shown in Fig. 7.1.

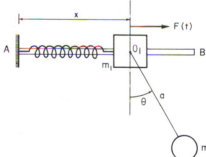

Figure 7.1   A system of two masses.

The equations of motion of this system were obtained in Chapter 3 by the direct application of Newton's laws, and in Chapter 5 by the application of Lagrange equations. We recall that these equations are given by

$$(m_1 + m_2)\ddot{x} + m_2 a\ddot{\theta} \cos \theta - m_2 a\dot{\theta}^2 \sin \theta + \mu N \operatorname{sgn} \dot{x} + kx = F \qquad (7.3)$$

$$m_2 a\ddot{x} \cos \theta + m_2 a^2\ddot{\theta} + c\dot{\theta} + m_2 ga \sin \theta = 0 \qquad (7.4)$$

where in (7.3), the normal force N between the mass m_1 and rod AB is given by

$$N = m_1 g + m_2 a\dot{\theta}^2 \cos \theta + m_2 g \cos^2 \theta - m_2 \ddot{x} \sin \theta \cos \theta \qquad (7.5)$$

---

*In nonlinear constant parameter systems, these matrices consist of constant parameters and functions of $\{q\}$ and $\{\dot{q}\}$ as elements. As $\{q\}$ and $\{\dot{q}\}$ are functions of time, the matrices must be updated to their current values at every integration step.

In Chapter 5 these equations were expressed as a set of first-order equations in the state-variable form by employing the Hamilton's canonic equations, where the generalized momenta are described, respectively, by

$$p_1 = \frac{\partial L}{\partial \dot{x}} = (m_1 + m_2)\dot{x} + m_2 a \dot{\theta} \cos \theta$$

and

$$p_2 = \frac{\partial L}{\partial \dot{\theta}} = m_2 a \dot{x} \cos \theta + m_2 a^2 \dot{\theta}$$

The state-variable vector $\{x\}$ has been defined as

$$\{x\} = \begin{Bmatrix} x \\ \theta \\ p_1 \\ p_2 \end{Bmatrix}$$

and the equations of motion expressed as

$$\{x\} = \{f(x_1, x_2, x_3, x_4, F(t)\} \tag{7.6}$$

Equation (7.6) is in the standard form of (7.1). In order to express the equations in the form of (7.2), we substitute for N from (7.5) in (7.3) and obtain the equations of motion as

$$(m_1 + m_2)\ddot{x} + m_2 a \ddot{\theta} \cos \theta - m_2 a \dot{\theta}^2 \sin \theta + kx$$
$$+ \mu(m_1 g + m_2 a \dot{\theta}^2 \cos \theta + m_2 g \cos^2 \theta - m_2 \ddot{x} \sin \theta \cos \theta) \operatorname{sgn} \dot{x} = F(t) \tag{7.7}$$
$$m_2 a \ddot{x} \cos \theta + m_2 a^2 \ddot{\theta} + c \dot{\theta} + m_2 g a \sin \theta = 0 \tag{7.8}$$

Equations (7.7) and (7.8) are now expressed as

$$\begin{bmatrix} m_1 + m_2 - m_2 \sin \theta \cos \theta \operatorname{sgn} \dot{x} & m_2 a \cos \theta \\ m_2 a \cos \theta & m_2 a^2 \end{bmatrix} \begin{Bmatrix} \ddot{x} \\ \ddot{\theta} \end{Bmatrix}$$

$$+ \begin{bmatrix} 0 & -m_2 a \dot{\theta} (\sin \theta + \cos \theta \operatorname{sgn} \dot{x}) \\ 0 & c \end{bmatrix} \begin{bmatrix} \dot{x} \\ \dot{\theta} \end{bmatrix} + \begin{bmatrix} k & 0 \\ 0 & m_2 g a \dfrac{\sin \theta}{\theta} \end{bmatrix} \begin{bmatrix} x \\ \theta \end{bmatrix}$$

$$= \begin{Bmatrix} \{F(t) - \mu g(m_1 + m_2) \operatorname{sgn} \dot{x}\} \\ 0 \end{Bmatrix} \tag{7.9}$$

In the foregoing equation, the $[m]$, $[c]$, and $[k]$ matrices are not constant but are considered as time varying and are updated at each integration step. Of course, care is required to ensure that $\dot{x}$ does not change sign during an integration step Δt. The matrices are not uniquely defined. For example, in the $[k]$ matrix, the element $m_2 g a(\sin \theta/\theta)$ could be replaced by zero and equivalently a term $m_2 g a \sin \theta$ added to the second row of the right-hand-side forcing vector.

In a direct integration method the equations in (7.2) are integrated successively using a step-by-step numerical integration procedure. The direct integration method implies that no transformation of the equations into a different form is carried out prior to integration. In direct integration methods, time derivatives are generally approximated using difference formulas involving one or more increments of time. There are two basic approaches used in the direct integration method: (1) explicit and (2) implicit. In an explicit formulation the response quantities are expressed in terms of previously determined values of displacement, velocity, or acceleration; whereas in an implicit formulation, the temporal difference equations are combined with the equations of motion, and displacements are calculated directly by solving these equations.

There are also certain semi-implicit methods which have been used to solve the partial differential equations encountered in thermodynamic and fluid dynamic problems. However, we do not consider them here because they offer no real advantage either over the implicit or explicit method for the ordinary differential equations of dynamic systems.

Three algorithms for dynamic analysis by explicit methods are presented here. They are the central difference predictor, two-cycle iteration with trapezoidal rule, and the fourth-order Runge–Kutta method. Four algorithms based on implicit procedures—the Newmark beta, Wilson theta, Houbolt, and Park stiffly stable methods—are also discussed. Algorithms for these solution techniques are implemented into computer programs.

7.3 EXPLICIT METHODS

7.3.1 Central Difference Predictor

We consider a displacement-time history curve as shown in Fig. 7.2. The velocity in the middle of the time interval Δt is given by

$$\dot{q}_{i+1/2} = \frac{q_{i+1} - q_i}{\Delta t} \tag{7.10}$$

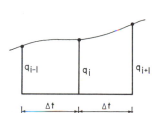

Figure 7.2   Displacement versus time.

The acceleration $\ddot{q}_i$ is obtained as

$$\ddot{q}_i = \frac{\dot{q}_{i+1/2} - \dot{q}_{i-1/2}}{\Delta t} \tag{7.11}$$

Substituting for $\dot{q}_{i+1/2}$ and $\dot{q}_{i-1/2}$ in (7.11), we obtain

$$\ddot{q}_i = \frac{1}{\Delta t^2}(q_{i+1} - 2q_i + q_{i-1}) \tag{7.12}$$

The difference formulas in the central difference predictor method will then be

$$\{\dot{q}_t\} = \frac{1}{2\,\Delta t}[\{q_{t+\Delta t}\} - \{q_{t-\Delta t}\}] \tag{7.13}$$

$$\{\ddot{q}_t\} = \frac{1}{\Delta t^2}[\{q_{t+\Delta t}\} - 2\{q_t\} + \{q_{t-\Delta t}\}] \tag{7.14}$$

Substituting the relations for $\{\dot{q}_t\}$ and $\{\ddot{q}_t\}$ from (7.13) and (7.14), respectively, into (7.2) we obtain

$$\left(\frac{1}{\Delta t^2}[m] + \frac{1}{2\,\Delta t}[c]\right)\{q_{t+\Delta t}\}$$

$$= \{Q_t\} - \left([k] - \frac{2}{\Delta t^2}[m]\right)\{q_t\} - \left(\frac{1}{\Delta t^2}[m] - \frac{1}{2\,\Delta t}[c]\right)\{q_{t-\Delta t}\} \tag{7.15}$$

Equation (7.15) can be rewritten as

$$[\bar{m}]\{q_{t+\Delta t}\} = \{\bar{Q}_t\} \tag{7.16}$$

where the effective mass matrix $[\bar{m}]$ and effective force vector $\{\bar{Q}_t\}$ are

$$[\bar{m}] = \frac{1}{\Delta t^2}[m] + \frac{1}{2\,\Delta t}[c] \tag{7.17a}$$

$$\{\bar{Q}_t\} = \{Q_t\} - \left([k] - \frac{2}{\Delta t^2}[m]\right)\{q_t\} - \left(\frac{1}{\Delta t^2}[m] - \frac{1}{2\,\Delta t}[c]\right)\{q_{t-\Delta t}\} \tag{7.17b}$$

Displacements $\{q_{t+\Delta t}\}$ at the time step $t + \Delta t$ can be calculated by solving (7.16), whereas the velocities and accelerations at time t are obtained by substituting these values of $\{q_{t+\Delta t}\}$ in (7.13) and (7.14). It can be observed that in the central difference predictor method, calculation of $\{q_{t+\Delta t}\}$ involves $\{q_t\}$ and $\{q_{t-\Delta t}\}$. Thus, to obtain the solution at time Δt, a special starting procedure is needed.

The local truncation error of the difference formulas used in the method is of the order Δt^2. The time step for linear dynamic analysis is limited by the highest frequency of the discrete system (i.e., ω_{max}) such that

$$\Delta t \leq \frac{2}{\omega_{max}} \tag{7.18}$$

When Δt does not satisfy the inequality (7.18), a spurious growth of the discrete solution occurs. This is known as a numerical instability.

For the linear dynamic analysis, (7.18) is the necessary and sufficient condition for the stability of the central difference predictor method. However, there is considerable empirical evidence that this equation is equally valid for stability in the nonlinear dynamic problems provided that Δt is reduced to account for the highest frequency during the computations.

7.3.2 Two-Cycle Iteration with Trapezoidal Rule

The incremental form of the equations of motion at any time t is expressed as

$$[m]\{\Delta\ddot{q}_t\} = \{\Delta Q_t\} - [k]\{\Delta q_t\} - [c]\{\Delta\dot{q}_t\} \tag{7.19}$$

In the first iteration cycle, increments in velocities and displacements are estimated using the following formulas:

For first time step:

$$\{\Delta\dot{q}_t\} = \Delta t\{\ddot{q}_{t-\Delta t}\} \tag{7.20a}$$

For other time step:

$$\{\Delta\dot{q}_t\} = 2\,\Delta t\{\ddot{q}_{t-\Delta t}\} - \{\Delta\dot{q}_{t-\Delta t}\} \tag{7.20b}$$

$$\{\dot{q}_t\} = \{\dot{q}_{t-\Delta t}\} + \{\Delta\dot{q}_t\} \tag{7.20c}$$

$$\{\Delta q_t\} = \frac{\Delta t}{2}(\{\dot{q}_{t-\Delta t}\} + \{\dot{q}_t\}) \tag{7.20d}$$

Increments in accelerations are evaluated, by substituting the relations for $\{\Delta\dot{q}_t\}$ and $\{\Delta q_t\}$ from (7.20a) or (7.20b) and (7.20d), respectively, into (7.19). These are then used to estimate the accelerations at time t as

$$\{\Delta\ddot{q}_t\} = [m]^{-1}(\{\Delta Q_t\} - [k]\{\Delta q_t\} - [c]\{\Delta\dot{q}_t\}) \tag{7.21}$$

$$\{\ddot{q}_t\} = \{\ddot{q}_{t-\Delta t}\} + \{\Delta\ddot{q}_t\} \tag{7.22}$$

In the second iteration cycle, increments in the velocities and accelerations are refined as follows:

$$\{\Delta\dot{q}_t\} = \frac{\Delta t}{2}(\{\ddot{q}_{t-\Delta t}\} + \{\ddot{q}_t\}) \tag{7.23a}$$

$$\{\dot{q}_t\} = \{\dot{q}_{t-\Delta t}\} + \{\Delta\dot{q}_t\} \tag{7.23b}$$

$$\{\Delta q_t\} = \frac{\Delta t}{2}(\{\dot{q}_{t-\Delta t}\} + \{\dot{q}_t\}) \tag{7.23c}$$

Finally, the relations for $\{\Delta\dot{q}_t\}$ and $\{\Delta q_t\}$ in (7.23a) and (7.23c) are substituted into (7.21) to calculate the new increments in the accelerations. These are then used in (7.22) to evaluate accelerations at time t.

7.3.3 Runge–Kutta Methods

In this method, the system equations are replaced in state-variables form; that is, both displacements and velocities are treated as unknowns defined by

$$\{x\} = \begin{Bmatrix} \{q\} \\ \{\dot{q}\} \end{Bmatrix} \tag{7.24}$$

Equation (7.2) is now rewritten as

$$\{\ddot{q}\} = -[m]^{-1}[k]\{q\} - [m]^{-1}[c]\{\dot{q}\} + [m]^{-1}\{Q(t)\} \tag{7.25}$$

Using the identity

$$\{\dot{q}\} = \{\dot{q}\} \tag{7.26}$$

equations (7.25) and (7.26) are written as

$$\{\dot{x}\} = \begin{Bmatrix} \{\dot{q}\} \\ \{\ddot{q}\} \end{Bmatrix} \left[\begin{array}{c|c} [0] & [I] \\ \hline -[m]^{-1}[k] & -[m]^{-1}[c] \end{array} \right] \begin{Bmatrix} \{q\} \\ \{\dot{q}\} \end{Bmatrix} + \begin{Bmatrix} \{0\} \\ [m]^{-1}\{Q(t)\} \end{Bmatrix} \tag{7.27a}$$

or

$$\{\dot{x}\} = [E]\{x\} + \{Q^*(t)\} \tag{7.27b}$$

or

$$\{\dot{x}\} = \{f(t, x)\} \tag{7.27c}$$

In the Runge–Kutta method, an approximation to $\{x_{t+\Delta t}\}$ is obtained from $\{x_t\}$ in such a way that the power series expansion of the approximation coincides, up to terms of a certain order $(\Delta t)^N$ in the time interval Δt, with the actual Taylor series expansion of $(t + \Delta t)$ in powers of Δt. However, the method is self-starting and also has the advantage that no initial values are needed beyond the prescribed values.

We first consider a scalar first-order differential equation described by

$$\dot{x} = f(x(t), t) \tag{7.28}$$

and later generalize to a set of first-order equations. It is assumed that conditions of Theorem 6.1 are satisfied about the point $(x(t), t)$ such that a solution of (7.28) exists and is unique in the interval of time Δt about that point. A Taylor series expansion of the solution yields

$$x(t + \Delta t) = x_{t+\Delta t} = x(t) + \Delta t \dot{x}(t) + \frac{(\Delta t)^2}{2!} \ddot{x}(t) + \frac{(\Delta t)^3}{3!} \dddot{x}(t) + \cdots \tag{7.29}$$

Since from (7.28), $\dot{x} = f(x(t), t) = f$ and further differentiation yields

$$\ddot{x}(t) = \frac{\partial f}{\partial t} + \frac{\partial f}{\partial x}\frac{dx}{dt} = f_t + ff_x.$$

Similarly,

$$\dddot{x}(t) = f_{tt} + 2ff_{tx} + f^2 f_{xx} + f_x(f_t + ff_x)$$

Substituting these results in (7.29), we obtain

$$x(t + \Delta t) = x(t) + \Delta t f + \frac{\Delta t^2}{2}(f_t + ff_x)$$

$$+ \frac{(\Delta t)^3}{6}[f_{tt} + 2ff_{tx} + f^2 f_{xx} + f_x(f_t + ff_x)] + \cdots \tag{7.30}$$

It has been also assumed in the foregoing that the higher derivatives and partial derivatives exist at points required. The simplest of the Runge–Kutta methods is the first-order method, also known as Euler method, which retains only the first two terms of the Taylor series expansion (7.30). Hence, in the Euler method, the approximation to the solution is given by

$$x(t + \Delta t) = x(t) + \Delta t f(x(t), t) \tag{7.31}$$

The results are reasonably accurate only for the first few time steps with small Δt; after that the approximation usually diverges from the actual solution. The general idea behind the higher-order Runge–Kutta methods is to retain the higher-order terms in (7.30). However, the method does not require evaluation of the derivatives of the function f. Instead, approximations are obtained at the expense of several evaluations of the function f at each time step.

As discussed in Chapter 6, the solution can also be written in the integral form

$$x(t + \Delta t) = x(t) + \int_{t}^{t+\Delta t} f(x(t'), t') \, dt' \tag{7.32}$$

Application of the mean value theorem of integral calculus to (7.32) yields

$$x(t + \Delta t) = x(t) + \Delta t \, f(x(t + \alpha \, \Delta t), t + \alpha \, \Delta t) \tag{7.33}$$

for some α such that $0 < \alpha < 1$. The problem now is to avoid the evaluation of explicit higher derivatives required in (7.30) and in the expansion of (7.33).

Second-order Runge–Kutta.   Here, α is chosen so that the Taylor series expansion of (7.33) agrees exactly with (7.30) up to terms of order $(\Delta t)^2$. Letting $x(t + \alpha \, \Delta t) = x(t) + \beta \, \Delta t + \cdots$ in (7.33), the Taylor series expansion of (7.33) up to orders of $(\Delta t)^2$ yields

$$x(t + \Delta t) = x(t) + \Delta t f + \alpha (\Delta t)^2 f_t + \beta (\Delta t)^2 f_x \tag{7.34}$$

Comparing (7.34) and (7.30) when only terms of order $(\Delta t)^2$ are retained in (7.30), we obtain

$$\alpha = \tfrac{1}{2} \quad \text{and} \quad \beta = \tfrac{1}{2} f$$

Hence, in the second-order Runge–Kutta method, the approximation to the solution is given by

$$x(t + \Delta t) = x(t) + \Delta t \, f\left(x(t) + \frac{\Delta t}{2} f(x(t), t), t + \frac{\Delta t}{2}\right) \tag{7.35}$$

Fourth-order Runge–Kutta method.   To obtain good accuracy, the commonly employed method is the fourth-order Runge–Kutta method. Again, to avoid the evaluation of explicit higher-order derivatives, we set

$$k_1 = f(t, x(t))$$
$$k_2 = f(t + \alpha_2 \, \Delta t, x + \beta_2 k_1 \, \Delta t)$$
$$k_3 = f(t + \alpha_3 \, \Delta t, x + \beta_3 k_1 \, \Delta t + r_3 k_2 \, \Delta t) \tag{7.36}$$
$$k_4 = f(t + \alpha_4 \, \Delta t, x + \beta_4 k_1 \, \Delta t + r_4 k_2 \, \Delta t + \delta_4 k_3 \, \Delta t)$$
$$x(t + \Delta t) = x(t) + \Delta t(\mu_1 k_1 + \mu_2 k_2 + \mu_3 k_3 + \mu_4 k_4)$$

The problem now is to determine the 13 parameters in (7.36)—3 α's, 3 β's, 2 r's, δ_4, and 4 μ's—such that the Taylor series expansion of (7.36) agrees exactly with (7.30) up to terms of order $(\Delta t)^4$. After expanding the k's in Taylor series up to terms of order $(\Delta t)^4$, substituting the result in (7.36), and comparing this expansion with (7.30) term by term, one obtains eight equations in 13 unknowns. Three further equations are obtained from the fact that the method must be independent of the function f. We omit the details here and the interested reader is referred to reference [14]. Hence, two parameters may be chosen arbitrarily. The choice of $\alpha_2 = \frac{1}{2}$ and $\delta_4 = 1$ leads to the commonly employed results

$$k_1 = f(t, x)$$

$$k_2 = f\left(t + \frac{\Delta t}{2}, x + k_1 \frac{\Delta t}{2}\right)$$

$$k_3 = f\left(t + \frac{\Delta t}{2}, x + k_2 \frac{\Delta t}{2}\right) \tag{7.37}$$

$$k_4 = f(t + \Delta t, x + k_3 \Delta t)$$

$$x(t + \Delta t) = x(t) + \frac{\Delta t}{6}(k_1 + 2k_2 + 2k_3 + k_4)$$

Now, in order to generalize the fourth-order Runge–Kutta method to a set of first-order equations

$$\dot{x}_i = f_i(x_i, \ldots, x_n, t); \qquad \text{that is, } \{\dot{x}\} = \{f(x, t)\}$$

we define the vector $\{k_1\} = \{f(t, x)\}$. The vectors $\{k_2\}$, $\{k_3\}$, and $\{k_4\}$ are defined similarly. In vector form, (7.37) is written as

$$\{x(t + \Delta t)\} = \{x(t)\} + \frac{\Delta t}{6}(\{k_1\} + 2\{k_2\} + 2\{k_3\} + \{k_4\}) \tag{7.38}$$

The first- and second-order Runge–Kutta methods are hardly ever employed because, as mentioned earlier, the results that they yield are not very accurate. Hence, if a Runge–Kutta method is chosen as the integration technique, it is usually the fourth-order method.

The truncation error e_t for the fourth-order Runge–Kutta method is of the form

$$e_t = k(\Delta t)^5 \tag{7.39}$$

where k depends on $f(t, x)$ and its higher-order partial derivatives.

Since the fourth-order Runge–Kutta method is an explicit method, the maximum time step is usually governed by stability considerations. The method can be considered as an inherently stable method, since the change in time step can be easily implemented at any stage of the advance of calculations. However, the method generates an artificial damping which unduly suppresses the amplitude of the response. The principal disadvantage consists in the fact that each forward step requires several evaluations of the functions. This increases considerably the cost of computation. Moreover, no simple expression is available to calculate precisely the truncation error for the Runge–Kutta method. This is also a source of inconvenience.

7.4 IMPLICIT METHODS

7.4.1 Houbolt Method

This method is based on a third-order interpolation of displacements. In the Houbolt integration scheme, multistep implicit formulas for velocity and acceleration are derived in terms of displacements using backward differences, as shown in the following with reference to Fig. 7.3.

$$q_t = q_{t+\Delta t} - \Delta t \dot{q}_{t+\Delta t} + \frac{\Delta t^2}{2} \ddot{q}_{t+\Delta t} - \frac{\Delta t^3}{6} \dddot{q}_{t+\Delta t} \tag{7.40a}$$

$$q_{t-\Delta t} = q_{t+\Delta t} - (2\,\Delta t)\dot{q}_{t+\Delta t} + \left(\frac{2\,\Delta t}{2}\right)^2 \ddot{q}_{t+\Delta t} - \left(\frac{2\,\Delta t}{6}\right)^3 \dddot{q}_{t+\Delta t} \tag{7.40b}$$

$$q_{t-2\Delta t} = q_{t+\Delta t} - (3\,\Delta t)\dot{q}_{t+\Delta t} + \left(\frac{3\,\Delta t}{2}\right)^2 \ddot{q}_{t+\Delta t} - \left(\frac{3\,\Delta t}{6}\right)^3 \dddot{q}_{t+\Delta t} \tag{7.40c}$$

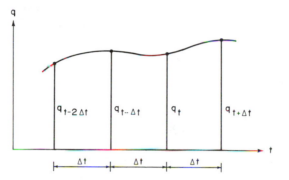

Figure 7.3   Displacement versus time.

Solving equations (7.40a), (7.40b), and (7.40c) for $\ddot{q}_{t+\Delta t}$ and $\dot{q}_{t+\Delta t}$ in terms of $q_{t+\Delta t}, q_t, q_{t-\Delta t},$ and $q_{t-2\Delta t},$ we obtain the following formulas:

$$\ddot{q}_{t+\Delta t} = \frac{1}{\Delta t^2}(2q_{t+\Delta t} - 5q_t + 4q_{t-\Delta t} - q_{t-2\Delta t}) \tag{7.41a}$$

$$\dot{q}_{t+\Delta t} = \frac{1}{6\,\Delta t}(11q_{t+\Delta t} - 18q_t + 9q_{t-\Delta t} - 2q_{t-2\Delta t}) \tag{7.41b}$$

The difference formulas in the Houbolt algorithms for a vector equation are then given by

$$\{\ddot{q}_{t+\Delta t}\} = \frac{1}{\Delta t^2}[2\{q_{t+\Delta t}\} - 5\{q_t\} + 4\{q_{t-\Delta t}\} - \{q_{t-2\Delta t}\}] \tag{7.42}$$

$$\{\dot{q}_{t+\Delta t}\} = \frac{1}{6\,\Delta t}[11\{q_{t+\Delta t}\} - 18\{q_t\} + 9\{q_{t-\Delta t}\} - 2\{q_{t-2\Delta t}\}] \tag{7.43}$$

Substituting the relations for $\{\ddot{q}_{t+\Delta t}\}$ and $\{\dot{q}_{t+\Delta t}\}$ from (7.42) and (7.43), respectively, into (7.2), we obtain

$$\left(\frac{2}{\Delta t^2}[m] + \frac{11}{6\,\Delta t}[c] + [k]\right)\{q_{t+\Delta t}\} = \{Q_{t+\Delta t}\} + \left(\frac{5}{\Delta t^2}[m] + \frac{3}{\Delta t}[c]\right)\{q_t\}$$

$$-\left(\frac{4}{\Delta t^2}[m] + \frac{3}{2\,\Delta t}[c]\right)\{q_{t-\Delta t}\}$$

$$+\left(\frac{1}{\Delta t^2}[m] + \frac{1}{3\Delta t}[c]\right)\{q_{t-2\Delta t}\} \quad (7.44)$$

Equation (7.44) is rewritten as

$$[\bar{m}]\{q_{t+\Delta t}\} = \{\bar{Q}_{t+\Delta t}\} \quad (7.45)$$

where the effective mass matrix $[\bar{m}]$ and effective force vector $\{\bar{Q}_{t+\Delta t}\}$ are

$$[\bar{m}] = \frac{2}{\Delta t^2}[m] + \frac{11}{6\,\Delta t}[c] + [k] \quad (7.46a)$$

$$\{\bar{Q}_{t+\Delta t}\} = \{Q_{t+\Delta t}\} + \left(\frac{5}{\Delta t^2}[m] + \frac{3}{\Delta t}[c]\right)\{q_t\}$$

$$-\left(\frac{4}{\Delta t^2}[m] + \frac{3}{2\,\Delta t}[c]\right)\{q_{t-\Delta t}\} + \left(\frac{1}{\Delta t^2}[m] + \frac{1}{3\,\Delta t}[c]\right)\{q_{t-2\Delta t}\} \quad (7.46b)$$

    Displacements $\{q_{t+\Delta t}\}$ at the time step $t + \Delta t$ can be calculated by solving (7.45), whereas the velocities and accelerations at time $t + \Delta t$ are obtained by substituting for $\{q_{t+\Delta t}\}$ in (7.41a) and (7.41b).

    It can be noticed that in the Houbolt method, calculation of $\{q_{t+\Delta t}\}$ involves displacements at t, $t - \Delta t$, and $t - 2\,\Delta t$. Therefore, a special starting procedure is required to obtain solution at time Δt and $2\Delta t$. This makes the method non-self-starting. The method also requires a large computer storage to store displacements for two previous time steps.

7.4.2 Wilson Theta Method

    In the Wilson theta method, it is assumed that the acceleration varies linearly over an increment of time $\theta\,\Delta t$, where $\theta \geq 1.0$ as shown in Fig. 7.4, whereas the properties of the dynamic system remain constant during this

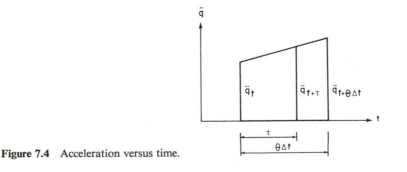

Figure 7.4 Acceleration versus time.

interval. If τ is the increase in time between t and $t + \theta \Delta t$ (i.e., $0 \leq \tau \leq \theta \Delta t$), then for the time interval t to $t + \theta \Delta t$, it is assumed that

$$\ddot{q}_{t+\tau} = \ddot{q}_t + \frac{\tau}{\theta \Delta t}(\ddot{q}_{t+\theta \Delta t} - \ddot{q}_t) \tag{7.47}$$

Integrating (7.47), we obtain the following expressions for $\dot{q}_{t+\tau}$ and $q_{t+\tau}$:

$$\dot{q}_{t+\tau} = \dot{q}_t + \ddot{q}_t\tau + \frac{\tau^2}{2\theta \Delta t}(\ddot{q}_{t+\theta \Delta t} - \ddot{q}_t) \tag{7.48a}$$

$$q_{t+\tau} = q_t + \dot{q}_t\tau + \frac{1}{2}\ddot{q}_t\tau + \frac{\tau^3}{6\theta \Delta t}(\ddot{q}_{t+\theta \Delta t} - \ddot{q}_t) \tag{7.48b}$$

Substituting $\tau = \theta \Delta t$ into (7.48a) and (7.48b), we obtain the following expressions at time $t + \theta \Delta t$:

$$\dot{q}_{t+\theta \Delta t} = \dot{q}_t + \frac{\theta \Delta t}{2}(\ddot{q}_t + \ddot{q}_{t+\theta \Delta t}) \tag{7.49a}$$

$$q_{t+\theta \Delta t} = q_t + \theta \Delta t\dot{q}_t + \frac{\theta^2 \Delta t^2}{6}(\ddot{q}_{t+\theta \Delta t} + 2\ddot{q}_t) \tag{7.49b}$$

Equations (7.49a) and (7.49b) are solved for $\ddot{q}_{t+\theta \Delta t}$ and $\dot{q}_{t+\theta \Delta t}$ in terms of $q_{t+\theta \Delta t}$ as

$$\ddot{q}_{t+\theta \Delta t} = \frac{6}{\theta^2 \Delta t^2}(q_{t+\theta \Delta t} - q_t) - \frac{6}{\theta \Delta t}\dot{q}_t - 2\ddot{q}_t \tag{7.50a}$$

$$\dot{q}_{t+\theta \Delta t} = \frac{3}{\theta \Delta t}(q_{t+\theta \Delta t} - q_t) - 2\dot{q}_t - \frac{\theta \Delta t}{2}\ddot{q}_t \tag{7.50b}$$

The difference formulas in the Wilson theta algorithm are then given by

$$\{\ddot{q}_{t+\theta \Delta t}\} = \frac{6}{\theta^2 \Delta t^2}(\{q_{t+\theta \Delta t}\} - \{q_t\}) - \frac{6}{\theta \Delta t}\{\dot{q}_t\} - 2\{\ddot{q}_t\} \tag{7.51}$$

$$\{\dot{q}_{t+\theta \Delta t}\} = \frac{3}{\theta \Delta t}(\{q_{t+\theta \Delta t}\} - \{q_t\}) - 2\{\dot{q}_t\} - \frac{\theta \Delta t}{2}\{\ddot{q}_t\} \tag{7.52}$$

We consider equation (7.2) at time $t + \theta \Delta t$ to obtain solution for the displacements, velocities, and accelerations at time $t + \Delta t$. Since the accelerations vary linearly, a linearly projected force vector is used such that

$$[m]\{\ddot{q}_{t+\theta \Delta t}\} + [c]\{\dot{q}_{t+\theta \Delta t}\} + [k]\{q_{t+\theta \Delta t}\} = \{Q_{t+\theta \Delta t}\} \tag{7.53}$$

where

$$\{Q_{t+\theta \Delta t}\} = \{Q_t\} + \theta(\{Q_{t+\Delta t}\} - \{Q_t\})$$

Substituting the relations for $\{\ddot{q}_{t+\theta \Delta t}\}$ and $\{\dot{q}_{t+\theta \Delta t}\}$ from (7.51) and (7.52), respectively, into (7.53), we obtain

$$\left(\frac{6}{\theta^2 \Delta t^2}[m] + \frac{3}{\theta \Delta t}[c] + [k]\right)\{q_{t+\theta \Delta t}\} = \{Q_{t+\theta \Delta t}\} + \left(\frac{6}{\theta^2 \Delta t^2}[m] + \frac{3}{\theta \Delta t}[c]\right)\{q_t\}$$

$$+ \left(\frac{6}{\theta \Delta t}[m] + 2[c]\right)\{\dot{q}_t\} + \left(2[m] + \frac{\theta \Delta t}{2}[c]\right)\{\ddot{q}_t\} \tag{7.54}$$

Equation (7.54) is rewritten as

$$[\bar{m}]\{q_{t+\theta\Delta t}\} = \{\bar{Q}_{t+\theta\Delta t}\} \tag{7.55}$$

where the effective mass matrix $[\bar{m}]$ and effective force vector $\{\bar{Q}_{t+\theta\Delta t}\}$ are given by

$$[\bar{m}] = \frac{6}{\theta^2 \, \Delta t^2}[m] + \frac{3}{\theta \, \Delta t}[c] + [k] \tag{7.56a}$$

$$\{\bar{Q}_{t+\theta\Delta t}\} = \{Q_{t+\theta\Delta t}\} + \left(\frac{6}{\theta^2 \, \Delta t^2}[m] + \frac{3}{\theta \, \Delta t}[c]\right)\{q_t\}$$

$$+ \left(\frac{6}{\theta \, \Delta t}[m] + 2[c]\right)\{\dot{q}_t\} + \left(2[m] + \frac{\theta \, \Delta t}{2}[c]\right)\{\ddot{q}_t\} \tag{7.56b}$$

The solution (7.55) yields $\{q_{t+\theta\Delta t}\}$, which is then substituted in the following expressions to obtain accelerations, velocities, and displacements at $t + \Delta t$:

$$\{\ddot{q}_{t+\Delta t}\} = \frac{6}{\theta^3 \, \Delta t^2}(\{q_{t+\theta\Delta t}\} - \{q_t\}) - \frac{6}{\theta^2 \, \Delta t}\{\dot{q}_t\} + \left(1 - \frac{3}{\theta}\right)\{\ddot{q}_t\} \tag{7.57a}$$

$$\{\dot{q}_{t+\Delta t}\} = \{\dot{q}_t\} + \frac{\Delta t}{2}(\{\ddot{q}_{t+\Delta t}\} + \{\ddot{q}_t\}) \tag{7.57b}$$

$$\{q_{t+\Delta t}\} = \{q_t\} + \Delta t\{\dot{q}_t\} + \frac{\Delta t^2}{6}(\{\ddot{q}_{t+\Delta t}\} + 2\{\ddot{q}_t\}) \tag{7.57c}$$

The overall method is proven to be unconditionally stable for values of $\theta \geq 1.37$ for linear dynamic systems, but a value of 1.5 is often used for nonlinear problems. An anomaly of this method is that equilibrium is never satisfied at time $t + \Delta t$.

7.4.3 Newmark Beta Method

The Newmark integration method can be treated as an extension of the linear integration scheme. The method uses parameters α and β, which can be changed to suit the requirements of the problem at hand. The equations used are given by

$$\dot{q}_{t+\Delta t} = \dot{q}_t + [(1 - \alpha)\ddot{q}_t + \alpha\ddot{q}_{t+\Delta t}] \, \Delta t \tag{7.58a}$$

$$q_{t+\Delta t} = q_t + \dot{q}_t\Delta_t + [(\tfrac{1}{2} - \beta)\ddot{q}_t + \beta\ddot{q}_{t+\Delta t}](\Delta t)^2 \tag{7.58b}$$

where α and β are parameters which are determined to obtain integration accuracy and stability. The net effect of these parameters is to change the form of the variation of an acceleration during the time interval Δt. By letting $\alpha = \frac{1}{2}$ and $\beta = 0$, the acceleration is constant and equal to $\ddot{q}_t$ during each time interval Δt. If $\alpha = \frac{1}{2}$ and $\beta = \frac{1}{8}$, the acceleration is constant from the beginning as $\ddot{q}_t$ and then changes to $\ddot{q}_{t+\Delta t}$ at the middle of the time interval Δt. With $\alpha = \frac{1}{2}$ and $\beta = \frac{1}{6}$, (7.58a) and (7.58b) imply that the acceleration varies linearly from $\ddot{q}_t$ to $\ddot{q}_{t+\Delta t}$, whereas values $\alpha = \frac{1}{2}$ and $\beta = \frac{1}{4}$ correspond to the assumption that acceleration remains constant at an average value of $(\ddot{q}_t + \ddot{q}_{t+\Delta t})/2$. The difference formulas in the Newmark beta algorithm are

$$\{\ddot{q}_{t+\Delta t}\} = \frac{1}{\beta\,\Delta t^2}(\{q_{t+\Delta t}\} - \{q_t\}) - \frac{1}{\beta\,\Delta t}\{\dot{q}_t\} - \left(\frac{1}{2\beta} - 1\right)\{\ddot{q}_t\} \tag{7.59}$$

$$\{\dot{q}_{t+\Delta t}\} = \frac{\alpha}{\beta\,\Delta t}(\{q_{t+\Delta t}\} - \{q_t\}) - \left(\frac{\alpha}{\beta} - 1\right)\{\dot{q}_t\} - \Delta t\left(\frac{\alpha}{2\beta} - 1\right)\{\ddot{q}_t\} \tag{7.60}$$

We consider (7.2) at time $t + \Delta t$ to obtain solution for the displacements, velocities, and accelerations. Substituting the relations for $\{\ddot{q}_{t+\Delta t}\}$ and $\{\dot{q}_{t+\Delta t}\}$ from (7.59) and (7.60), respectively, into (7.2), we obtain

$$\left(\frac{1}{\beta\,\Delta t^2}[m] + \frac{\alpha}{\beta\,\Delta t}[c] + [k]\right)\{q_{t+\Delta t}\}$$

$$= \{Q_{t+\Delta t}\} + \left[\left(\frac{1}{2\beta} - 1\right)[m] + \Delta t\left(\frac{\alpha}{2\beta} - 1\right)[c]\right]\{\ddot{q}_t\}$$

$$+ \left[\frac{1}{\beta\,\Delta t}[m] + \left(\frac{\alpha}{\beta} - 1\right)[c]\right]\{\dot{q}_t\}$$

$$+ \left[\frac{1}{\beta\,\Delta t^2}[m] + \frac{\alpha}{\beta\,\Delta t}[c]\right]\{q_t\} \tag{7.61}$$

Equation (7.61) is rewritten as

$$[\bar{m}]\{q_{t+\Delta t}\} = \{\bar{Q}_{t+\Delta t}\} \tag{7.62}$$

where the effective mass matrix $[\bar{m}]$ and effective force vector $\{\bar{Q}_{t+\Delta t}\}$ are

$$[\bar{m}] = \frac{1}{\beta\,\Delta t^2}[m] + \frac{\alpha}{\beta\,\Delta t}[c] + [k] \tag{7.63a}$$

$$\{\bar{Q}_{t+\Delta t}\} = \{Q_{t+\Delta t}\} + \left[\left(\frac{1}{2\beta} - 1\right)[m] + \Delta t\left(\frac{\alpha}{2\beta} - 1\right)[c]\right]\{\ddot{q}_t\}$$

$$+ \left[\frac{1}{\beta\,\Delta t}[m] + \left(\frac{\alpha}{\beta} - 1\right)[c]\right]\{\dot{q}_t\}$$

$$+ \left[\frac{1}{\beta\,\Delta t^2}[m] + \frac{\alpha}{\beta\,\Delta t}[c]\right]\{q_t\} \tag{7.63b}$$

Solution of (7.62) yields $\{q_{t+\Delta t}\}$, which is then substituted in (7.59) and (7.60) to obtain velocities and displacements at $t + \Delta t$.

The important features of this method are that for linear systems the amplitude of a mode is conserved, and the response is unconditionally stable provided that $\alpha \geq \frac{1}{2}$ and $\beta \geq 0.25\,(\alpha + 0.5)^2$. However, the $\alpha = \frac{1}{2}$ and $\beta = \frac{1}{4}$ values give the largest truncation error in the frequency of the response as opposed to other β values. For a multiple-degree-of-freedom system in which a number of modes constitute the total response, the peak amplitude may not be correct.

7.4.4 Park Stiffly Stable Method

The Gear two-step and three-step methods are based on the second- and third-order interpolation of displacements using backward difference formulas, respectively. The velocity formula at time $t + \Delta t$ in the Gear two-step method is

$$\dot{q}_{t+\Delta t} = \frac{1}{2\,\Delta t}(2q_{t+\Delta t} - 4q_t + q_{t-\Delta t}) \tag{7.64}$$

The difference formula for velocity at time $t + \Delta t$ in the Gear three-step method is given by (7.41b). The Gear two-step method introduces high numerical damping in the solution, whereas the Gear three-step method is unstable for the frequency ranges ($\omega\,\Delta t \le 2$) of interest. The Park stable method is the combination of the Gear two- and three-step methods to achieve an accurate and stable method for the low-frequency range and stable for all higher-frequency components. The velocity formula in the Park stiffly stable method is derived using a linear combination of (7.41b) and (7.64) as

$$\dot{q}_{t+\Delta t} = \frac{1}{2}\bigg[\frac{1}{2\,\Delta t}(3q_{t+\Delta t} - 4q_t + q_{t-\Delta t})$$
$$+ \frac{1}{6\,\Delta t}(11q_{t+\Delta t} - 18q_t + 9q_{t-\Delta t} - 2q_{t-2\Delta t})\bigg] \tag{7.65a}$$

or

$$\dot{q}_{t+\Delta t} = \frac{1}{6\,\Delta t}(10q_{t+\Delta t} - 15q_t + 6q_{t-\Delta t} - q_{t-2\Delta t}) \tag{7.65b}$$

Similarly,

$$\ddot{q}_{t+\Delta t} = \frac{1}{6\,\Delta t}(10\dot{q}_{t+\Delta t} - 15\dot{q}_t + 6\dot{q}_{t-\Delta t} - \dot{q}_{t-2\Delta t}) \tag{7.66}$$

The difference formulas in the Park algorithm will then be given by

$$\{\ddot{q}_{t+\Delta t}\} = \frac{1}{6\,\Delta t}[10\{\dot{q}_{t+\Delta t}\} - 15\{\dot{q}_t\} + 6\{\dot{q}_{t-\Delta t}\} - \{\dot{q}_{t-2\Delta t}\}] \tag{7.67}$$

$$\{\dot{q}_{t+\Delta t}\} = \frac{1}{6\,\Delta t}[10\{q_{t+\Delta t}\} - 15\{q_t\} + 6\{q_{t-\Delta t}\} - \{q_{t-2\Delta t}\}] \tag{7.68}$$

We consider (7.2) at time $t + \Delta t$ to obtain solution for the displacements, velocities, and accelerations. Substituting the relations for $\{\ddot{q}_{t+\Delta t}\}$ and $\{\dot{q}_{t+\Delta t}\}$ from (7.67) and (7.68), respectively, into (7.2), we obtain

$$\left(\frac{100}{36\,\Delta t^2}[m] + \frac{10}{6\,\Delta t}[c] + [k]\right)\{q_{t+\Delta t}\}$$
$$= \{Q_{t+\Delta t}\} + \frac{15}{6\,\Delta t}[m]\{\dot{q}_t\} - \frac{1}{\Delta t}[m]\{\dot{q}_{t-\Delta t}\} + \frac{1}{6\,\Delta t}[m]\{\dot{q}_{t-2\Delta t}\}$$
$$+ \left(\frac{150}{36\,\Delta t^2}[m] + \frac{15}{6\,\Delta t}[c]\right)\{q_t\} - \left(\frac{10}{6\,\Delta t^2}[m] + \frac{1}{\Delta t}[c]\right)\{q_{t-\Delta t}\}$$
$$+ \left(\frac{1}{36\,\Delta t^2}[m] + \frac{1}{6\,\Delta t}[c]\right)\{q_{t-2\Delta t}\} \tag{7.69}$$

Equation (7.69) is rewritten as

$$[\bar{m}]\{q_{t+\Delta t}\} = \{\bar{Q}_{t+\Delta t}\} \tag{7.70}$$

where the effective mass matrix $[\bar{m}]$ and effective force vector $\{\bar{Q}_{t+\Delta t}\}$ are given by

$$[\bar{m}] = \frac{100}{36\,\Delta t^2}[m] + \frac{10}{6\,\Delta t}[c] + [k] \tag{7.71a}$$

$$\{\bar{Q}_{t+\Delta t}\} = \{Q_{t+\Delta t}\} + \frac{15}{6\,\Delta t}[m]\{\dot{q}_t\} - \frac{1}{\Delta t}[m]\{\dot{q}_{t-\Delta t}\}$$

$$+ \frac{1}{6\,\Delta t}[m]\{\dot{q}_{t-2\Delta t}\} + \left(\frac{150}{36\,\Delta t^2}[m] + \frac{15}{6\,\Delta t}[c]\right)\{q_t\}$$

$$- \left(\frac{10}{6\,\Delta t^2}[m] + \frac{1}{\Delta t}[c]\right)\{q_{t-\Delta t}\}$$

$$+ \left(\frac{1}{36\,\Delta t^2}[m] + \frac{1}{6\,\Delta t}[c]\right)\{q_{t-2\Delta t}\} \tag{7.71b}$$

Solution of (7.70) yields $\{q_{t+\Delta t}\}$, which is substituted in (7.68) to obtain velocities. Then $\{\ddot{q}_{t+\Delta t}\}$ are obtained by substituting the calculated values of $\{\dot{q}_{t+\Delta t}\}$ into (7.67).

It can be observed that in the Park stiffly stable method, calculation of $\{q_{t+\Delta t}\}$ involves displacements and velocities at t, $t - \Delta t$, and $t - 2\Delta t$. Thus, to obtain the solution at time Δt and $2\,\Delta t$, a special starting procedure is required. This makes the method not self-starting. The method also requires a large computer memory to store velocities and displacements for two previous time steps.

7.5 CASE STUDY

7.5.1 Linear Dynamic System

In order to compare economy, efficiency, stability, and accuracy of various integration schemes, the response is obtained here for a system with two degrees of freedom as shown in Fig. 7.5. We first consider linear springs and viscous damping and obtain the equations of motion as

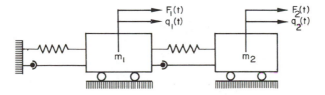

Figure 7.5   Two-degree-of-freedom system.

$$\begin{bmatrix} m_1 & 0 \\ 0 & m_2 \end{bmatrix} \begin{Bmatrix} \ddot{q}_1 \\ \ddot{q}_2 \end{Bmatrix} + \begin{bmatrix} c_1 + c_2 & -c_2 \\ -c_2 & c_2 \end{bmatrix} \begin{Bmatrix} \dot{q}_1 \\ \dot{q}_2 \end{Bmatrix} + \begin{bmatrix} k_1 + k_2 & -k_2 \\ -k_2 & k_2 \end{bmatrix} \begin{Bmatrix} q_1 \\ q_2 \end{Bmatrix} = \begin{Bmatrix} F_1(t) \\ F_2(t) \end{Bmatrix}$$

$$(7.72)$$

The numerical values of the mass, damping, and stiffness matrices are chosen as

$$[M] = \begin{bmatrix} 1 & 0 \\ 0 & 10 \end{bmatrix}, \quad [C] = \begin{bmatrix} 0.2 & -0.1 \\ -0.1 & 0.1 \end{bmatrix}, \quad [K] = \begin{bmatrix} 21 & -1 \\ -1 & 1 \end{bmatrix} \quad (7.73)$$

All the initial conditions are selected as zero and the forcing function vector as

$$\begin{Bmatrix} F_1(t) \\ F_2(t) \end{Bmatrix} = \begin{Bmatrix} 0 \\ 4 \end{Bmatrix} \quad \text{for } t > 0 \quad \text{and} \quad \begin{Bmatrix} F_1(t) \\ F_2(t) \end{Bmatrix} = \begin{Bmatrix} 0 \\ 0 \end{Bmatrix} \quad \text{for } t < 0 \quad (7.74)$$

Since this system is linear, first we obtain an analytic solution which is later used to compare with the numerical solutions yielded by the different integration methods. Laplace-transforming (7.72) with zero initial conditions, we obtain

$$[Z(s)]\{\hat{q}(s)\} = \{\hat{F}(s)\} \quad (7.75)$$

where the impedance matrix $Z(s)$ is given by

$$[Z(s)] = \begin{bmatrix} m_1 s^2 + (c_1 + c_2)s + (k_1 + k_2) & -(c_2 s + k_2) \\ -(c_2 s + k_2) & m_2 s^2 + c_2 s + k_2 \end{bmatrix} \quad (7.76)$$

The system characteristic equation becomes

$$\begin{aligned} \Delta(s) &= \det[Z(s)] \\ &= [s^2 m_1 + s(c_1 + c_2) + (k_1 + k_2)](s^2 m_2 + c_2 s + k_2) - (c_2 s + k_2)^2 \\ &= m_1 m_2 s^4 + [m_1 c_2 + m_2(c_1 + c_2)]s^3 + [m_1 k_2 + m_2(k_1 + k_2) + c_1 c_2]s^2 \\ &\quad + (c_1 k_2 + c_2 k_1)s + k_1 k_2 = 0 \end{aligned} \quad (7.77)$$

Inverting the impedance matrix (7.76), it follows that

$$\{\hat{q}(s)\} = [G(s)]\{\hat{F}(s)\} \quad (7.78)$$

where the transfer function matrix is given by

$$[G(s)] = \frac{1}{\Delta(s)} \begin{bmatrix} m_2 s^2 + c_2 s + k_2 & c_2 s + k_2 \\ c_2 s + k_2 & m_1 s^2 + (c_1 + c_2)s + (k_1 + k_2) \end{bmatrix} \quad (7.79)$$

The displacements in the Laplace domain are

$$\hat{q}_1(s) = \frac{m_2 s^2 + c_2 s + k_2}{\Delta(s)} \hat{F}_1(s) + \frac{c_2 s + k_2}{\Delta(s)} \hat{F}_2(s) \quad (7.80)$$

$$\hat{q}_2(s) = \frac{c_2 s + k_2}{\Delta(s)} \hat{F}_1(s) + \frac{m_1 s^2 + (c_1 + c_2)s + (k_1 + k_2)}{\Delta(s)} \hat{F}_2(s) \quad (7.81)$$

Since the initial conditions and F_1 are all zero, the displacements in the time domain may be obtained from (7.80) and (7.81) by employing the convolution integral as

$$q_1(t) = \int_0^t G_{12}(t - t')F_2(t')\,dt' \tag{7.82}$$

$$q_2(t) = \int_0^t G_{22}(t - t')F_2(t')\,dt' \tag{7.83}$$

where

$$G_{12}(t) = L^{-1}\left[\frac{c_2 s + k_2}{\Delta(s)}\right] \tag{7.84}$$

$$G_{22}(t) = L^{-1}\left[\frac{m_1 s^2 + (c_1 + c_2)s + (k_1 + k_2)}{\Delta(s)}\right] \tag{7.85}$$

For the parameter values given by (7.73), the characteristic equation (7.77) becomes

$$\Delta(s) = 10s^4 + 2.1s^3 + 211.01s^2 + 2.1s + 20 = 0$$

and its roots are $\lambda_1, \lambda_2 = -0.10045635 \pm j4.581906$ and $\lambda_3, \lambda_4 = -0.0045436458 \pm j0.3085442$. Obtaining the inverse Laplace transformation indicated by (7.84) and (7.85) and carrying out the convolution integrals of (7.82) and (7.83), we get

$$\begin{aligned}
q_1(t) = {} & 0.2 + (0.46337973 + j0.17946622)10^{-3}e^{\lambda_1 t} \\
& + (0.46337973 - j0.17946622)10^{-3}e^{\lambda_2 t} \\
& + (-0.1004633779 - j0.00133652411)e^{\lambda_3 t} \\
& + (-0.1004633779 + j0.00133652411)e^{\lambda_4 t}
\end{aligned} \tag{7.86}$$

$$\begin{aligned}
q_2(t) = {} & 4.2 + (-0.187764 - j0.178863)10^{-5}e^{\lambda_1 t} \\
& + (-0.187764 + j0.178863)10^{-5}e^{\lambda_2 t} \\
& + (-2.099998084 + j0.03095190715)e^{\lambda_3 t} \\
& + (-2.099998084 - j0.03095190715)e^{\lambda_4 t}
\end{aligned} \tag{7.87}$$

The displacements $q_1(t)$ and $q_2(t)$ given by (7.86) and (7.87) are shown plotted in Fig. 7.6. The responses show decaying oscillations which, as time increases to a sufficiently high value, will reach the constant values of 0.2 and 4.2, respectively.

In this numerical study, the value of the constant θ in the Wilson theta scheme was taken as 1.5, whereas the parameters α and β in the Newmark beta scheme were chosen to be 1/2 and 1/6, respectively. Those methods that are not self-starting were started by using the Runge–Kutta method for the initial time steps. The results for the displacements obtained using the various integration schemes with a time step of 0.01 s are shown in Fig. 7.7. For this time step all the integration schemes yield the same results, which are therefore shown in a single Fig. 7.7. On comparing Figs. 7.6 and 7.7, it is seen that by choosing a sufficiently small time step, all the integration schemes can yield accurate results, but at the expense of computational time.

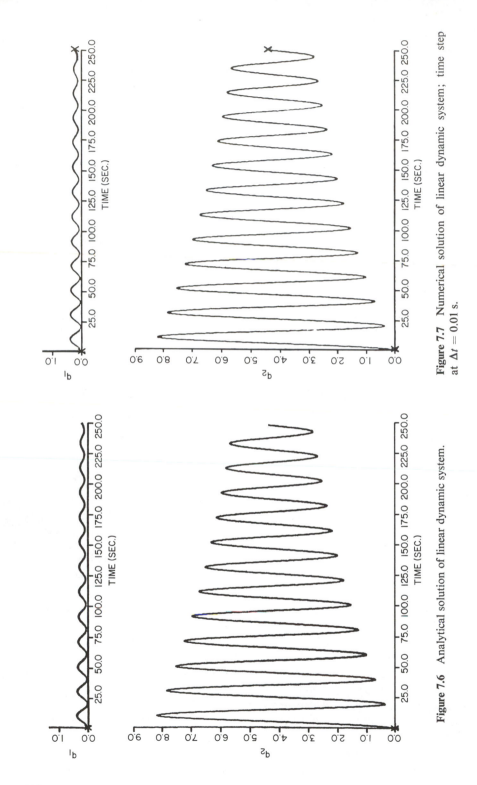

Figure 7.7 Numerical solution of linear dynamic system; time step at $\Delta t = 0.01$ s.

Figure 7.6 Analytical solution of linear dynamic system.

The results obtained when the time step is increased to 0.05 s are shown in Fig. 7.8(a) for the displacement q_1 and in Fig. 7.8(b) for the displacement q_2. The Newmark beta method becomes unstable at this time step and the spurious growth in q_1 is seen after two cycles in Fig. 7.8(a), whereas the spurious growth

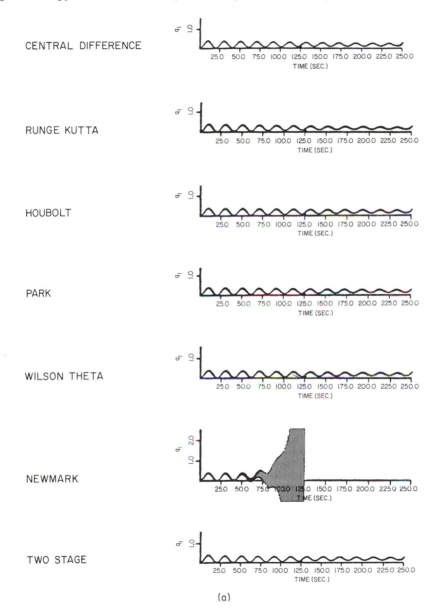

(a)

Figure 7.8   (a) Displacement q_1 for numerical solution of linear dynamic system; time step $\Delta t = 0.05$ s.

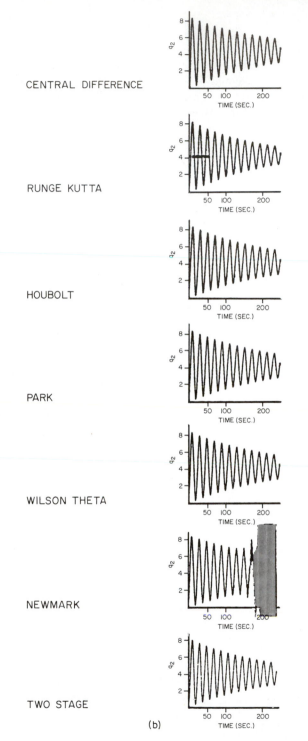

CENTRAL DIFFERENCE

RUNGE KUTTA

HOUBOLT

PARK

WILSON THETA

NEWMARK

TWO STAGE

(b)

Figure 7.8 (b) displacement q_2 for numerical solution of linear dynamic system; time step $\Delta t = 0.05$ s.

in q_2 occurs after seven cycles, as seen in Fig. 7.8(b). The other integration schemes remain stable for this time step.

As the value of time step is increased further to 0.1 s, the instability exhibited by the Newmark beta method appears after a lesser number of time increments, as seen in Fig. 7.9. The other schemes continue to show stable

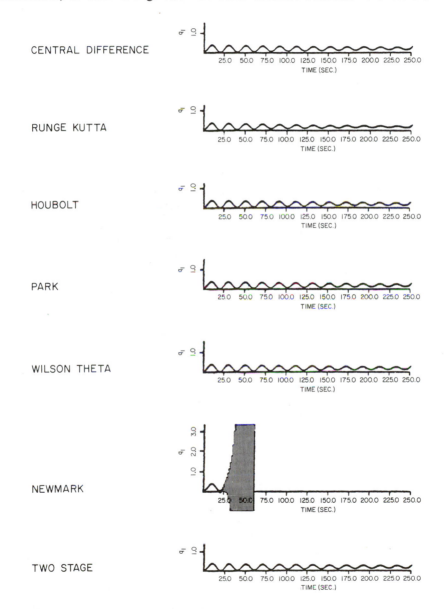

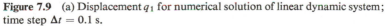

Figure 7.9   (a) Displacement q_1 for numerical solution of linear dynamic system; time step $\Delta t = 0.1$ s.

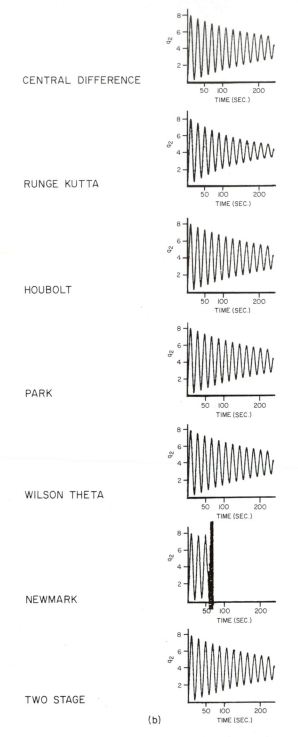

CENTRAL DIFFERENCE

RUNGE KUTTA

HOUBOLT

PARK

WILSON THETA

NEWMARK

TWO STAGE

(b)

Figure 7.9 (b) displacement q_2 for numerical solution of linear dynamic system; time step $\Delta t = 0.1$ s.

behavior; however, the response obtained from the fourth-order Runge–Kutta scheme is highly damped, as seen in Fig. 7.9. When the time step is increased to 0.5 s, the Newmark beta, central difference predictor, two-cycle iteration with trapezoidal rule, and the fourth-order Runge–Kutta schemes give unstable solutions, whereas the Park, Houbolt, and Wilson theta schemes remain stable.

A comparison of the CPU time used on the DEC-20 system for 5500 time steps (275 s) is given in Table 7.1 for time increment $\Delta t = 0.05$ s. It can be seen

TABLE 7.1   Comparison of Integration Scheme for Linear Problem[a]

Integration Method	DEC-20, CPU Time (s)
1. Central difference	26.32
2. Newmark beta ($\alpha = 0.5$, $\beta = 1/6$)	27.95 (unstable)
3. Wilson theta ($\theta = 1.5$)	27.21
4. Houbolt	26.04
5. Park	26.69
6. Runge–Kutta	27.41
7. Two-cycle interation	25.20

[a]Number of time steps = 5500; time increment = 0.05 s.

that the two-cycle iteration with trapezoidal rule requires the least CPU time; however, the scheme becomes unstable for a larger time step. Next are the Houbolt, central difference predictor, Park, Wilson theta, fourth-order Runge–Kutta, and Newmark beta schemes, in that order. The Houbolt scheme has the advantage of using less CPU time and being unconditionally stable. The CPU time for the Newmark beta method in Table 7.1 does not have much significance since the scheme is unstable for this time step. The differences in the CPU time shown in Table 7.1 are not large because this example considers only a two-degree-of-freedom system.

7.5.2 Nonlinear Dynamic System

To compare the performance of each integration scheme further, a non-linearity is introduced in the two-degree-of-freedom system of Fig. 7.5. Specifically, the force F_s in the spring connecting the masses m_1 and m_2 is assumed to be related to its displacement x_s by $F_s = k_2(x_s + 0.5x_s^3)$. The equations of motion are given by

$$m_1\ddot{q}_1 + c_1\dot{q}_1 + k_1q_1 - k_2[(q_2 - q_1) + 0.5(q_2 - q_1)^3] - c_2(\dot{q}_2 - \dot{q}_1) = F_1$$
$$m_2\ddot{q}_2 + c_2(\dot{q}_2 - \dot{q}_1) + k_2[(q_2 - q_1) + 0.5(q_2 - q_1)^3] = F_2$$

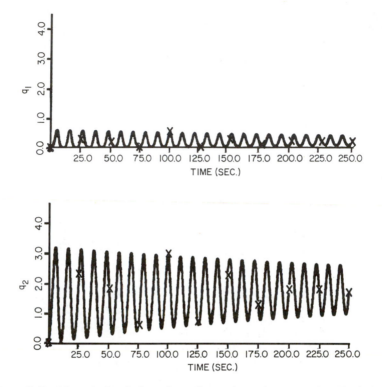

Figure 7.10 Numerical solution of nonlinear dynamic system by Houbolt scheme; time step $\Delta t = 0.01$ s.

In matrix notation, these equations may be written as

$$\begin{bmatrix} m_1 & 0 \\ 0 & m_2 \end{bmatrix} \begin{Bmatrix} \ddot{q}_1 \\ \ddot{q}_2 \end{Bmatrix} + \begin{bmatrix} c_1 + c_2 & -c_2 \\ -c_2 & c_2 \end{bmatrix} \begin{Bmatrix} \dot{q}_1 \\ \dot{q}_2 \end{Bmatrix}$$
$$+ \begin{bmatrix} k_1 + k_2 + 0.5k_2(q_2 - q_1)^2 & -k_2 - 0.5k_2(q_2 - q_1)^2 \\ -k_2 - 0.5k_2(q_2 - q_1)^2 & k_2 + 0.5k_2(q_2 - q_1)^2 \end{bmatrix} \begin{Bmatrix} q_1 \\ q_2 \end{Bmatrix} = \begin{Bmatrix} F_1 \\ F_2 \end{Bmatrix}$$

(7.88)

In the foregoing equation, the mass and damping matrices $[m]$ and $[c]$ are as given by (7.73) and are constant. The stiffness matrix $[k]$ in (7.88) is computed and updated at each integration step. The values of k_1 and k_2 are as given by (7.73). All the initial conditions and the forcing function F_1 are assumed to be zero and the forcing function F_2 is a step function of magnitude 4 as in (7.74).

The results by the Houbolt scheme by using a time step of 0.01 s are shown in Fig. 7.10. All the other integration schemes, except the two-cycle iteration method, yielded identical results for this time step and hence are not shown here. The two-cycle iteration scheme yielded an unstable solution for time step $\Delta t \geq 0.01$ s, and those results are omitted here.

   The results obtained when the time step is increased to 0.05 s are shown in
Fig. 7.11(a) for the displacement q_1 and in Fig. 7.11(b) for the displacement q_2.
For this time step, the Newmark beta method becomes unstable after two
cycles. The Wilson theta method is also unstable, whereas the Houbolt and Park
schemes are on the verge of instability. The fourth-order Runge–Kutta method
yields a solution that is highly damped, whereas the central difference scheme
yields accurate results.

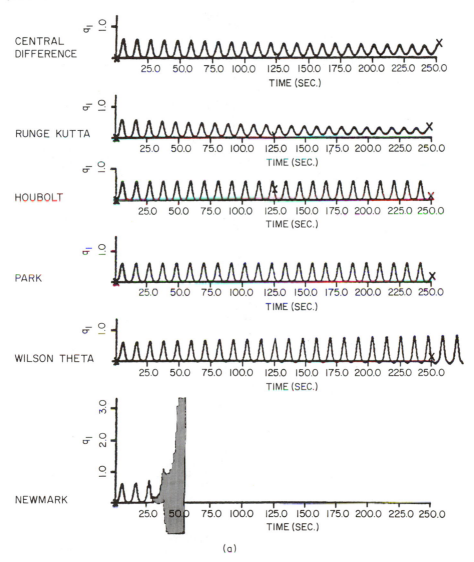

(a)

Figure 7.11   (a) Displacement q_1 for numerical solution of nonlinear dynamic
system; time step $\Delta t = 0.05$ s.

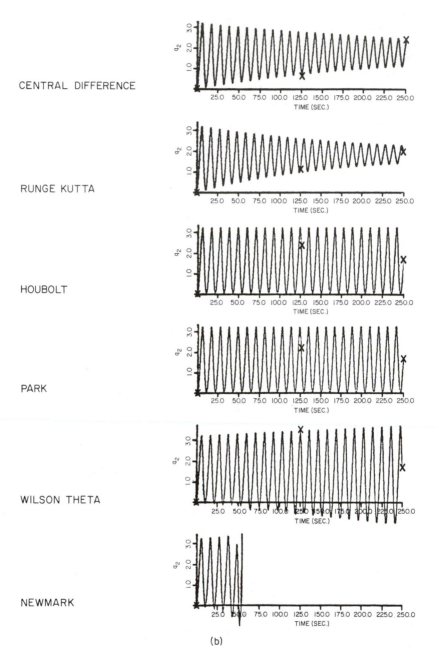

CENTRAL DIFFERENCE

RUNGE KUTTA

HOUBOLT

PARK

WILSON THETA

NEWMARK

(b)

Figure 7.11   (b) Displacement q_2 for numerical solution of nonlinear dynamic system; time step $\Delta t = 0.05$ s.

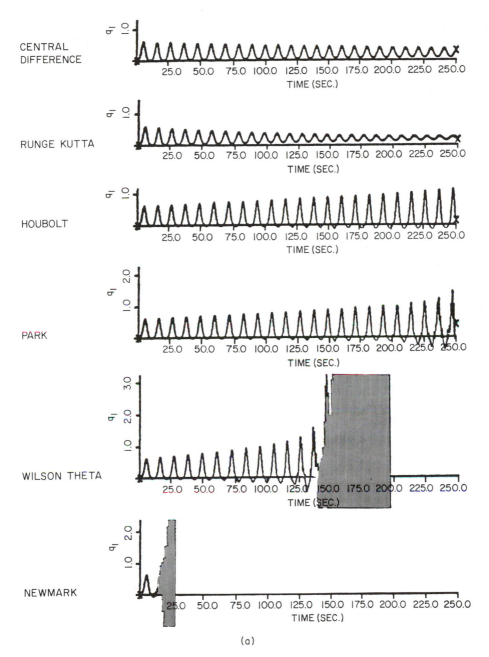

(a)

Figure 7.12 (a) Displacement q_1 for numerical solution of nonlinear dynamic system; time step $\Delta t = 0.1$ s.

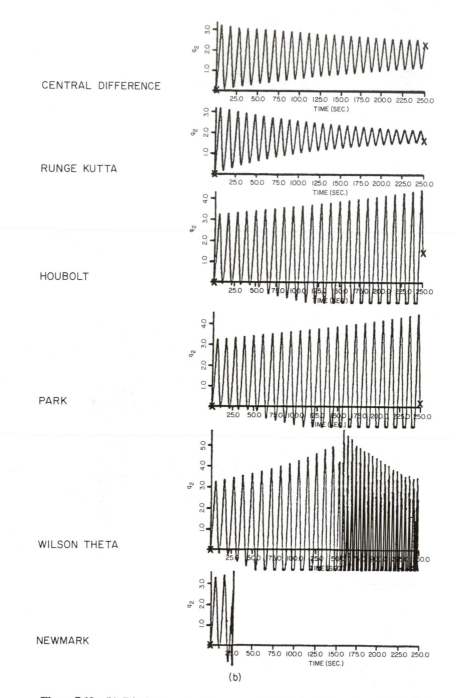

CENTRAL DIFFERENCE

RUNGE KUTTA

HOUBOLT

PARK

WILSON THETA

NEWMARK

Figure 7.12 (b) Displacement q_2 for numerical solution of nonlinear dynamic system; time step $\Delta t = 0.1$ s.

As the time step is increased to 0.1 s, as seen from Fig. 7.12, the Newmark beta, two-cycle iteration, Wilson theta, Park, and Houbolt schemes all yield unstable solutions. Only the fourth-order Runge–Kutta and the central difference predictor schemes are stable. As seen from Fig. 7.12, the Runge–Kutta scheme introduces damping, whereas the central difference predictor method yields results that are quite accurate compared to those of Fig. 7.10.

A comparison of the CPU time used on the DEC-20 computer for 5500 time steps (55 s) is given in Table 7.2 for the time increment $\Delta t = 0.01$ s. It is seen that the central difference predictor scheme uses the least CPU time. The Park method is a close second and it is followed by the Runge–Kutta and Houbolt methods. The CPU time listed for the two-cycle iteration scheme has no significance since the method is unstable.

TABLE 7.2   Comparison of Numerical Schemes for
Nonlinear Problem[a]

Integration Method	DEC-20 CPU Time (s)
1. Central difference	27.76
2. Newmark beta ($\alpha = 0.5$, $\beta = 1/6$)	28.83
3. Wilson theta ($\theta = 1.5$)	28.51
4. Houbolt	28.33
5. Park	27.93
6. Runge–Kutta	28.32
7. Two-cycle iteration	85.12 (unstable)

[a]Number of time steps $= 5500$; time increment $= 0.01$ s.

7.6 SUMMARY

Three explicit and four implicit methods have been studied in detail. Each integration scheme was used to obtain the response of a system with two degrees of freedom for different time steps and a sufficient number of time increments to show whether the scheme becomes unstable, if it introduces damping, and the amount of CPU time that is required. For the linear problem studied, it was found that when small time increments are used, the CPU time is the least for the two-cycle iteration with trapezoidal rule, but the method becomes unstable as the time step is increased. The Houbolt method requires a little more CPU time than the two-cycle iteration scheme. However, it has the advantage of being unconditionally stable. The central difference and Park schemes are not far behind.

When these integration schemes are used to obtain the response of a nonlinear problem, it is found that all the schemes that are stable for the linear example do not remain stable for the nonlinear example. It is found that the two-cycle iteration, Newmark beta, and Wilson theta methods are unstable even with a relatively small time step, while the response from the fourth-order Runge–Kutta scheme is highly damped.

From the linear and nonlinear examples considered in this study, it appears that the Houbolt, Park, and central difference predictor methods exhibit superior stability characteristics for large time steps and the CPU time requirements of these three methods are not significantly different for a system with few degrees of freedom.

PROBLEMS

7.1. Obtain a digital computer simulation of the system described by (7.9) with the following parameter values: $F(t) = 0$, $\theta(0) = 1$, $\dot{\theta}(0) = 0$, $m_2 = 1$, $k = 1$, $c = 1$, $x(0) = 1$, $\dot{x}(0) = 1$, $m_1 = 1$, $a = 1$, $\mu = 0.3$, $g = 9.81$. Use the following integration techniques and compare the stability and cost of computations:
   (a) Fourth-order Runge–Kutta method.
   (b) Park stiffly stable method.
   (c) Houbolt method.

7.2. An elasto-plastic spring with a force versus displacement curve as shown in Fig. P7.2(a) supports a mass of 37,500 kg [Fig. P7.2(b)]. A dynamic force linearly

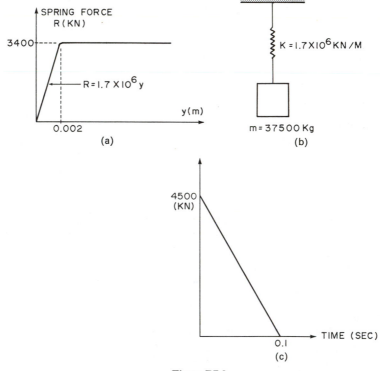

Figure P7.2

varying from 4500 kN at $t = 0$ to zero kN at 0.1 s is applied to the mass [Fig. P7.2(c)]. Find the maximum deflection attained and the time required to attain this maximum deflection. Assume the initial conditions as $\dot{y}(0) = y(0) = 0$.

7.3. An undamped spring–mass system with mass $m = 8$ kg has a natural period of 0.5 s. The system is subjected to an impulse of 9 N-s which has a triangular shape with a time duration of 0.4 s. Determine the maximum displacement of the mass. Use the following numerical methods:
 (a) Fourth-order Runge–Kutta method.
 (b) Central difference method.

7.4. Figure P7.4(a) shows a wheel–axle set. The nonlinear equations of motion for this wheel–axle set, for the lateral and yaw degrees of freedom, are as follows:

$$m\ddot{y} + \frac{2f_{11}}{V}(\dot{y} - V\psi) + \frac{2f_{12}}{V}\dot{\psi} - \frac{2f_{12}}{r_0}\Delta_2(y) + W_A\Delta_L(y) + k_y y + c_y \dot{y} = F_y(t)$$

$$I_\omega\ddot{\psi} + \frac{2a^2}{V}f_{33}\dot{\psi} + \frac{2af_{33}}{r_0}\left(\frac{r_L - r_R}{2}\right) + \frac{2f_{22}}{V}\dot{\psi}$$

$$- \frac{2f_{12}}{V}(\dot{y} - V\psi) - \frac{2f_{22}}{r_0}\Delta_1(y) - a\psi W_A\delta_0 + k_\psi\psi + c_\psi\dot{\psi} = F_\psi(t)$$

where m = mass of the wheel–axle set = 30 lb-sec^2/in.
    I_ω = yaw moment of inertia of the wheel = 16,500 lb-in.-sec^2
    a = 30 in.
    r_0 = radius of the wheel = 20 in.

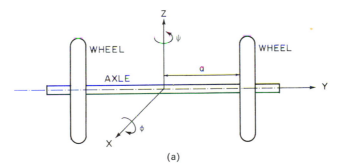

(a)

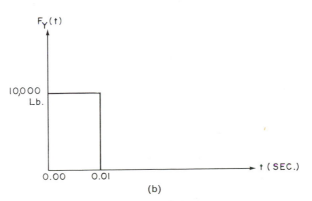

(b)

Figure P7.4

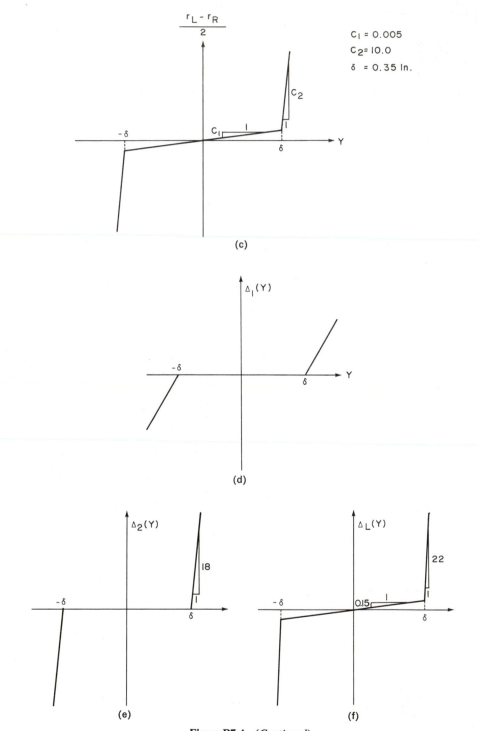

Figure P7.4 (*Continued*)

f_{11} = longitudinal creep coefficient = 3.6×10^6 lb
f_{12} = 0.46×10^6 in.-lb
f_{22} = spin creep coefficient = $66,000$ in^2-lb
f_{33} = lateral creep coefficient = 3.9×10^6 lb
W_A = axle load = $66,000$ lb
k_y = lateral stiffness = 5000 lb/in.
c_y = lateral damping = 100 lb-sec/in.
k_ψ = yaw stiffness = $187,200,000$ in.-lb/rad
c_ψ = yaw damping = $31,200$ lb-in.-sec/rad
V = axle speed = $15,056$ in./sec.
δ_0 = initial taper angle = 0.05
$F_y(t)$ = lateral input force
$F_\psi(t)$ = yaw input moment = 0.0

Figure P7.4(b) shows the lateral impact force versus time relationship, whereas Figs. P7.4(c)–(f) represent variations of $\Delta_1(y)$, $\Delta_2(y)$, and $\Delta_L(y)$. Solve the equations of motion numerically, using the following schemes:

(a) Newmark beta ($\alpha = 0.5$, $\beta = \frac{1}{6}$).
(b) Wilson theta ($\theta = 1.5$).
(c) Houbolt.
(d) Park stiffly stable.
(e) Central difference.
(f) Runge–Kutta.
(g) Two-cycle interation with trapezoidal rule.

    Plot the time histories for parameters y and ψ and compare the results with respect to:
(1) Computing cost.
(2) Stability of solution.
(3) Accuracy.

7.5. Resolve Problem 7.4 using an initial lateral displacement of 0.36 in. instead of the initial lateral force. Use the following numerical schemes:
(a) Newmark beta ($\alpha = 0.5$, $\beta = \frac{1}{6}$).
(b) Park stiffly stable.
(c) Central difference.
(d) Wilson theta ($\theta = 1.5$).
(e) Runge–Kutta.
(f) Two-cycle integration with trapezoidal rule.

    Plot the time histories for y and ψ and compare the results with respect to:
(1) Stability of solution.
(2) Accuracy.
(3) Computing cost.

REFERENCES

1. Bathe, K. J., and Wilson, E. L., *Numerical Methods in Finite Element Analysis*, Prentice-Hall, Inc., Englewood Cliffs, N.J., 1976.

2. Clough, R. W., and Penzien, J., *Dynamics of Structures*, McGraw-Hill Book Company, New York, 1975.

3. Hurty, W. C., and Rubinstein, M. F., *Dynamics of Structures*, Prentice-Hall, Inc., Englewood Cliffs, N.J., 1970.

4. Froberg, C. E., *Introduction to Numerical Analysis*, Addison-Wesley Publishing Company, Inc., Reading, Mass. 1969.

5. Brice, C., Luther, H. A., and Wilkes, J. O., *Applied Numerical Methods*, John Wiley & Sons, Inc., New York, 1969.

6. Hildebrand, F. B., *Introduction to Numerical Analysis*, McGraw-Hill Book Company, New York, 1956.

7. McNamara, J. F., "Solution Schemes for Problems of Nonlinear Structural Dynamics," *Journal of Pressure Vessel Technology*, ASME, May 1974, pp. 96–102.

8. Park, K. C., "An Improved Stiffly Stable Method for Direct Integration of Nonlinear Structural Dynamic Equations," *Journal of Applied Mechanics*, ASME, June 1975, pp. 464–470.

9. Hojjat, A., Gere, J. M., and Weaver, W., "Algorithms for Nonlinear Structural Dynamics," *Journal of the Structural Division*, ASCE, Feb. 1978, pp. 263–279.

10. Belytschko, T., and Schoeberle, D. F., "On the Unconditional Stability of an Implicit Algorithm for Nonlinear Structural Dynamics," *Journal of Applied Mechanics*, Vol. 42, 1975, pp. 865–869.

11. Belytschko, T., Holmes, N., and Mullen, R., "Explicit Integration—Stability, Solution Properties, Cost," *Finite-Element Analysis of Transient Nonlinear Structural Behavior*, ASME, AMD Vol. 14, 1975.

12. Tillerson. J. R., Stricklin, J. A., and Haisler, W. E., "Numerical Methods for the Solution of Nonlinear Problems in Structural Analysis," Winter Annual Meeting of ASME, Detroit, Mich., Nov. 11–15, 1973.

13. Wang, P. C., *Numerical and Matrix Methods in Structural Mechanics*, John Wiley & Sons, Inc., New York, 1966.

14. Romanelli, M. J. "Runge–Kutta Method for the Solution of Ordinary Differential Equations," in *Mathematical Methods for Digital Computers*, ed. A. Ralston and H. S. Wilf, John Wiley & Sons, Inc., New York, 1965.

8

LINEAR VIBRATIONS

8.1 INTRODUCTION

This chapter deals with the study of linear vibrations of dynamic systems. In many applications, vibrations occur about an equilibrium state or about a stationary motion. Assuming that the equations of motion contain all the non-linearities that are analytic functions of their arguments, the equations that describe small perturbations can be linearized as discussed in Chapter 6. The vibrations studied in this chapter may then be considered as perturbations about an equilibrium state or stationary motion and are governed by linear, time-invariant ordinary differential equations.

After a discussion of the preliminary concepts in vibration analysis, we begin with the study of vibrations of single-degree-of-freedom systems. Two methods are presented for the analysis of linear vibrations. The first method is a time-domain solution and employs the state transition matrix developed in Chapter 6. The second method is a frequency-domain solution and employs the harmonic response function for the analysis of steady-state forced vibrations.

The analysis techniques for a single-degree-of-freedom systems are then generalized to multiple-degree-of-freedom systems. The time-domain method is based on matrix diagonalization and normal-mode solution techniques that were developed in Chapter 6. The frequency-domain method employs the harmonic response function matrix for the study of steady-state forced vibrations. Continuous systems with distributed mass and elasticity have infinite degrees of freedom and are not included in this study.

8.2 CLASSIFICATION OF VIBRATIONS

Vibration is in general a motion periodic in time and is used to describe oscillation in mechanical systems. In most cases, the general purpose is to prevent or attenuate the vibrations, because of their detrimental effects, such as fatigue failure of components, failure of bearings, and generation of noise. However, there are some applications where vibrations are desirable and are usefully employed, as in vibratory conveyors and hair-cutting shears.

Vibrations may be classified into three categories: free vibrations, forced vibrations, and self-excited vibrations.

Free vibrations.   Free vibrations can occur only in conservative systems where there is no friction or damping and any external exciting force is absent. Here, the total mechanical energy, which is due to the initial conditions, is conserved and exchange takes place between the kinetic and potential energies. Since almost all mechanical systems exhibit some form of damping, the only applications where free vibrations can exist belong to the area of celestial mechanics, space dynamics, and orbits of satellites, where the orbit lies outside the atmosphere of the body around which it translates such that there is no drag.

Forced vibrations.   The vibrations in this case are caused by an external force that excites the system. In forced vibrations, in contrast to free vibrations, the exciting force supplies energy continuously to the system to compensate for that dissipated by damping. Forced vibrations may be either deterministic or random. The differential equations of motion of the dynamic systems that we consider in this book are deterministic; that is, the parameters are not randomly varying with time. But the exciting force may be either a deterministic or a random function of time; that is, its amplitude and period may either be deterministic or randomly varying. In deterministic vibrations, the amplitude and frequency at any designated future time can be completely predicted from the past history. Random forced vibrations are defined in statistical terms and only the probability of occurrence of designated magnitudes and frequencies can be predicted.

Self-excited vibrations.   Self-excited vibrations are periodic and deterministic oscillations of the limit-cycle type and are caused by some nonlinear phenomenon. Under certain conditions, the equilibrium state or stationary motion is unstable and any disturbance causes the perturbations to grow until some nonlinear effects limit any further growth. The energy required to maintain the vibrations is obtained from a nonalternating power source. In self-excited vibrations, the periodic force that excites the vibrations is created by the vibrations themselves. If the system is prevented from vibrating, the exciting force disappears. By contrast, in forced vibrations the exciting force is independent of the vibrations and can persist when the system is prevented from vibrating.

In this chapter we study linear forced vibrations of the deterministic type. Study of linear free vibrations is also included. In many vibrating systems, the only damping is the so-called structural damping, which is very small, and in such cases it will be seen that the damped natural frequencies are very close to the natural frequencies. Hence, even though free vibrations do not occur in practice, their study is important for the purpose of determining the natural frequencies. The values of the damped natural frequencies are then known approximately and can be chosen such that they are not close to the forcing frequencies in order to prevent large-amplitude forced vibrations and near-resonance.

8.3 UNDAMPED SINGLE-DEGREE-OF-FREEDOM SYSTEMS

As discussed in Chapter 5, a single-degree-of-freedom system is represented by a single Lagrange equation of motion in the form

$$\frac{d}{dt}\left(\frac{\partial L}{\partial \dot{q}}\right) - \frac{\partial L}{\partial q} = Q \tag{8.1}$$

where L is the Lagrangian, q the generalized coordinate, and Q the generalized force. Assuming that the nonlinearities are analytic functions of their arguments, we linearize the equations that represent small perturbations about an equilibrium state or a stationary motion. In Example 6.7, which represents a single-degree-of-freedom translational system, the linearized equation for perturbations about the equilibrium state ($q = 0$, $\dot{q} = 0$) can be represented as given by (6.62) in the form

$$m(\Delta\ddot{q}) + c(\Delta\dot{q}) + k(\Delta q) = \Delta F$$

For simplicity of notation, we represent the deviation Δq by q and rewrite this equation as

$$m\ddot{q} + c\dot{q} + kq = F \tag{8.2}$$

This system, which represents a mass m attached to a linear spring with spring constant k and a linear dash pot with coefficient c as shown in Fig. 8.1, is one of the simplest dynamic systems in which elastic and inertia forces interact. In many applications, the mass–spring–damper of Fig. 8.1 is a conceptual model for a more complicated physical system such as liquid sloshing in a propellant tank. For example, reference [8] shows how to determine the slosh mass, associated spring constant, and damping constant for the study of the fundamental mode of oscillation of liquid sloshing in a cylindrical tank. A single-degree-of-freedom torsional system where q represents an angular displacement is described by

$$I\ddot{q} + c\dot{q} + kq = M \tag{8.3}$$

where I is the mass moment of inertia, c the torsional damper constant, k the

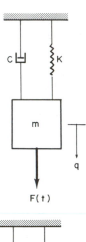

Figure 8.1 Mass, spring, and damper system.

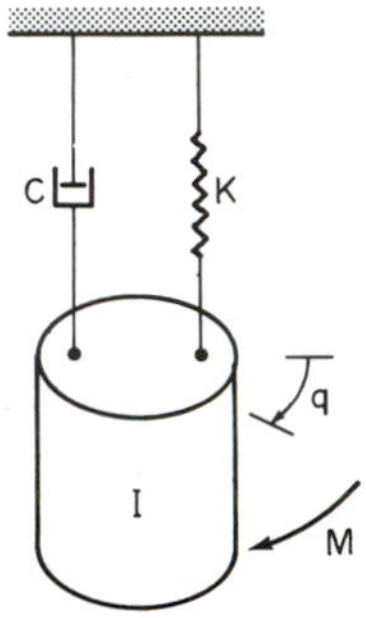

c⊐ K

m

q

F(t)

c⊐ K

I M

q

Figure 8.2 Inertia, torsional spring, and torsional damper system.

torsional spring constant, and M the applied moment or torque as indicated in Fig. 8.2.

Dividing (8.2) throughout by m and defining natural frequency ω_n and damping ratio ζ as

$$\omega_n = \sqrt{\frac{k}{m}} \quad \text{and} \quad \zeta = \frac{1}{2}\frac{c}{\sqrt{mk}}$$

we obtain

$$\ddot{q} + 2\zeta\omega_n\dot{q} + \omega_n^2 q = \frac{1}{m}F \tag{8.4}$$

The reason for the definition of the natural frequency in this manner would become clear in the following when we consider the free vibrations of the undamped system. The frequency $\omega_n\sqrt{1-\zeta^2}$ is called the damped natural frequency, for reasons that would become obvious when we consider the vibrations of damped systems in the next section. In a similar manner, dividing (8.3) throughout by I and defining the natural frequency and damping ratio as

$$\omega_n = \sqrt{\frac{k}{I}} \quad \text{and} \quad \zeta = \frac{1}{2}\frac{c}{\sqrt{Ik}}$$

we get

$$\ddot{q} + 2\zeta\omega_n\dot{q} + \omega_n^2 q = \frac{1}{I}M \tag{8.5}$$

In this section we consider the undamped case where $c = 0$ and let $\zeta = 0$ in (8.4) and (8.5). As discussed earlier, such conservative systems do not really exist except in the area of space dynamics. However, there are many applications where the damping ratio ζ has a small value around 0.05 to 0.02 and the damped natural frequency $\omega_n\sqrt{1 - \zeta^2}$ is very close to the natural frequency ω_n. In such cases, when the system is subjected to a periodic force whose frequency ω is close to the natural frequency, near-resonance occurs. It is a good first approximation to consider the system as undamped and determine its natural frequency. The damped natural frequency is then known approximately and care can be taken to ensure that it is not close to the forcing frequency.

Example 8.1

A torsion pendulum consists of a vertical shaft AB, assumed massless, rigidly attached at the top and supporting a rigid rod CD of uniform cross section and having length L and mass m_1. (a) Knowing the dimensions and material of shaft AB, determine the natural frequency of the pendulum. (b) Determine the distance b on the rod at which two equal weights, each of mass m_2, should be clamped such that the natural frequency is reduced to one-half of the value obtained in part (a).

(a) Assuming that there is no damping, the equation of motion of the torsional pendulum is

$$I\ddot{\theta} + k\theta = 0 \tag{8.6}$$

where θ is the angular displacement about the z axis as shown in Fig. 8.3. The natural frequency is given by $\omega_n = \sqrt{k/I}$. The torsional stiffness k is obtained from a knowledge of strength of materials. If a twisting moment M is applied to end B of shaft AB with build-in end A, within the elastic limit the angle of twist θ of end B is given by

$$\theta = \frac{ML_1}{GI_p}$$

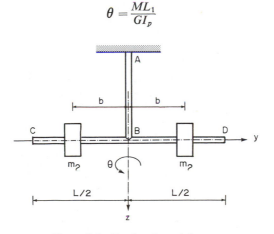

Figure 8.3 Torsional pendulum.

where L_1 is the length AB, I_p the polar moment of inertia of its cross-sectional area, and G its shear modulus. Hence, the torsional stiffness becomes

$$k = \frac{M}{\theta} = \frac{GI_p}{L_1}$$

The mass moment of inertia of rod CD about the z axis can be obtained as

$$I = \int_{-L/2}^{L/2} \frac{m_1}{L} y^2 \, dy$$

$$= \tfrac{1}{12} m_1 L^2$$

Hence, in the absence of the attached masses, the natural frequency becomes

$$\omega_n = \left[12 \frac{GI_p}{L_1 m_1 L^2}\right]^{1/2} \tag{8.7}$$

It should be noted that it is assumed that shaft AB does not contribute to the mass moment of inertia about the z axis; otherwise, the natural frequency will be smaller than that given by (8.7).

(b) When two equal masses, each of mass m_2, are attached to rod CD, the combined mass moment of inertia about the z axis becomes

$$I' = \tfrac{1}{12} m_1 L^2 + 2 m_2 b^2$$

and the natural frequency is given by

$$\omega_n' = \left[\frac{GI_p}{L_1(\tfrac{1}{12}m_1 L^2 + 2m_2 b^2)}\right]^{1/2} \tag{8.8}$$

Since, $\omega_n' = 0.5\omega_n$, from (8.7) and (8.8) we obtain

$$\frac{1}{\tfrac{1}{12}m_1 L^2 + 2m_2 b^2} = \frac{1}{4}\frac{1}{\tfrac{1}{12}m_1 L^2}$$

It follows that

$$b = \left[\frac{1}{8}\frac{m_1}{m_2}\right]^{1/2} L \tag{8.9}$$

8.3.1 Free Vibrations

We now consider free vibrations of a single-degree-of-freedom system whose equation of motion is obtained by omitting the damping and external forcing function is (8.4) or (8.5):

$$\ddot{q} + \omega_n^2 q = 0 \tag{8.10}$$

Choosing state variables as $x_1 = q$ and $x_2 = \dot{q}$, the state representation of (8.10) may be written in the form

$$\{\dot{x}\} = [A]\{x\} \tag{8.11}$$

where the $[A]$ matrix is

$$[A] = \begin{bmatrix} 0 & 1 \\ -\omega_n^2 & 0 \end{bmatrix} \tag{8.11a}$$

The solution of (8.11) is given by

$$\begin{Bmatrix} x_1 \\ x_2 \end{Bmatrix} = \Phi(t) \begin{Bmatrix} x_1(0) \\ x_2(0) \end{Bmatrix} \tag{8.11b}$$

where, as discussed in Chapter 6, the Laplace transform of the state transition matrix is

$$\hat{\Phi}(s) = (s\mathbf{I} - \mathbf{A})^{-1} = \frac{1}{s^2 + \omega_n^2} \begin{bmatrix} s & 1 \\ -\omega_n^2 & s \end{bmatrix} \tag{8.12}$$

The characteristic equation of matrix $\mathbf{A}$ is given by $s^2 + \omega_n^2 = 0$ and its eigenvalues are $s = \pm j\omega_n$. Employing partial-fraction expansion of each element of (8.12) and then the inverse Laplace transformation, we obtain

$$\Phi(t) = \begin{bmatrix} \dfrac{e^{j\omega_n t} + e^{-j\omega_n t}}{2} & \dfrac{e^{-j\omega_n t} - e^{-j\omega_n t}}{2j\omega_n} \\[2ex] -\omega_n \dfrac{e^{j\omega_n t} - e^{-j\omega_n t}}{2j} & \dfrac{e^{j\omega_n t} + e^{-j\omega_n t}}{2} \end{bmatrix}$$

$$= \begin{bmatrix} \cos \omega_n t & \dfrac{1}{\omega_n} \sin \omega_n t \\[2ex] -\omega_n \sin \omega_n t & \cos \omega_n t \end{bmatrix} \tag{8.13}$$

Substituting (8.13) in (8.11b), we get

$$q(t) = q(0) \cos \omega_n t + \dot{q}(0) \frac{1}{\omega_n} \sin \omega_n t \tag{8.14}$$

$$\dot{q}(t) = -q(0)\omega_n \sin \omega_n t + \dot{q}(0) \cos \omega_n t \tag{8.15}$$

The displacement (8.14) and velocity (8.15) may also be written as

$$q(t) = u \sin (\omega_n t + \psi) \tag{8.16}$$

$$\dot{q}(t) = u\omega_n \cos (\omega_n t + \psi) \tag{8.17}$$

where the amplitude u and phase angle ψ are defined by

$$u = \left\{ q^2(0) + \left[\frac{\dot{q}(0)}{\omega_n} \right]^2 \right\}^{1/2} \tag{8.18}$$

$$\psi = \tan^{-1} \frac{\omega_n q(0)}{\dot{q}(0)} \tag{8.19}$$

The total mechanical energy at any time is the sum of the kinetic and potential energies and is expressed as

$$E = \tfrac{1}{2} m \dot{q}^2(t) + \tfrac{1}{2} k q^2(t)$$

$$= \frac{m}{2} [\dot{q}^2(t) + \omega_n^2 q^2(t)] \tag{8.20}$$

At the initial time, we have

$$E = \frac{m}{2} [\dot{q}^2(0) + \omega_n^2 q^2(0)] \tag{8.21}$$

Also, substituting from (8.16) and (8.17) in (8.20), the mechanical energy at any instant of time becomes

$$E = \frac{m}{2}[u^2\omega_n^2]$$

and after employing (8.18), we obtain

$$E = \frac{m}{2}[\dot{q}^2(0) + \omega_n^2 q^2(0)] \tag{8.22}$$

The system here is conservative and, as expected, the mechanical energy is conserved and at any instant of time, it is equal to the initial mechanical energy.

8.3.2 Forced Vibrations

Let the undamped system be excited by a sinusoidal force $F(t) = f_0 \sin \omega t$ having amplitude f_0 and circular frequency ω. The equation of motion becomes

$$\ddot{q} + \omega_n^2 q = \frac{1}{m} f_0 \sin \omega t \tag{8.23}$$

Choosing the state variables as $x_1 = q$ and $x_2 = \dot{q}$ as done earlier, it follows that

$$\{\dot{x}\} = \mathbf{A}\{x\} + \{b\}f_0 \sin \omega t \tag{8.24}$$

where the matrix $\mathbf{A}$ has been defined by (8.11a) and

$$\{b\} = \left\{ \begin{matrix} 0 \\ \frac{1}{m} \end{matrix} \right\} \tag{8.25}$$

From Chapter 6 it follows that the solution of (8.24) is given by

$$\{x\} = \mathbf{\Phi}(t)\{x(0)\} + \int_0^t \mathbf{\Phi}(t - t')\{b\}f_0 \sin \omega t' \, dt' \tag{8.26}$$

where the state transition matrix $\mathbf{\Phi}(t)$ is given by (8.13).

The convolution integral, which is the second term on the right-hand side of (8.26), may be written as

$$\int_0^t \begin{bmatrix} \cos \omega_n(t - t') & \frac{1}{\omega_n} \sin \omega_n(t - t') \\ -\omega_n \sin \omega_n(t - t') & \cos \omega_n(t - t') \end{bmatrix} \left\{ \begin{matrix} 0 \\ \frac{1}{m} \end{matrix} \right\} f_0 \sin \omega t' \, dt'$$

$$= \int_0^t \begin{bmatrix} \frac{f_0}{m\omega_n} \sin \omega_n(t - t') \sin \omega t' \, dt' \\ \frac{f_0}{m} \cos \omega_n(t - t') \sin \omega t' \, dt' \end{bmatrix}$$

$$= \left\{ \begin{matrix} \frac{f_0}{m} \frac{1}{\omega_n^2 - \omega^2} \sin \omega t - \frac{\omega}{\omega_n} \sin \omega_n t \\ \frac{f_0}{m} \frac{\omega}{\omega_n^2 - \omega^2} (\cos \omega t - \cos \omega_n t) \end{matrix} \right\} \tag{8.27}$$

Letting $m = k/\omega_n^2$ and the frequency ratio $\alpha = \omega/\omega_n$ in (8.27) and then employing this equation and (8.14) in (8.26), we obtain

$$q(t) = q(0) \cos \omega_n t + \dot{q}(0) \frac{1}{\omega_n} \sin \omega_n t + \frac{f_0}{k} \frac{1}{1 - \alpha^2} (\sin \omega t - \alpha \sin \omega_n t) \quad (8.28)$$

Hence, the response contains two frequencies: the natural frequency ω_n and forcing frequency ω. When ω_n and ω are close to each other, the response exhibits a beat phenomenon and when $\alpha = \omega/\omega_n = 1$, it is obvious from (8.28) that $q(t)$ is infinite (i.e., there is resonance). When the initial conditions $q(0) = \dot{q}(0) = 0$, the response ratio, which is defined as the ratio between the dynamic displacement and static displacement is given by

$$\frac{q(t)}{f_0/k} = \frac{1}{1 - \alpha^2} (\sin \omega t - \alpha \sin \omega_n t) \quad (8.29)$$

8.4 DAMPED SINGLE-DEGREE-OF-FREEDOM SYSTEMS

8.4.1 Free Vibrations

Since undamped systems are rarely encountered in practice, almost all single-degree-of-freedom systems are described by (8.4) or (8.5), where the damping ratio ζ is nonzero. We first consider the unforced system and letting F or M be zero in (8.4) or (8.5), we get

$$\ddot{q} + 2\zeta\omega_n\dot{q} + \omega_n^2 q = 0 \quad (8.30)$$

Again choosing state variables as $x_1 = q$ and $x_2 = \dot{q}$, the state representation of (8.30) becomes

$$\begin{Bmatrix} \dot{x}_1 \\ \dot{x}_2 \end{Bmatrix} = \begin{bmatrix} 0 & 1 \\ -\omega_n^2 & -2\zeta\omega_n \end{bmatrix} \begin{Bmatrix} x_1 \\ x_2 \end{Bmatrix} \quad (8.31)$$

and its solution is given by

$$\{x\} = \mathbf{\Phi}(t)\{x(0)\} \quad (8.32)$$

It is recalled that the state transition matrix for this system was obtained in Chapter 6 and is given by (6.81). When $\zeta > 1$, the system is overdamped and the eigenvalues of matrix $\mathbf{A}$ which are the roots of the characteristic equation are real and are given by

$$\lambda_1, \lambda_2 = -\zeta\omega_n \pm \omega_n\sqrt{\zeta^2 - 1} \quad (8.33)$$

Employing (6.81), the displacement for the overdamped case can be obtained as

$$q(t) = q(0)\left[\frac{\lambda_1 + 2\zeta\omega_n}{\lambda_1 - \lambda_2} e^{\lambda_1 t} + \frac{\lambda_2 + 2\zeta\omega_n}{-\lambda_1 + \lambda_2} e^{\lambda_2 t}\right]$$

$$+ \dot{q}(0)\left[\frac{1}{\lambda_1 - \lambda_2} e^{\lambda_1 t} + \frac{1}{-\lambda_1 + \lambda_2} e^{\lambda_2 t}\right] \quad (8.34)$$

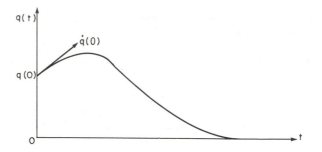

Figure 8.4   Free response of overdamped system.

This response is shown in Fig. 8.4. It is seen that the free response of an overdamped system is not oscillatory and decays to zero as time increases.

When $0 < \zeta < 1$, the system is underdamped and the eigenvalues of matrix **A** are complex conjugates with negative real part and are given by

$$\lambda_1, \lambda_2 = -\zeta\omega_n \pm j\omega_n\sqrt{1 - \zeta^2} \tag{8.35}$$

Again employing (6.81) and substituting for λ_1 and λ_2 from (8.35), the displacement for the underdamped case becomes

$$q(t) = e^{-\zeta\omega_n t}\left[\frac{\dot{q}(0) + \zeta\omega_n q(0)}{\omega_n\sqrt{1 - \zeta^2}}\sin\omega_n\sqrt{1 - \zeta^2}\,t + q(0)\cos\omega_n\sqrt{1 - \zeta^2}\,t\right] \tag{8.36}$$

As mentioned earlier, the damped natural frequency is defined by $\omega_d = \omega_n\sqrt{1 - \zeta^2}$. Now letting

$$u = \left\{\left[\frac{\dot{q}(0) + \zeta\omega_n q(0)}{\omega_d}\right]^2 + [q(0)]^2\right\}^{1/2}$$

$$\psi = \tan^{-1}\left[\frac{\omega_d q(0)}{\dot{q}(0) + \zeta\omega_n q(0)}\right]$$

equation (8.36) is written as

$$q(t) = ue^{-\zeta\omega_n t}\sin(\omega_d t + \psi) \tag{8.37}$$

This response is shown in Fig. 8.5. It is seen that the free response of an underdamped system is a damped sinusoid and decays to zero with time as the initial mechanical energy is continuously dissipated per each cycle. The period T is related to the damped natural frequency by $\omega_d = 2\pi/T$. The amplitudes a and b, which are one period apart, are related by logarithmic decrement to the damping ratio as discussed in Chapter 6 by

$$\ln\frac{a}{b} = \frac{2\pi\zeta}{\sqrt{1 - \zeta^2}} \tag{8.38}$$

Even though free vibrations are not sustained in damped systems, knowledge of the decaying free response is important because it is often employed in practice for the experimental determination of the natural frequency and damping ratio of complicated systems.

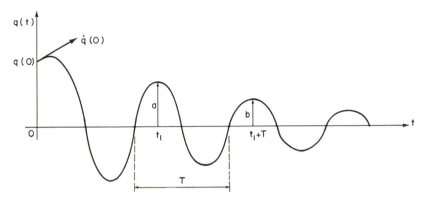

Figure 8.5   Free response of underdamped system.

Example 8.2

We consider a setup for experimental determination of the weight of a tracked vehicle. The vehicle approaches a massless bumper and couples with it. The bumper displacement $q(t)$ from its equilibrium position is recorded versus time as shown in Fig. 8.6. It is known that the spring constant $k = 180,000$ N/m. Determine the mass of the vehicle in kilograms and the value of the damping ratio ζ.

After the vehicle couples with the bumper, the differential equation that describes the motion is given by

$$m\ddot{q} + c\dot{q} + kq = 0 \tag{8.39}$$

with initial conditions $q(0) = 0$ and $\dot{q}(0) \neq 0$. Hence, in (8.37), we have $u = \dot{q}(0)/\omega_d$ and $\psi = 0$. It follows that the response is

$$q(t) = \frac{\dot{q}(0)}{\omega_d} e^{-\zeta\omega_n t} \sin \omega_d t \tag{8.40}$$

From the experimental results of Fig. 8.6, the period of the damped oscillations is $T = 0.8$ s and hence the damped natural frequency is $1/T = 1.25$ Hz. It follows that

$$\omega_n\sqrt{1 - \zeta^2} = (1.25)2\pi = 7.854 \text{ rad/s} \tag{8.41}$$

Also, from the response of Fig. 8.6, we get

$$\ln\frac{0.15}{0.01} = \frac{2\pi\zeta}{\sqrt{1 - \zeta^2}} \tag{8.42}$$

Solving for ζ from (8.42), we obtain $\zeta = 0.396$ and after substituting this value in (8.41), we get $\omega_n = 8.553$ rad/s. Then it follows that $m = k/\omega_n^2 = 2460.4$ kg.

This method of determining the natural frequency and damping ratio from the experimental impulse response is often employed for complicated structures such as machine tools where a simple analytical model is not available. Another method of experimentally determining the natural frequency and damping ratio is based on the frequency response, and this method will be discussed later.

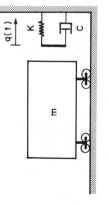

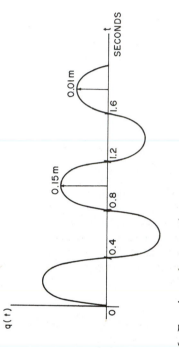

Figure 8.6 Experimental setup and response.

Example 8.3

The disk shown in Fig. 8.7 has a total mass m distributed uniformly throughout. It contacts the point A on the flat surface without slipping. Assuming small displacements, (a) write the equation of motion for the disk, (b) neglect damping and calculate the circular frequency of the system in terms of k, and m, and (c) for $c = 0.1\sqrt{km}$, calculate damping ratio, damped frequency, logarithmic decrement, and amplitude ratio after five cycles of free vibration.

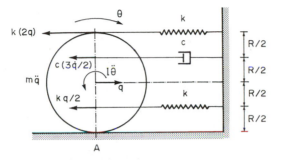

Figure 8.7   Free vibrations of a disk rolling without slipping.

(a) As the disk moves along the flat surface, its center of gravity (c.g.) translates horizontally and the disk rotates about the axis normal to the disk. The inertia force acting at the center of mass due to translation is $m\ddot{q}$ and the inertia moment due to rotation is $I\ddot{\theta}$, where I, the mass moment of inertia about the axis normal to the disk, is given by $I = mR^2/2$. Other forces acting on the disk are shown in Fig. 8.7. Since the disk rolls without slipping, there is only one degree of freedom, and the angular displacement θ can be expressed in terms of q as $\theta = q/R$.

The point A may be taken as the instantaneous center of rotation of the disk. Summing the moments about A, we obtain

$$m\ddot{q}R + I\ddot{\theta} + \frac{kq}{2}\frac{R}{2} + c\frac{3\dot{q}}{2}\frac{3R}{2} + k(2q)(2R) = 0 \tag{8.43}$$

After substituting for I and $\ddot{\theta}$ in (8.43) and simplifying, we obtain

$$1.5m\ddot{q} + \tfrac{9}{4}c\dot{q} + \tfrac{17}{4}kq = 0$$

or

$$\ddot{q} + \frac{3}{2}\frac{c}{m}\dot{q} + \frac{17}{6}\frac{k}{m}q = 0 \tag{8.44}$$

(b) To obtain the undamped natural frequency, we compare the foregoing equation with the standard form given by (8.30) and get

$$\omega_n = \sqrt{\frac{17k}{6m}}$$

(c) Again, comparing the foregoing equation with the standard form (8.30), we obtain

$$2\zeta\omega_n = \frac{3}{2}\frac{c}{m}$$

or

$$\zeta = \frac{3}{4} \frac{c}{m} \frac{1}{\omega_n} = \frac{3}{4} \frac{c}{m} \frac{\sqrt{6m}}{\sqrt{17k}}$$

For $c = 0.1 \sqrt{km}$, it follows that

$$\zeta = \frac{3}{4}(0.1)\frac{\sqrt{6}}{\sqrt{17}} = 0.0445$$

The damped natural frequency is now given by

$$\omega_d = \omega_n\sqrt{1 - \zeta^2}$$

$$= 0.999\omega_n$$

$$= 1.683\sqrt{\frac{k}{m}}$$

If q_n and q_{n+1} are the amplitudes of the free vibration $q(t)$ one cycle apart, from Fig. 8.5 and equation (8.38) it follows that

$$\ln\frac{q_n}{q_{n+1}} = \frac{2\pi\zeta}{\sqrt{1 - \zeta^2}} = 0.278$$

The amplitudes of $q(t)$ that are five cycles apart are related by

$$\ln\frac{q_n}{q_{n+5}} = 5(0.278) = 1.39$$

or

$$\frac{q_n}{q_{n+5}} = 0.249$$

8.4.2 Forced Vibrations: Time-Domain Method

We consider the forced vibrations of a damped single-degree-of-freedom system of (8.4), where the exciting force $F = f_0 \sin \omega t$ (i.e., simple harmonic with amplitude f_0 and circular frequency ω). The equation of motion, therefore, becomes

$$\ddot{q} + 2\zeta\omega_n\dot{q} + \omega_n^2 q = \frac{f_0}{m}\sin \omega t \qquad (8.45)$$

We first discuss a time-domain method of solution based on the state transition matrix and later consider another method which is a frequency-domain technique and is based on the harmonic response function. Choosing state variables $x_1 = q$ and $x_2 = \dot{q}$, (8.45) may be expressed in the form

$$\{\dot{x}\} = \mathbf{A}\{x\} + \{b\}f_0 \sin \omega t \qquad (8.46)$$

where

$$\mathbf{A} = \begin{bmatrix} 0 & 1 \\ -\omega_n^2 & -2\zeta\omega_n \end{bmatrix}, \qquad \{b\} = \begin{Bmatrix} 0 \\ \dfrac{1}{m} \end{Bmatrix}$$

and the solution in the form

$$\{x(t)\} = \mathbf{\Phi}(t)\{x(0)\} + \int_0^t \mathbf{\Phi}(t - t') \begin{Bmatrix} 0 \\ \dfrac{1}{m} \end{Bmatrix} f_0 \sin \omega t' \, dt' \tag{8.47}$$

where the state transition matrix $\mathbf{\Phi}(t)$, as obtained in Chapter 6, is given by (6.81). The part of the response $\mathbf{\Phi}(t)\{x(0)\}$ due to initial conditions has been discussed earlier and the displacement given by (8.34) for the overdamped case and by (8.37) for the underdamped case. The part of the velocity response due to initial conditions may be obtained by differentiating (8.34) or (8.37) with respect to time. The part of the response due to the forcing function is obtained by evaluating the convolution integral in (8.47). Employing the state transition matrix of (6.81) for the underdamped case, we obtain

$$\int_0^t \mathbf{\Phi}(t - t') \begin{Bmatrix} 0 \\ \dfrac{1}{m} \end{Bmatrix} f_0 \sin \omega t' \, dt'$$

$$= \int_0^t \begin{Bmatrix} \dfrac{f_0}{m\omega_d} [e^{-\zeta\omega_n(t-t')} \sin \omega_d(t - t')] \sin \omega t' \, dt' \\ \dfrac{f_0}{m} \left[e^{-\zeta\omega_n(t-t')} \left(\dfrac{-\zeta\omega_n}{\omega_d} \sin \omega_d(t - t') + \cos \omega_d(t - t') \right) \right] \sin \omega t' \, dt' \end{Bmatrix} \tag{8.48}$$

A similar expression can also be obtained for the overdamped case. Our main interest is to obtain the steady-state forced vibrations after the transients have decayed to zero. We denote $\lim_{t \to \infty} q(t) = q_{ss}(t)$, which is the steady-state forced vibrations. For the underdamped case, performing the integration in (8.48) and employing it and (8.37) in (8.47), after taking limit as $t \to \infty$, we obtain

$$q_{ss}(t) = \frac{f_0/k}{[(1 - \omega^2/\omega_n^2)^2 + (2\zeta\omega/\omega_n)^2]^{1/2}} \sin(\omega t + \psi) \tag{8.49}$$

where

$$\text{phase angle } \psi = -\tan^{-1} \left(\frac{2\zeta\omega/\omega_n}{1 - \omega^2/\omega_n^2} \right) \tag{8.50}$$

Hence, for steady-state vibrations, the displacement is sinusoidal with the same frequency as the forcing frequency, but it has a different amplitude and there is a phase angle. It should be noted that the phase angle as defined by (8.50) is a negative angle and the displacement lags behind the exciting force. The velocity for steady-state vibrations obtained from (8.48) is the same expression that is obtained by differentiating (8.49) with respect to time. By considering the overdamped system, it can be shown that expressions (8.49) and (8.50) are also valid for this case. The requirement for the transients to decay to zero with time, so that steady-state vibrations exist, is that both eigenvalues of matrix $\mathbf{A}$ have negative real part. This condition is satisfied whether the system is over-damped or underdamped.

Denoting the amplitude of q_{ss} in (8.49) by u, the dynamic magnification factor, which is defined as the ratio of the resultant response amplitude to the static displacement f_0/k, is given by

$$\frac{u}{f_0/k} = \frac{1}{[(1 - \omega^2/\omega_n^2)^2 + (2\zeta\omega/\omega_n)^2]^{1/2}} \tag{8.51}$$

It may be noticed from (8.51) and (8.50) that both the dynamic magnification factor and the phase angle vary with the frequency ratio ω/ω_n and the damping ratio ζ. Figures 8.8 and 8.9 show the plots of these relationships. Another manner of plotting these relationships in the form of a Bode diagram will be discussed later when we consider the frequency-response method.

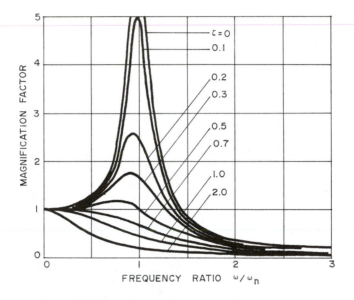

Figure 8.8   Magnification factor versus frequency ratio.

When $\zeta = 0$, the result is obtained from the term $(f_0/k)(1 - \omega^2/\omega_n^2)^{-1} \sin \omega t$ in (8.29), which is the expression for the steady-state vibrations of undamped system. It is noted that the steady-state vibration of undamped system is not obtained from (8.49). For an undamped system, both eigenvalues of matrix $\mathbf{A}$ have zero real part and the terms containing the initial conditions and sinusoidal terms with natural frequency do not decay to zero with time as happens in a damped system.

Example 8.4

A machine of mass m is resting on a foundation whose spring constant is k and viscous damping coefficient is c (Fig. 8.10). It it subjected to a sinusoidal force $F = f_0 \sin \omega t$. Determine the amplitude of the force transmitted to the foundation under steady-state vibrations when $\omega/\omega_n = 1$ and damping ratio $\zeta = 0.05$.

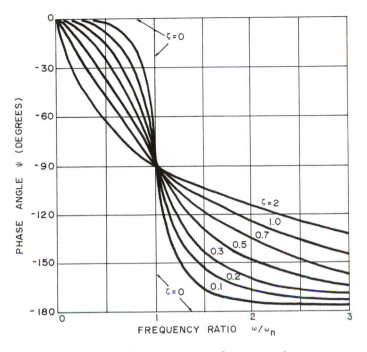

Figure 8.9   Phase angle versus frequency ratio.

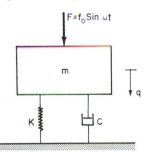

Figure 8.10   Machine on foundation.

The equation of motion of the vibrating system is

$$m\ddot{q} + c\dot{q} + kq = F$$

which may be written as

$$\ddot{q} + 2\zeta\omega_n\dot{q} + \omega_n^2 q = \frac{f_0}{m}\sin \omega t$$

This equation is identical to (8.45). Hence, under steady-state vibrations, q_{ss} is given by (8.49) and (8.50). The force transmitted to the foundation is

$$F_T = c\dot{q}_{ss} + kq_{ss}$$

$$= k\left(\frac{2\zeta}{\omega_n}\dot{q}_{ss} + q_{ss}\right) \tag{8.52}$$

Substituting for q_{ss} and its time derivative from (8.49), we obtain

$$F_T = \frac{f_0}{[(1 - \omega^2/\omega_n^2)^2 + (2\zeta\omega/\omega_n)^2]^{1/2}} \left[\frac{2\zeta\omega}{\omega_n} \cos{(\omega t + \psi)} + \sin{(\omega t + \psi)}\right]$$

Now,

$$\frac{2\zeta\omega}{\omega_n} \cos{(\omega t + \psi)} + \sin{(\omega t + \psi)}$$

$$= \left[1 + \left(\frac{2\zeta\omega}{\omega_n}\right)^2\right]^{1/2} \sin{(\omega t + \psi + \psi_1)}$$

where $\psi_1 = \tan^{-1}(2\zeta\omega/\omega_n)$. Hence, we obtain

$$F_T = \frac{f_0[1 + (2\zeta\omega/\omega_n)^2]^{1/2}}{[(1 - \omega^2/\omega_n^2)^2 + (2\zeta\omega/\omega_n)^2]^{1/2}} \sin{(\omega t + \psi + \psi_1)} \qquad (8.53)$$

The amplitude of F_T for $\omega/\omega_n = 1$ and $\zeta = 0.05$ becomes $10.05f_0$. The ratio of the steady-state amplitude of the force transmitted to the foundation to the amplitude of the exciting force is called the transmissibility, which here has the value of 10.05.

8.4.3 Energy Balance in Forced Vibrations

We now consider the energy supplied by the exciting force per cycle and show that it is exactly balanced by the energy dissipated per cycle by damping under steady-state forced vibrations. Firstly, we restrict ourselves to the case where the exciting force is simple harmonic (i.e., $F = f_0 \sin{\omega t}$). The work done or energy supplied by the exciting force per cycle is given by

$$W_1 = \int F \, dq = \int_0^T F\dot{q} \, dt$$

$$= \frac{1}{\omega} \int_0^{2\pi} F\dot{q} \, d(\omega t)$$

Since for steady-state vibration, $q_{ss}(t) = u \sin{(\omega t + \psi)}$, where u and ψ are defined by (8.51) and (8.50), respectively, we get

$$W_1 = f_0 u \int_0^{2\pi} \sin{\omega t} \cos{(\omega t + \psi)} \, d(\omega t)$$

$$= f_0 u \left[\cos{\psi} \int_0^{2\pi} \sin{\omega t} \cos{\omega t} \, d(\omega t) - \sin{\psi} \int_0^{2\pi} \sin^2{\omega t} \, d(\omega t)\right]$$

$$= f_0 u [0 - \pi \sin{\psi}]$$

After substituting for u and $\sin{\psi}$ in the expression above from (8.51) and (8.50), respectively, it follows that

$$W_1 = \pi \frac{f_0^2}{k} \frac{2\zeta\omega/\omega_n}{(1 - \omega^2/\omega_n^2)^2 + (2\zeta\omega/\omega_n)^2} \qquad (8.54)$$

On the other hand, the work done by the damper force per cycle is obtained as

$$W_2 = \int c\dot{q}\, dq = \int_0^T c\dot{q}^2\, dt$$

$$= \frac{c}{\omega} \int_0^{2\pi} \dot{q}^2\, d(\omega t)$$

$$= cu^2\omega \int_0^{2\pi} \cos^2(\omega t + \psi)\, d(\omega t)$$

$$= \pi cu^2\omega$$

After substituting for u from (8.51) and with $c = k2\zeta/\omega_n$, we obtain

$$W_2 = \pi \frac{f_0^2}{k} \frac{2\zeta\omega/\omega_n}{(1 - \omega^2/\omega_n^2)^2 + (2\zeta\omega/\omega_n)^2} \tag{8.55}$$

On comparing (8.54) and (8.55), we conclude that the energy supplied by the exciting force per cycle is exactly balanced by the energy dissipated by damping force per cycle.

8.4.4 Forced Vibrations under Periodic Force

The exciting force considered so far has been simple harmonic. Now, we generalize the results when the exciting force is periodic as shown in Fig. 8.11.

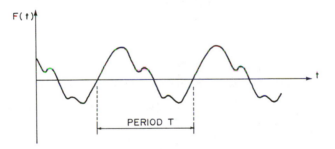

Figure 8.11   Periodic force.

Employing Fourier series expansion for this force, we get

$$F(t) = a_1 \sin \omega t + b_1 \cos \omega t + a_2 \sin 2\omega t + b_2 \cos 2\omega t$$
$$+ \cdots + a_m \sin m\omega t + b_m \cos m\omega t + \cdots \tag{8.56}$$

where a_m and b_m are the coefficients of the Fourier series expansion and it has been assumed that the constant term $b_0 = 0$. Here, ω is the fundamental frequency. Since,

$$a_m \sin m\omega t + b_m \cos m\omega t = f_m \sin(m\omega t + \alpha_m) \tag{8.57}$$

where $f_m = [a_m^2 + b_m^2]^{1/2}$ and $\alpha_m = \tan^{-1}(b_m/a_m)$, it follows that

$$F(t) = f_1 \sin(\omega t + \alpha_1) + f_2 \sin(2\omega t + \alpha_2) + \cdots + f_m \sin(m\omega t + \alpha_m) + \cdots \tag{8.58}$$

Since we are dealing with the vibrations of a linearized system, superposition is valid and we can consider each term on the right-hand side of (8.58) as a separate forcing function. The steady-state vibration displacement is obtained by adding the responses due to each term of (8.58) acting separately. Hence, it follows that

$$q_{ss}(t) = u_1 \sin(\omega t + \alpha_1 + \psi_1) + u_2 \sin(2\omega t + \alpha_2 + \psi_2) + \cdots$$
$$+ u_m \sin(m\omega t + \alpha_m + \psi_m) + \cdots \qquad (8.59)$$

where from (8.49) and (8.50), we have

$$u_m = \frac{f_m/k}{[(1 - (m\omega)^2/\omega_n^2)^2 + (2\zeta m\omega/\omega_n)^2]^{1/2}} \qquad (8.60)$$

$$\psi_m = -\tan^{-1}\frac{2\zeta m\omega/\omega_n}{1 - (m\omega/\omega_n)^2}, \qquad m = 1, 2, \ldots \qquad (8.61)$$

Hence, the steady-state vibration displacement is also periodic with the same period as the force but with a different amplitude and there exits a phase lag. The energy supplied by the periodic force per cycle can be determined as follows. The work done by the component $f_m \sin(m\omega t + \alpha_m)$ on the component $u_n \sin(n\omega t + \alpha_n + \psi_n)$ per cycle can be expressed as

$$W_{n,m} = nf_m u_n \int_0^{2\pi} \sin(m\omega t + \alpha_m)\cos(n\omega t + \alpha_n + \psi_n)\,d(\omega t)$$

Expanding the sine and cosine terms in the expression above and integrating, it can be shown that

$$W_{n,m} = 0, \qquad n \neq m$$
$$W_{m,m} = mf_m u_m \pi \sin(\alpha_m - \psi_m) \qquad n = m \qquad (8.62)$$

Hence, the work done by the force component $f_m \sin(m\omega t + \alpha_m)$ on the displacement component $u_n \sin(n\omega t + \alpha_n + \psi_n)$ per cycle is zero when $n \neq m$. The total work done per cycle is

$$W = \sum_{m=1}^{\infty} mf_m u_m \pi \sin(\alpha_m - \psi_m) \qquad (8.63)$$

By considering the work done per cycle by the damping force, it can be shown that the work dissipated per cycle is also given by (8.63) and the energy per cycle is therefore balanced.

8.4.5 Forced Vibrations: Frequency-Domain Method

We now discuss an alternative method for the analysis of forced vibrations of a single-degree-of-freedom damped system. This method, which is closely related to the time-domain method discussed in the foregoing, is a frequency-domain method and is based on the concept of transfer function and harmonic response function. We define an ordinary differential operator D as $D = d/dt$. Then the equation

$$m\ddot{q} + c\dot{q} + kq = F(t) \qquad (8.2)$$

may be written as $(mD^2 + cD + k)q = F(t)$, which may also be expressed as

$$q(t) = \frac{1}{mD^2 + cD + k} F(t) \tag{8.64}$$

In (8.64), $q(t)$ is called the output, $F(t)$ the input, and a transfer operator $G(D)$ is defined by

$$G(D) = \frac{1}{mD^2 + cD + k} \tag{8.65}$$

Taking the Laplace transformation of (8.2) and letting the symbol ($\hat{\ }$) over a variable denote its Laplace transform, we obtain

$$ms^2\hat{q} + cs\hat{q} + k\hat{q} = m[sq(0) + \dot{q}(0)] + cq(0) + \hat{F}(s)$$

and it follows that

$$\hat{q}(s) = \frac{smq(0) + m\dot{q}(0) + cq(0)}{ms^2 + cs + k} + \frac{1}{ms^2 + cs + k} \hat{F}(s) \tag{8.66}$$

If the initial conditions are zero [i.e., $q(0) = \dot{q}(0) = 0$], we get

$$\hat{q}(s) = \frac{1}{ms^2 + cs + k} \hat{F}(s) \tag{8.67}$$

where a transfer function $G(s)$ is defined by

$$G(s) = \frac{1}{ms^2 + cs + k} \tag{8.68}$$

On comparing (8.65) and (8.68), we note that a transfer function is obtained by replacing the operator D by s in the transfer operator and it relates the input and output in the Laplace domain as in (8.67). This equation implies that all the initial conditions are zero or that the initial conditions have not been accounted for. Quite often, the transfer operator of (8.65) is called the transfer function even though operator D has not been replaced by s. Of course, a transfer function can be defined only for linear time-invariant equations of motion. There is a close relationship between the state transition matrix and the transfer function as shown in the following. Choosing state variables as $x_1 = q$ and $x_2 = \dot{q}$, (8.2) is written as

$$\begin{Bmatrix} \dot{x}_1 \\ \dot{x}_2 \end{Bmatrix} = \begin{bmatrix} 0 & 1 \\ -\dfrac{k}{m} & -\dfrac{c}{m} \end{bmatrix} \begin{Bmatrix} x_1 \\ x_2 \end{Bmatrix} + \begin{Bmatrix} 0 \\ \dfrac{1}{m} \end{Bmatrix} F(t) \tag{8.69}$$

or

$$\{\dot{x}\} = A\{x\} + \{B\}F$$

The solution of (8.69) may be written as

$$\{x(t)\} = \boldsymbol{\Phi}(t)\{x(0)\} + \int_0^t \boldsymbol{\Phi}(t - t')\{B\}F(t')\, dt'$$

and in the Laplace domain this equation becomes

$$\{\hat{x}(s)\} = \hat{\boldsymbol{\Phi}}(s)\{x(0)\} + \hat{\boldsymbol{\Phi}}(s)\{B\}\hat{F}(s) \tag{8.70}$$

If we are interested in observing only the displacement, then we define a matrix E such that

$$q = \lfloor E \rfloor \{x\} = \lfloor 1 \quad 0 \rfloor \begin{Bmatrix} x_1 \\ x_2 \end{Bmatrix} \tag{8.71}$$

and from (8.70), we obtain

$$\hat{q}(s) = \lfloor E \rfloor \hat{\boldsymbol{\Phi}}(s)\{x(0)\} + \lfloor E \rfloor \hat{\boldsymbol{\Phi}}(s)\{B\}\hat{F}(s)$$

In case all initial conditions are zero, it follows that

$$\hat{q}(s) = \lfloor E \rfloor \hat{\boldsymbol{\Phi}}(s)\{B\}\hat{F}(s)$$

Now,

$$\lfloor E \rfloor \hat{\boldsymbol{\Phi}}(s)\{B\} = \lfloor 1 \quad 0 \rfloor (s\mathbf{I} - \mathbf{A})^{-1} \begin{Bmatrix} 0 \\ 1 \\ \overline{m} \end{Bmatrix}$$

$$= \frac{1}{ms^2 + cs + k} \tag{8.72}$$

It is seen that (8.72) is identical to (8.68) and hence for a single input–single output system, the transfer function can also be defined as $G(s) = \lfloor E \rfloor \hat{\boldsymbol{\Phi}}(s)\{B\}$. The input–output relationship is represented in the form of a block diagram as shown in Fig. 8.12. Steady-state forced vibrations are now analyzed by employing the transfer function. However, in order to be able to generalize the results to multiple-degree-of-freedom systems at a later stage, here we consider a general scalar differential equation

$$(D^n + b_{n-1}D^{n-1} + \cdots + b_0)q(t) = (a_m D^m + a_{m-1}D^{m-1} + \cdots + a_0)F(t) \tag{8.73}$$

$\hat{F}(s)$ → [G(s)] → $\hat{q}(s)$

Figure 8.12 Block diagram of input–output relationship.

where the coefficients a's and b's are constants and $n > m$. Here, the transfer function becomes

$$G(s) = \frac{a_m s^m + a_{m-1}s^{m-1} + \cdots + a_0}{s^n + b_{n-1}s^{n-1} + \cdots + b_0} \tag{8.74}$$

and the characteristic equation is

$$s^n + b_{n-1}s^{n-1} + \cdots + b_0 = 0 \tag{8.75}$$

We now prove that when the exciting force $F(t) = f_0 \sin \omega t$ and all the roots of the characteristic equation (8.75), which are in fact the eigenvalues of matrix $\mathbf{A}$ in the state-variable formulation, have negative real part, then for steady-state vibration,

$$q_{ss} = u \sin (\omega t + \psi)$$

where $u = f_0 |G(j\omega)|$ and $\psi = \angle G(j\omega)$. Taking the Laplace transformation of (8.73) with $F(t) = f_0 \sin \omega t$ [i.e., $F(s) = f_0 \omega / (s^2 + \omega^2)$], we obtain

$$\hat{q}(s) = \frac{I(s)}{s^n + b_{n-1}s^{n-1} + \cdots + b_0} + \frac{a_m s^m + \cdots + a_0}{s^n + b_{n-1}s^{n-1} + \cdots + b_0} \frac{f_0 \omega}{s^2 + \omega^2}$$

where $I(s)$ is a polynomial in s due to the initial conditions. Letting $(s^n + b_{n-1}s^{n-1} + \cdots + b_0) = (s + r_1)(s + r_2) \ldots (s + r_n)$, where $-r_1$, $-r_2$, $\ldots$, $-r_n$ are the roots of the characteristic equation and employing partial-fraction expansion, we get

$$\hat{q}(s) = \left[\frac{c_1}{s + r_1} + \cdots + \frac{c_n}{s + r_n} \right]$$

$$+ \left[\frac{k_1}{s + r_1} + \cdots + \frac{k_n}{s + r_n} + \frac{k_{n+1}}{s - j\omega} + \frac{k_{n+2}}{s + j\omega} \right]$$

where c's and k's are constants of the partial-fraction expansion. Since all the roots of the characteristic equation have negative real parts, we obtain $\lim_{t \to \infty} c_i e^{-r_i t} = 0$. Hence, it follows that

$$q_{ss}(t) = \lim_{t \to \infty} q(t) = L^{-1} \left(\frac{k_{n+1}}{s - j\omega} + \frac{k_{n+2}}{s + j\omega} \right) \tag{8.76}$$

where the symbol L^{-1} denotes the inverse Laplace transformation. Now,

$$k_{n+1} = \lim_{s \to j\omega} \left[(s - j\omega) G(s) \frac{f_0 \omega}{(s - j\omega)(s + j\omega)} \right]$$

$$= f_0 \frac{G(j\omega)}{2j}$$

$$= \frac{f_0}{2j} |G(j\omega)| e^{j\psi}$$

where $\psi = \angle G(j\omega)$. Also,

$$k_{n+2} = \frac{f_0}{-2j} |G(j\omega)| e^{-j\psi}$$

Hence, (8.76) yields

$$q_{ss} = f_0 |G(j\omega)| \left[\frac{e^{j(\omega t + \psi)}}{2j} - \frac{e^{-j(\omega t + \psi)}}{2j} \right]$$

$$= f_0 |G(j\omega)| \sin (\omega t + \psi) \tag{8.77}$$

Hence, the amplitude of steady-state vibration $u = f_0 |G(j\omega)|$ and the phase angle between the vibration and the exciting force is $\angle G(j\omega)$. The function $G(j\omega)$ which is obtained by replacing the operator D by $j\omega$ in $G(D)$ or by replacing s by $j\omega$ in $G(s)$ is called the harmonic response function. Of course, a condition for the harmonic response function to exist such that it may be used as in (8.77) is that all the roots of the characteristic equation have negative real part. A method for checking whether this condition has been satisfied, without actually determining the eigenvalues, is the Routh criterion, which is discussed in Chapter 9.

Now, for the damped single-degree-of-freedom system that we have been considering, the transfer function is defined by (8.68), which may be expressed as

$$G(s) = \frac{1/k}{(1/\omega_n^2)s^2 + (2\zeta/\omega_n)s + 1} \tag{8.78}$$

Both roots of the characteristic equation of (8.78) have negative real parts. Hence, substituting $j\omega$ for s, the harmonic response function is obtained as

$$G(j\omega) = \frac{1/k}{(1 - \omega^2/\omega_n^2) + (j2\zeta\omega/\omega_n)} \tag{8.79}$$

The steady-state vibration of this system given by (8.45) may be expressed as

$$q_{ss} = f_0 |G(j\omega)| \sin(\omega t + \measuredangle G(j\omega))$$

$$= \frac{f_0/k}{[(1 - \omega^2/\omega_n^2)^2 + (2\zeta\omega/\omega_n)^2]^{1/2}} \sin(\omega t + \psi)$$

where

$$\psi = -\tan^{-1}\left(\frac{2\zeta\omega/\omega_n}{1 - \omega^2/\omega_n^2}\right) = \measuredangle G(j\omega)$$

This result as expected is the same as that given by (8.49) and (8.50). When an exciting force is periodic but not simple harmonic, the result expressed by (8.60) and (8.61) also follows from the harmonic response function $G(jm\omega)$.

8.5 BODE DIAGRAM FOR FREQUENCY RESPONSE OF DAMPED SINGLE-DEGREE-OF-FREEDOM SYSTEMS

In the foregoing, it has been shown that for the steady-state vibrations of stable dynamic systems, the amplitude ratio $u/f_0 = |G(j\omega)|$, where u is the amplitude of the vibration displacement, f_0 the amplitude of the sinusoidal exciting force, and $G(j\omega)$ the harmonic response function. Also, the phase angle ψ between the vibration and the force is given by $\psi = \measuredangle G(j\omega)$. Hence, a plot of $|G(j\omega)|$ and $\measuredangle G(j\omega)$ versus the frequency ω is very useful for the analysis of steady-state vibrations. One method of representing this information is shown in Figs. 8.8 and 8.9. We now present an alternative form called the Bode diagram.

It consists of two plots, $\log_{10}|G(j\omega)|$ and the phase angle $\measuredangle G(j\omega)$, both plotted versus ω, on a log scale, using semilog graph paper. The Bode diagram has several advantages for multiple-degree-of-freedom systems. Since the frequency scale is logarithmic, a larger range of frequencies can be represented than that which would be possible by using a linear scale. The plotting of $\log_{10}|G(j\omega)|$ can be done very simply by using straight-line asymptotes, as shown shortly. It is common practice to plot $20\log_{10}|G(j\omega)|$ in decibels (dB) instead of $\log_{10}|G(j\omega)|$. The procedure will be clarified by considering the following examples.

Example 8.5

We consider the single-degree-of-freedom damped system defined by (8.2). It has been shown that the transfer function relating the displacement to the force is given by (8.68), and the harmonic response function by (8.79) for steady-state vibrations. Hence, we have

$$20 \log_{10} |G(j\omega)| = 20 \log \frac{1}{k} - \frac{20}{2} \log \left[\left(1 - \frac{\omega^2}{\omega_n^2} \right)^2 + \left(\frac{2\zeta\omega}{\omega_n} \right)^2 \right] \qquad (8.80)$$

$$\angle G(j\omega) = \angle \frac{1}{k} - \angle \left[\left(1 - \frac{\omega^2}{\omega_n^2} \right) + j \frac{2\zeta\omega}{\omega_n} \right]$$

$$= 0 - \tan^{-1} \left(\frac{2\zeta\omega/\omega_n}{1 - \omega^2/\omega_n^2} \right) \qquad (8.81)$$

It is seen that both $20 \log |G(j\omega)|$ and $\angle G(j\omega)$ are obtained by adding the contributions of the factors that constitute $G(j\omega)$. In both expressions, the contribution of a term on the denominator of $G(j\omega)$ has a negative sign. First, we consider the contribution of the constant term, $20 \log 1/k$ and $\angle 1/k$. The plot is shown in Fig. 8.13 for $k = 0.1$, where $20 \log (1/0.1) = 20$ dB and $\angle 1/0.1 = 0$ for all frequencies.

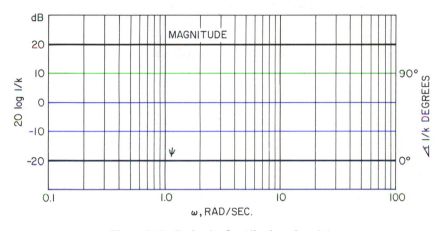

Figure 8.13   Bode plot for $1/k$ where $k = 0.1$.

Both the magnitude and phase angle curves are straight lines. The slope of the magnitude curve is 0 dB/decade. A decade is the horizontal distance on the frequency scale from any value of ω to 10 times ω. Thus, $\omega = 3$ to $\omega = 30$ is a decade. Now, we consider the contribution of the second term, namely,

$$\frac{-20}{2} \log \left[\left(1 - \frac{\omega^2}{\omega_n^2} \right)^2 + \left(\frac{2\zeta\omega}{\omega_n} \right)^2 \right]$$

$$-\tan^{-1} \left(\frac{2\zeta\omega/\omega_n}{\omega^2/\omega_n^2} \right)$$

When $\omega/\omega_n \ll 1$, the magnitude expression becomes $(-20/2) \log 1 = 0$. This is the equation for the low-frequency asymptote, which is a horizontal straight line whose slope is 0 dB/decade. When $\omega/\omega_n \gg 1$, the magnitude expression becomes $(-20/2) \log (\omega/\omega_n)^4 = -40 \log (\omega/\omega_n)$. This is the high-frequency asymptote, which is a straight line whose slope is -40 dB/decade. The low- and high-frequency asymptotes intersect at $\omega/\omega_n = 1$, where the exact value of the magnitude expression becomes $-20 \log 2\zeta$. When $\omega/\omega_n \rightarrow 0$, the phase angle tends to zero, and when $\omega/\omega_n \rightarrow \infty$, the phase angle tends to $-180°$. For $\omega/\omega_n = 1$, the phase angle is $-90°$. The plot is shown in Fig. 8.14 for various values of the damping ratio ζ and $\omega_n = 10$ rad/s. The frequency $\omega = \omega_n$, where the asymptotes intersect, is called the corner frequency. The Bode diagram for expressions (8.80) and (8.81) can now be completed by adding the diagrams of the two individual factors shown in Figs. 8.13 and 8.14.

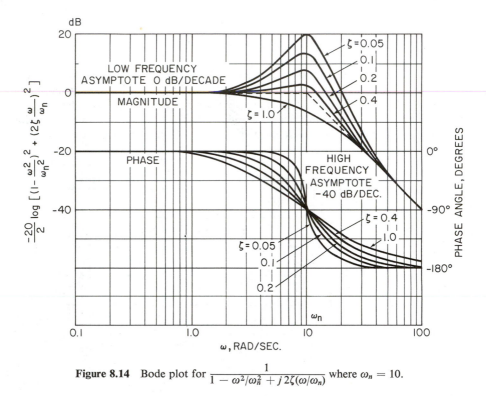

Figure 8.14   Bode plot for $\dfrac{1}{1 - \omega^2/\omega_n^2 + j\,2\zeta(\omega/\omega_n)}$ where $\omega_n = 10$.

This diagram is shown in Fig. 8.15 for $k = 0.1$, $\omega_n = 10$ rad/s, and $\zeta = 0.05$. In practice, it is not necessary to first draw the Bode diagram for the individual factors and then obtain the overall Bode diagram by adding the individual diagrams. The overall Bode diagram is drawn by utilizing the asymptotes, their slopes, and the corner frequencies.

8.5.1 Identification from Experimental Frequency Response

In the earlier part of this chapter, we have discussed a method for the identification of natural frequency and damping ratio of a single-degree-of-freedom system from the experimental response to a step or impulse input. Another method of identification is from the experimental frequency response,

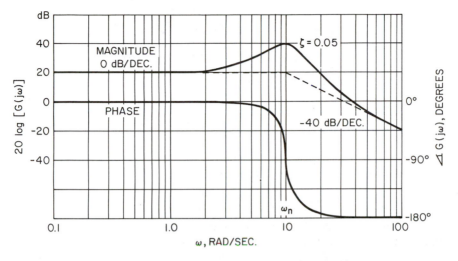

Figure 8.15   Bode diagram of $G(j\omega)$ of Eq. (8.79).

which is applicable to multiple-degree-of-freedom systems. The experimental procedure employs a force generator to provide sinusoidal excitation to the system. The amplitude ratio u/f_0 and the phase angle ψ between the displacement and force waveforms are measured for different frequencies. The information is then plotted in the form of a Bode diagram. For example, let us suppose that Fig. 8.15 represents experimental frequency-response data. The actual magnitude curve would have no corner. Hence, the corner frequency is determined by fitting asymptotes to the experimental data. From the corner frequency, we can identify the natural frequency as $\omega_n = 10$ rad/s. The overshoot of the magnitude curve from the corner is $+20$ dB. By comparing this to a standard plot given in Fig. 8.14, we identify the damping ratio as $\zeta = 0.05$. Also, at $\omega = 0.1$ rad/s, $20 \log |G(j\omega)| = 20$ dB. Now, at $\omega = 0.1$ rad/s, $\omega/\omega_n = 0.01$ and the contribution of the second term in (8.80) is negligible. Hence, at $\omega = 0.1$ rad/s,

$$20 \log |G(j\omega)| \simeq 20 \log \frac{1}{k} = 20 \text{ dB}$$

It follows that $1/k = 10$ and the spring constant is identified as $k = 0.1$.

Example 8.6

In Example 8.4, the relationship between the exciting force and displacement is given by

$$m\ddot{q} + c\dot{q} + kq = F$$

and the force transmitted to the foundation by

$$F_T = c\dot{q} + kq$$

These two equations can be expressed in the form of a block diagram as shown in Fig. 8.16. The transfer function relating the exciting force to the transmitted force

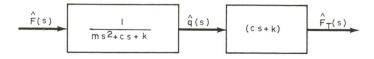

Figure 8.16 Transfer function between the exciting and transmitted forces.

is obtained as

$$G(s) = \frac{cs + k}{ms^2 + cs + k} = \frac{(2\zeta/\omega_n)s + 1}{(1/\omega_n^2)s^2 + (2\zeta/\omega_n)s + 1} \tag{8.82}$$

The roots of the characteristic equation of this transfer function have negative real parts and the harmonic response function becomes

$$G(j\omega) = \frac{(j2\zeta\omega/\omega_n) + 1}{(1 - \omega^2/\omega_n^2) + j2\zeta\omega/\omega_n} \tag{8.83}$$

We now draw a Bode diagram for this harmonic response function for the damping ratio $\zeta = 0.05$. Hence,

$$20 \log |G(j\omega)| = \frac{20}{2} \log \left[\left(0.1 \frac{\omega}{\omega_n} \right)^2 + 1 \right] - \frac{20}{2} \log \left[\left(1 - \frac{\omega^2}{\omega_n^2} \right)^2 + \left(0.1 \frac{\omega}{\omega_n} \right)^2 \right] \tag{8.84}$$

$$\angle G(j\omega) = \tan^{-1} \left(0.1 \frac{\omega}{\omega_n} \right) - \tan^{-1} \left(\frac{0.1(\omega/\omega_n)}{1 - \omega^2/\omega_n^2} \right) \tag{8.85}$$

First, we consider the contribution of the first term in (8.84) and (8.85). For $0.1 (\omega/\omega_n) \ll 1$, we have

$$\frac{20}{2} \log \left[\left(0.1 \frac{\omega}{\omega_n} \right)^2 + 1 \right] \simeq \frac{20}{2} \log 1 = 0$$

This is the low-frequency asymptote which is a horizontal straight line whose slope is 0 dB/decade. For $0.1(\omega/\omega_n) \gg 1$, we get

$$\frac{20}{2} \log \left[\left(0.1 \frac{\omega}{\omega_n} \right)^2 + 1 \right] \simeq \frac{20}{2} \log \left(0.1 \frac{\omega}{\omega_n} \right)^2 = 20 \log 0.1 \frac{\omega}{\omega_n}$$

This is a high-frequency asymptote whose slope is 20 dB/decade. The low- and high-frequency asymptotes intersect at the corner frequency $0.1(\omega/\omega_n) = 1$, where the exact value of the amplitude is $20/2 \log (1 + 1) = 3$ dB. The Bode diagram for this first term is shown in Fig. 8.17, where normalized frequency ω/ω_n is employed. In case this first-order term was on the denominator of the transfer function, the slope of the high-frequency asymptote would be -20 dB/decade and the phase angle would be negative, varying from zero to $-90°$. The Bode plot of the second term in (8.84) and (8.85) is similar to that shown in Fig. 8.14, the corner frequency being $\omega/\omega_n = 1$. Combining these individual plots, the overall Bode diagram for (8.84) and (8.85) is shown in Fig. 8.18, where normalized frequency ω/ω_n is employed. It is seen that for $\omega/\omega_n \gg 1$, the amplitude of the transmitted force becomes very small and the exciting force is filtered out. In order to study the effect of damping ratio ζ on the transmissibility and filter characteristics, now let $\zeta = 0.25$. From (8.83), we then obtain

$$20 \log |G(j\omega)| = \frac{20}{2} \log \left[\left(0.5 \frac{\omega}{\omega_n} \right)^2 + 1 \right] - \frac{20}{2} \log \left[\left(1 - \frac{\omega^2}{\omega_n^2} \right)^2 + \left(0.5 \frac{\omega}{\omega_n} \right)^2 \right] \tag{8.86}$$

$$\angle G(j\omega) = \tan^{-1} 0.5 \frac{\omega}{\omega_n} - \tan^{-1} \left(\frac{0.5 (\omega/\omega_n)}{1 - \omega^2/\omega_n^2} \right) \tag{8.87}$$

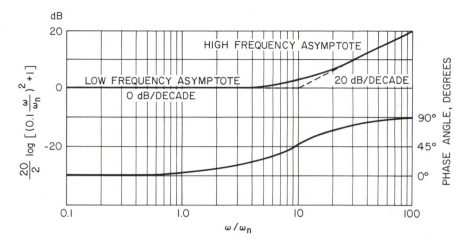

Figure 8.17   Bode diagram for $j0.1(\omega/\omega_n) + 1$.

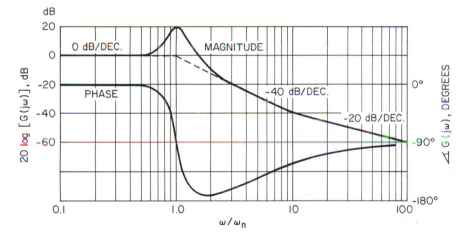

Figure 8.18   Bode diagram for Eqs. (8.84) and (8.85).

In order to draw the Bode diagram for (8.86) and (8.87), we need not first draw the diagrams for the individual factors. The overall Bode diagram is simply drawn by utilizing the asymptotes, their slopes, and the corner frequencies of (8.86). The first term in (8.86) is approximated by two asymptotes: the low-frequency asymptote with a slope of 0 dB/decade and the high-frequency asymptote with a slope of $+20$ dB/decade, and the corner frequency is $0.5\,(\omega/\omega_n) = 1$ (i.e., $\omega/\omega_n = 2$). The second term is also approximated by two asymptotes with slopes of 0 dB/decade and -40 dB/decade, respectively, and the corner frequency is $\omega/\omega_n = 1$. At frequencies below the first corner frequency, we have $20 \log |G(j\omega)| = 0$. The Bode diagram for (8.86) and (8.87) is shown in Fig. 8.19.

The effect of the damping ratio on the amplitude of the force transmitted to the foundation can be seen on comparing Fig. 8.18 with Fig. 8.19. For $\omega \ll \omega_n$, the damp-

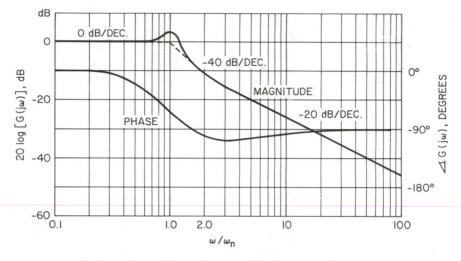

Figure 8.19 Bode diagram for Eqs. (8.86) and (8.87).

ing ratio has no effect on the transmissibility. When ω is in the neighborhood of ω_n, increased damping causes less amplification. For $\omega \gg \omega_n$, increased damping has a disadvantage since it causes less attenuation or less filtering.

Example 8.7

Figure 8.20 shows a seismic mass which is mounted on a frame with linear spring and damper as a displacement transducer. The purpose is to measure the vibration displacement $y(t)$ of the frame from the relative displacement $q(t)$ of the mass.

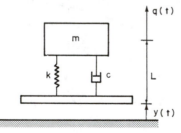

Figure 8.20 Seismic mass displacement transducer.

The displacement $q(t)$ is measured relative to the frame. Hence, the equation of motion of the mass becomes

$$m(\ddot{q} + \ddot{y}) + c\dot{q} + kq = 0$$

or

$$m\ddot{q} + c\dot{q} + kq = -m\ddot{y} \tag{8.88}$$

The transfer function relating the displacement y to the relative displacement q is shown in Fig. 8.21 and is

$$G(s) = \frac{-s^2/\omega_n^2}{(1/\omega_n^2)s^2 + (2\zeta/\omega_n)s + 1} \tag{8.89}$$

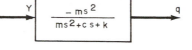

Figure 8.21   Block diagram of displacement transducer.

After checking that the roots of the characteristic equation have negative real parts, the harmonic response function is obtained by substituting $j\omega$ for s and is

$$G(j\omega) = \frac{\omega^2/\omega_n^2}{(1 - \omega^2/\omega_n^2) + j2\zeta\omega/\omega_n} \tag{8.90}$$

From (8.90), we obtain

$$20 \log |G(j\omega)| = 40 \log\frac{\omega}{\omega_n} - \frac{20}{2} \log\left[\left(1 - \frac{\omega^2}{\omega_n^2}\right)^2 + \left(2\zeta\frac{\omega}{\omega_n}\right)^2\right] \tag{8.91}$$

$$\triangle G(j\omega) = -\tan^{-1}\left(\frac{2\zeta\omega/\omega_n}{1 - \omega^2/\omega_n^2}\right) \tag{8.92}$$

The Bode diagram of (8.91) and (8.92) is shown in Fig. 8.22. The first term in (8.91) is a straight line with slope of $+40$ dB/decade, and for $\omega/\omega_n = 0.1$, its value is -40 dB. For $\omega/\omega_n \ll 1$, the value of the second term in (8.91) is zero and it is approximated by two asymptotes with corner frequency $\omega = \omega_n$. It is seen from Fig. 8.22 that

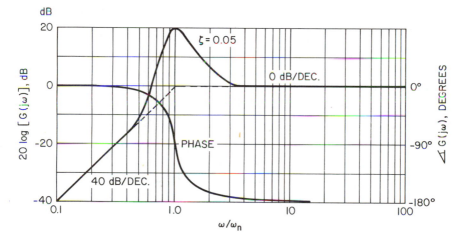

Figure 8.22   Bode diagram for Eqs. (8.91) and (8.92).

for $\omega > \omega_n$, the frequency response becomes flat. Hence, a seismic mass can be used as a vibration displacement transducer only for frequencies greater than its natural frequency. The natural frequency may be decreased by increasing the mass and decreasing the spring constant. But then the transducer would become very bulky. A better approach might be to measure the acceleration and integrate it twice to obtain the displacement.

In this example, for steady-state vibrations where $y = y_0 \sin \omega t$, we have $q = u \sin(\omega t + \psi)$. It is seen from Fig. 8.22 that the amplitude ratio $u/y_0 = 1$ for $\omega/\omega_n \gg 1$ and $u/y_0 = \omega^2/\omega_n^2$ for $\omega/\omega_n \ll 1$ (i.e., the amplitude u is a function of the frequency).

Example 8.8

This example illustrates the use of experimental frequency-response data for system identification. The system of Fig. 8.10 was externally excited by a sinusoidal force $F = f_0 \sin \omega t$ and the steady-state response observed was $q = u \sin(\omega t + \psi)$. The experimental frequency response data are given in the following table. Identify the transfer function of the system.

Frequency, ω (rad/s)	0.2	0.25	0.32	0.4	0.5	0.55	0.65	0.8	1	2
$20 \log \dfrac{u}{f_0}$ (dB)	-25.2	-24.4	-23.9	-23.4	-24.1	-25.2	-28.0	-32.0	-35.6	-49.1
Phase angle, ψ (deg)	-20	-30	-41	-62	-90	-104	-123	-140	-152	-166

The Bode diagram of the response data is shown in Fig. 8.23. The magnitude curve exhibits an initial slope of zero and a final slope of -40 dB/decade. The initial phase angle tends to zero and the final phase angle to $-180°$. It is apparent that the break in the initial zero slope is due to a quadratic term on the denominator. Thus the transfer function is of the form

$$G(s) = \frac{1/k}{(1/\omega_n^2)s^2 + (2\zeta/\omega_n)s + 1} \tag{8.93}$$

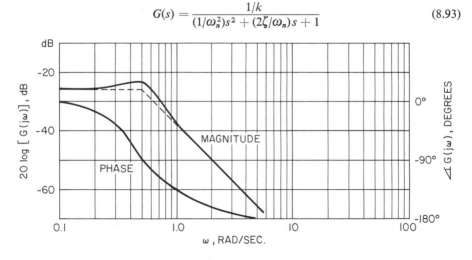

Figure 8.23 Bode diagram of experimental data of Example 8.8.

The phase-angle curve also confirms this model. Drawing the best asymptotes to the experimental magnitude curve, the corner frequency is obtained as $\omega = 0.5$ rad/s. Hence, the natural frequency is $\omega_n = 0.5$ rad/s. The phase angle of $-90°$ at this frequency also confirms this value of the natural frequency. On comparing the overshoot of the magnitude plot at the corner frequency to that of the standard plot shown in Fig. 8.14, it is seen that the damping ratio $\zeta = 0.4$. Now, the value of the stiffness k

in (8.93) is to be determined. From (8.93), we obtain

$$20 \log |G(j\omega)| = 20 \log \frac{1}{k} - \frac{20}{2} \log \left[\left(1 - \frac{\omega^2}{0.5^2}\right)^2 + \left(\frac{0.8\omega}{0.5}\right)^2 \right] \qquad (8.94)$$

At $\omega = 0.1$ rad/s, from Fig. 8.23 we obtain $20 \log |G(j\omega)| = -25.8$ dB. Substituting these values in (8.94), it follows that $20 \log 1/k = -26.03$ dB (i.e., $k = 20$).

8.6 MULTIPLE-DEGREE-OF-FREEDOM SYSTEMS

When the number of degrees of freedom for a dynamic system is more than 1, we obtain as many natural frequencies and modes of vibration as there are degrees of freedom. In the remainder of this chapter, we discuss the general procedure of analysis for multiple-degree-of-freedom systems. The equations of motion can be formulated by employing the procedures discussed in Chapters 3 to 5. For holonomic systems with n degrees of freedom, the Lagrange equations are

$$\frac{d}{dt}\left(\frac{\partial L}{\partial \dot{q}_i}\right) - \frac{\partial L}{\partial q_i} = Q_i, \qquad i = 1, \ldots, n \qquad (8.95)$$

where L is the Lagrangian and Q_i is the generalized force in the ith direction due to the work done by the nonconservatives forces, both frictional and externally applied. We assume that the nonlinearities in (8.95) are analytic functions of their arguments and that the frictional forces are viscous, for which a Rayleigh dissipation function $\frac{1}{2}\{\dot{q}\}^T[C]\{\dot{q}\}$ can be defined. Linearization of the equations for small displacements about an equilibrium then yields

$$[M]\{\ddot{q}\} + [C]\{\dot{q}\} + [K]\{q\} = \{Q\} \qquad (8.96)$$

where $\{Q\}$ now denotes only the externally applied forces. In (8.96), $[M]$, $[C]$, and $[K]$ are the $n \times n$ mass, damping, and stiffness matrices, respectively. For the time-invariant systems considered in this chapter, these matrices are constant. Another formulation that is convenient in some cases is the state-variables formulation, where (8.96) is expressed as a set of $2n$ first-order differential equations. For this purpose, we note that (8.96) is unchanged when it is expressed as

$$\{\ddot{q}\} = -[M]^{-1}[C]\{\dot{q}\} - [M]^{-1}[K]\{q\} + [M]^{-1}\{Q\} \qquad (8.97)$$

We rewrite this equation in the form

$$\left\{ \begin{matrix} \{\dot{q}\} \\ \{\ddot{q}\} \end{matrix} \right\} = \left[\begin{array}{c|c} [0] & [I] \\ \hline -[M]^{-1}[K] & -[M]^{-1}[C] \end{array} \right] \left\{ \begin{matrix} \{q\} \\ \{\dot{q}\} \end{matrix} \right\} + \left\{ \begin{matrix} \{0\} \\ [M]^{-1}\{Q\} \end{matrix} \right\} \qquad (8.98)$$

The first set of n equations in the foregoing represent an identity (i.e., $\{\dot{q}\} = \{\dot{q}\}$) and the second set of n equations represent (8.97). We now choose phase variables as state variables and define $(2n \times 1)$ vector $\{x\}$, $(2n \times 2n)$ matrix $[A]$, and $[2n \times n]$ matrix $[B]$ as

$$\{x\} = \left\{\begin{array}{c} \{q\} \\ \hline \{\dot{q}\} \end{array}\right\}, \qquad \left\{\begin{array}{c} \{0\} \\ \hline [M]^{-1}\{Q\} \end{array}\right\} = [B]\{Q\},$$

$$[A] = \left[\begin{array}{c:c} [0] & [I] \\ \hline -[M]^{-1}[K] & -[M]^{-1}[C] \end{array}\right]$$

Equation (8.98) is now expressed as a set of $2n$ first-order equations in state-variable form as

$$\{\dot{x}\} = [A]\{x\} + [B]\{Q\} \tag{8.99}$$

Considering the set of generalized coordinates as output, we define a $(n \times 2n)$ matrix E such that

$$\{q\} = [E]\{x\} \tag{8.100}$$

where

$$\underset{(n \times 2n)}{[E]} = \begin{bmatrix} 1 & 0 & 0 & \cdots & 0 & \cdots & 0 & 0 \\ 0 & 1 & 0 & \cdots & 0 & \cdots & 0 & 0 \\ 0 & 0 & 1 & \cdots & 0 & \cdots & 0 & 0 \\ \cdot & & & & & & & \\ \cdot & & & & & & & \\ \cdot & & & & & & & \\ 0 & 0 & 0 & \cdots & 1 & \cdots & 0 & 0 \end{bmatrix}$$

It is noted that (8.99) and (8.100) represent the generalization of (8.69) and (8.71) from a single-degree- to a multiple-degree-of-freedom system. For the analysis of steady-state vibrations of damped multiple-degree-of-freedom systems, we also use the harmonic response-function matrix, which is obtained from the transfer function matrix by substituting $j\omega$ for s or for the operator d/dt in stable systems. Since in stable systems, the response due to initial conditions decays to zero and is not reflected in steady-state vibrations, we Laplace-transform (8.96) assuming zero initial conditions on $\{q\}$ and $\{\dot{q}\}$. It follows that

$$[s^2[M] + s[C] + [K]]\{\hat{q}(s)\} = \{\hat{Q}(s)\} \tag{8.101a}$$

$$\{\hat{q}(s)\} = [s^2[M] + s[C] + [K]]^{-1}\{\hat{Q}(s)\} \tag{8.101b}$$

Hence, the n outputs $\{\hat{q}(s)\}$ are related to the n inputs $\{\hat{Q}(s)\}$ by the $n \times n$ transfer function matrix

$$[G(s)] = [s^2[M] + s[C] + [K]]^{-1} \tag{8.102}$$

and the relationship is represented in the form of the block diagram shown in Fig. 8.24. In case the state-variable formulation has been employed, this same transfer function matrix can be obtained from (8.99) and (8.100) as follows. It

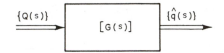

Figure 8.24 Block diagram for input–output relationship.

has been shown in Chapter 6 that the solution of (8.99) is given by

$$\{x(t)\} = \mathbf{\Phi}(t)\{x_0\} + \int_0^t \mathbf{\Phi}(t - t')[B]\{Q(t')\} \, dt' \tag{8.103}$$

In the Laplace transform domain, the foregoing equation may be written as

$$\{\hat{x}(s)\} = \hat{\mathbf{\Phi}}(s)\{x_0\} + \hat{\mathbf{\Phi}}(s)[B]\{\hat{Q}(s)\} \tag{8.104}$$

and from (8.100) it follows that

$$\{\hat{q}(s)\} = [E]\hat{\mathbf{\Phi}}(s)\{x_0\} + [E]\hat{\mathbf{\Phi}}(s)[B]\{\hat{Q}(s)\} \tag{8.105}$$

For steady-state vibrations of stable systems, the first term on the right-hand side of (8.105) will decay to zero with time and we would obtain

$$\{\hat{q}(s)\} = [E]\hat{\mathbf{\Phi}}(s)[B]\{\hat{Q}(s)\} \tag{8.106}$$

Since $\hat{\mathbf{\Phi}}(s) = (s\mathbf{I} - \mathbf{A})^{-1}$, on comparing (8.106) with (8.101b), it is seen that the transfer function matrix can also be expressed as

$$[G(s)] = [s[M] + s[C] + [K]]^{-1} = [E](s\mathbf{I} - \mathbf{A})^{-1}[B] \tag{8.107}$$

The formulations expressed by (8.96), (8.101b) and (8.99) will be employed in the following two sections for the analysis of vibrations of multiple-degree-of-freedom systems.

8.7 UNDAMPED MULTIPLE-DEGREE-OF-FREEDOM SYSTEMS

8.7.1 Analysis of Free Vibrations by Modal Decomposition

It was pointed out earlier in this chapter that undamped conservative systems are not encountered in practice, with the exception of the area of celestial mechanics. However, it is seen from previous sections that in lightly damped systems, the natural frequencies of the undamped system do not differ appreciably from the ones of the damped system. It is shown in this section that the values of the natural frequencies can be easily determined knowing the values of the mass and stiffness matrices. In practice, the values of the mass matrix are usually known and the values of the stiffness matrix can be approximated from strength of materials. The determination of the values of the damping matrix usually requires time-consuming experimentation such as experimental frequency response. Hence, our purpose in analyzing free vibrations of undamped systems in this section is to determine the natural frequencies by neglecting the damping matrix in lightly damped systems.

The dynamic response analysis of a freely vibrating system consists of determining the natural frequencies (i.e., the eigenvalues and the corresponding free vibration mode shapes). The mode shapes represent n independent displacement patterns. Either the formulation of (8.96) or the state-variable formulation of (8.99) and (8.100) could be employed for the analysis after setting matrix

$[C] = 0$ and $\{Q\} = 0$. The formulation of (8.99) leads to a complex eigenvalue problem (in this case the eigenvalues are in fact purely imaginary since there is no damping). Since the coefficients of matrix $[A]$ are real, there exist complex-conjugate pairs. The formulation of (8.96) with $[C] = 0$ leads to a real eigenvalue problem when certain restrictions are satisfied and hence is preferable. The restrictions are that matrices $[M]$ and $[K]$ be symmetric and positive definite. This requires that in the linearized system, the kinetic and potential energies be quadratic functions and that

$$T = \tfrac{1}{2}\lfloor \dot{q} \rfloor [M]\{\dot{q}\} > 0 \qquad \text{for } \{q\} \neq 0$$
$$U = \tfrac{1}{2}\lfloor q \rfloor [K]\{q\} > 0 \qquad \text{for } \{q\} \neq 0 \tag{8.108}$$

It follows from Maxwell's reciprocal theorem that matrix $[K]$ is symmetric but matrix $[M]$ need not always be symmetric. When these restrictions are not satisfied, the state-variable formulation (8.99) can be employed. Here, we employ the formulation of (8.96), and the equations of motion for the free vibrations of a conservative system are represented by

$$[M]\{\ddot{q}\} + [K]\{q\} = \{0\} \tag{8.109}$$

Letting $[H] = [M]^{-1}[K]$, where $[H]$ is called the dynamical matrix, (8.109) is represented as

$$\{\ddot{q}\} + [H]\{q\} = \{0\} \tag{8.110}$$

We seek a harmonic solution of the form

$$\{q\} = \{v\} \sin(\omega_n t + \psi) \tag{8.111}$$

where $\{v\}$ is an eigenvector or modal vector, ω_n a natural frequency, and ψ a phase angle. Substituting for $\{q\}$ from (8.111) in (8.110) and noting that for a nontrivial solution $\sin(\omega_n t + \psi) \neq 0$, we obtain

$$[\omega_n^2[I] - [H]]\{v\} = \{0\} \tag{8.112}$$

Letting $\lambda = \omega_n^2$, where λ is an eigenvalue, (8.112) is expressed in the form of an eigenvalue problem as

$$[\lambda[I] - [H]]\{v\} = \{0\} \tag{8.113}$$

For a nontrivial solution of this equation, it follows that

$$\det[\lambda[I] - [H]] = 0 \tag{8.114}$$

This is the characteristic equation which is also called the frequency equation when ω_n^2 is used instead of λ. Since matrices $[M]$ and $[K]$ are assumed to be positive definite, matrix $[H]$ is positive definite and its n eigenvalues are all real and positive. The n natural frequencies are given by $\omega_{ni} = \sqrt{\lambda_i}$ for $i = 1, 2, \ldots, n$. The eigenvector $\{v_i\}$ corresponding to an eigenvalue λ_i is also called a modal vector and corresponds to a particular mode shape of vibration. We now show that the eigenvectors corresponding to two distinct eigenvalues are orthogonal with respect to the mass and stiffness matrices. For two distinct eigenvalues, from (8.113), we obtain

$$\lambda_i[M]\{v_i\} = [K]\{v_i\} \tag{8.115}$$

$$\lambda_j[M]\{v_j\} = [K]\{v_j\} \tag{8.116}$$

Premultiplying (8.115) by $\lfloor v_j \rfloor$ and (8.116) by $\lfloor v_i \rfloor$, it follows that

$$\lambda_i\lfloor v_j \rfloor[M]\{v_i\} = \lfloor v_j \rfloor[K]\{v_i\} \tag{8.117}$$

$$\lambda_j\lfloor v_i \rfloor[M]\{v_j\} = \lfloor v_i \rfloor[K]\{v_j\} \tag{8.118}$$

Since matrices $[M]$ and $[K]$ are assumed to be symmetric, we subtract the transpose of (8.118) from (8.117) and obtain

$$(\lambda_i - \lambda_j)\lfloor v_j \rfloor[M]\{v_i\} = 0, \qquad i \neq j$$

Since $\lambda_i \neq \lambda_j$, it follows that

$$\lfloor v_j \rfloor[M]\{v_i\} = 0, \qquad i \neq j \tag{8.119}$$

By substituting the result of (8.119) in (8.117), it can be easily verified that

$$\lfloor v_j \rfloor[K]\{v_i\} = 0, \qquad i \neq j \tag{8.120}$$

It is seen in Chapter 6 that only the direction of the eigenvector can be determined from (8.114) and its length is arbitrary. Each eigenvector can be normalized such that

$$\lfloor v_i \rfloor[M]\{v_i\} = 1 \tag{8.121}$$

The eigenvectors belonging to two different eigenvalues are then mutually orthonormal. We now assume further that the eigenvalues of (8.113) are distinct. It rarely happens that any two or more natural frequencies are identical. It is then seen from Chapter 6 that the n eigenvectors of (8.113) are linearly independent and we can define a nonsingular similarity transformation matrix $[P]$ as

$$[P] = [\{v_1\}, \{v_2\}, \ldots, \{v_n\}] \tag{8.122}$$

This similarity transformation matrix $[P]$ is now employed to uncouple the equations of motion (8.109) by defining a new set of generalized coordinates such that

$$\{q\} = [P]\{y\} \tag{8.123}$$

Equation (8.109) is then converted to

$$[M][P]\{\ddot{y}\} + [K][P]\{y\} = \{0\} \tag{8.124a}$$

Premultiplying (8.124a) by $[P]^T$, we obtain

$$[P]^T[M][P]\{\ddot{y}\} + [P]^T[K][P]\{y\} = \{0\} \tag{8.124b}$$

It can be seen from (8.121) and (8.115) that $[P]^T[M][P] = [I]$, where $[I]$ is an identity matrix, and that $[P]^T[K][P] = \Lambda$, where Λ is a diagonal matrix with $\lambda_i = \omega_{ni}^2$ along the main diagonal. Hence, in the new generalized coordinates the equations of motion (8.124b) are uncoupled and we obtain

$$\ddot{y}_i + \omega_{ni}^2 y_i = 0, \qquad i = 1, \ldots, n \tag{8.125}$$

The solution of (8.125) can be written easily as

$$y_i(t) = b_i \sin(\omega_{ni} t + \psi_i) \qquad (8.126)$$

where the constants b_i and ψ_i are evaluated from the two initial conditions $y_i(0)$ and $\dot{y}_i(0)$ which can be obtained from the initial conditions $\{q(0)\}$ and $\{\dot{q}(0)\}$ from (8.123). The new set of coordinates $\{y\}$ are called normal generalized coordinates, a name that is derived from the Jordan normal form. The solutions given by (8.126) represent normal modes of vibration. If free vibration in the generalized coordinates $\{q\}$ is required, it can be obtained from $\{q\} = [P]\{y\}$. The similarity transformation matrix $[P]$ is also called the modal matrix. Since free vibrations of conservative systems are not encountered in practice, it is not necessary to obtain the solution. It is sufficient to determine only the natural frequencies from the characteristic equation (8.114).

Example 8.9

In an overhead crane, a truck of mass m_2 is resting at the center of a beam with stiffness k_2. The truck is lifting a mass m_1 through a cable of stiffness k_1. We wish to determine the natural frequencies of this system. We assume a two-degree-of-freedom system.

The system is shown in Fig. 8.25 and its conceptual model in Fig. 8.26. The equations of motion can be derived by employing Newton's law or Lagrange equations. The free-body diagram is shown in Fig. 8.27 and the equations of motion of the unforced system are given by

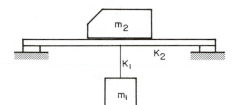

Figure 8.25 Overhead crane lifting a mass.

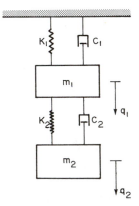

Figure 8.26 Conceptual model of overhead crane lifting a mass.

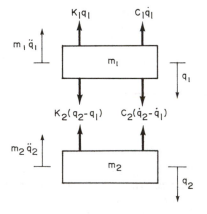

Figure 8.27 Free-body diagram of system of Fig. 8.25.

$$m_1\ddot{q}_1 + (c_1 + c_2)\dot{q}_1 + (k_1 + k_2)q_1 - c_2\dot{q}_2 - k_2q_2 = 0$$

$$m_2\ddot{q}_2 + c_2\dot{q}_2 + k_2q_2 - c_2\dot{q}_1 - k_2q_1 = 0$$

which can be represented in the matrix notation as

$$\begin{bmatrix} m_1 & 0 \\ 0 & m_2 \end{bmatrix}\begin{Bmatrix} \ddot{q}_1 \\ \ddot{q}_2 \end{Bmatrix} + \begin{bmatrix} c_1 + c_2 & -c_2 \\ -c_2 & c_2 \end{bmatrix}\begin{Bmatrix} \dot{q}_1 \\ \dot{q}_2 \end{Bmatrix} + \begin{bmatrix} k_1 + k_2 & -k_2 \\ -k_2 & k_2 \end{bmatrix}\begin{Bmatrix} q_1 \\ q_2 \end{Bmatrix} = \begin{Bmatrix} 0 \\ 0 \end{Bmatrix} \tag{8.127}$$

The mass, damping, and stiffness matrices are obvious from (8.127). The values of the mass matrix can be determined easily. The values of the stiffness matrix can be evaluated from strength of materials. For a simply supported beam loaded at the center within elastic limits, we obtain $k_2 = 48E_bI_b/L_b^3$ and for the cable loaded in tension, $k_1 = A_cE_c/L_c$. The subscripts b and c denote the beam and cable, respectively. The evaluation of the damping matrix would require experimental data. However, it is known that for this system the damping is very small. Hence, in (8.127), we set the damping matrix to zero and obtain approximation to the damped natural frequencies from the natural frequencies. From (8.114) the characteristic equation is obtained as

$$\begin{vmatrix} k_1 + k_2 + \lambda m_1 & -k_2 \\ -k_2 & k_2 + \lambda m_2 \end{vmatrix} = 0$$

or

$$m_1m_2\lambda^2 + (m_1k_2 + m_2k_1 + m_2k_2)\lambda + k_1k_2 = 0 \tag{8.128}$$

In order to obtain numerical values, let $m_1 = 28.49$ lb-sec²/in., $m_2 = 6.83$ lb-sec²/in., $k_1 = 2000$ lb/in., $k_2 = 42.66$ lb/in. The numerical values of λ_1 and λ_2 that satisfy (8.128) can now be determined numerically and the two natural frequencies are $\omega_{n1} = \sqrt{\lambda_1} = 1.25$ rad/s and $\omega_{n2} = \sqrt{\lambda_2} = 19.17$ rad/s. In vibration analysis, the lowest natural frequency is called the fundamental frequency. In the normal generalized coordinates the equations are

$$\ddot{y}_1 + \omega_{n1}^2 y_1 = 0$$

$$\ddot{y}_2 + \omega_{n2}^2 y = 0$$

The normal modes of vibration are $y_1 = b_1 \sin(\omega_{n1}t + \psi_1)$ and $y_2 = b_2 \sin(\omega_{n2}t + \psi_2)$, where b_1, b_2, ψ_1, and ψ_2 are obtained from initial conditions $y_1(0), y_2(0), \dot{y}_1(0)$, and $\dot{y}_2(0)$.

Example 8.10

An automobile suspension system is shown in Fig. 8.28. If m and I are the mass and moment of inertia of the sprung mass, develop the equations of motion and calculate the natural frequencies of free vibration.

The natural frequency of vertical motion of the unsprung mass is much higher than the natural frequency of the sprung mass. At sufficiently small frequencies, the disturbances are transmitted directly to the sprung mass and the degrees of freedom of the unsprung mass can be neglected.

Assuming the sprung mass to be rigid, two degrees of freedom are assigned: (a) vertical bounce q and (b) pitch θ (Fig. 8.29).

Considering small displacements, the translation at the front and rear suspension points are $(q - L_1\theta)$ and $(q + L_2\theta)$.

The free-body diagram for the system is shown in Fig. 8.30. Summing up the forces and moments, we obtain

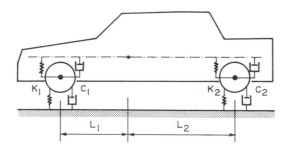

Figure 8.28 Automobile suspension system.

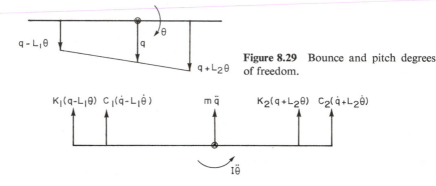

Figure 8.29 Bounce and pitch degrees of freedom.

Figure 8.30 Free-body diagram.

$$m\ddot{q} + c_1(\dot{q} - L_1\dot{\theta}) + c_2(\dot{q} + L_2\dot{\theta}) + k_1(q - L_1\theta) + k_2(q + L_2\theta) = 0$$

$$I\ddot{\theta} + c_2(\dot{q} + L_2\dot{\theta})L_2 - c_1(\dot{q} - L_1\dot{\theta})L_1 + k_2(q + L_2\theta)L_2 - k_1(q - L_1\theta)L_1 = 0$$

Rewriting the foregoing equations of motion in matrix form, it follows that

$$\begin{bmatrix} m & 0 \\ 0 & I \end{bmatrix} \begin{Bmatrix} \ddot{q} \\ \ddot{\theta} \end{Bmatrix} + \begin{bmatrix} c_1 + c_2 & -c_1L_1 + c_2L_2 \\ -c_1L_1 + c_2L_2 & c_1L_1^2 + c_2L_2^2 \end{bmatrix} \begin{Bmatrix} \dot{q} \\ \dot{\theta} \end{Bmatrix}$$

$$+ \begin{bmatrix} k_1 + k_2 & -k_1L_1 + k_2L_2 \\ -k_1L_1 + k_2L_2 & k_1L_1^2 + k_2L_2^2 \end{bmatrix} \begin{Bmatrix} q \\ \theta \end{Bmatrix} = \begin{Bmatrix} 0 \\ 0 \end{Bmatrix} \qquad (8.129)$$

The $[M]$, $[C]$, and $[K]$ matrices are defined from the foregoing equation. In order to determine the natural frequencies, we let $[C] = [0]$ and obtain

$$[M] \begin{Bmatrix} \ddot{q} \\ \ddot{\theta} \end{Bmatrix} + [K] \begin{Bmatrix} q \\ \theta \end{Bmatrix} = \begin{Bmatrix} 0 \\ 0 \end{Bmatrix}$$

The characteristic equation (8.114) is obtained as

$$\det \left| \lambda \begin{bmatrix} m & 0 \\ 0 & I \end{bmatrix} - \begin{bmatrix} k_1 + k_2 & -k_1L_1 + k_2L_2 \\ -k_1L_1 + k_2L_2 & k_1L_1^2 + k_2L_2^2 \end{bmatrix} \right| = 0$$

Expanding the determinant in the foregoing equation, we obtain

$$mI\lambda^2 - I(k_1 + k_2)\lambda - m(k_1L_1^2 + k_2L_2^2)\lambda + (k_1 + k_2)(k_1L_1^2 + k_2L_2^2)$$

$$- (k_2L_2 - k_1L_1)^2 = 0$$

The two roots λ_1 and λ_2 correspond to ω_{n1}^2 and ω_{n2}^2, respectively, and are given by

$$\omega_{n1}^2, \omega_{n2}^2 = \frac{1}{2}\left[\left(\frac{k_1 + k_2}{m} + \frac{k_1 L_1^2 + k_2 L_2^2}{I}\right)\right.$$

$$\left.\pm \left\{\left(\frac{k_1 + k_2}{m} - \frac{k_1 L_1^2 + k_2 L_2^2}{I}\right)^2 + \frac{4(k_2 L_2 - k_1 L_1)^2}{mI}\right\}^{1/2}\right]$$

In practice, the natural frequencies ω_{n1} and ω_{n2} are in the range 1 to 2 Hz. However, in this example, intentional damping is introduced by means of the shock absorbers and hence we can expect the damped natural frequencies to differ appreciably from the undamped natural frequencies that have been evaluated. For a damped system, the characteristic equation and the determination of the damped natural frequencies are discussed in the following section.

8.8 FORCED VIBRATIONS OF DAMPED MULTIPLE-DEGREE-OF-FREEDOM SYSTEMS

8.8.1 Forced Vibration Analysis by Modal Decomposition

In this section we discuss two methods for the analysis of forced vibrations of damped multiple-degree-of-freedom systems. The first method, which is discussed next, involves modal analysis; the second method, which is discussed later, employs the harmonic response function. The method of modal analysis described in the preceding section does not generally apply to damped systems. The equations of motion cannot be uncoupled by the modal matrix of undamped systems except when the damping is proportional; that is, the damping matrix $[C] = \alpha[M] + \beta[K]$, where α and β are constants. The case of proportional damping is rarely encountered. For this reason, we do not employ the formulation of (8.96) for the modal analysis of damped system but use the state-variable formulation of (8.99).

The method of analysis is similar to the one employed in the preceding section. It involves the solution of the eigenvalue problem and the determination of the eigenvectors and the similarity transformation to uncouple the equations of motion. The eigenvalues and eigenvectors, however, are complex quantities. These techniques have been discussed in Chapter 6 in connection with the determination of the state transition matrix for linear time-invariant systems. Here, we merely apply the method for the analysis of vibrations. The equations of motion are described in the form of (8.99) and it is assumed that the exciting forces are harmonic; that is, $Q_i = f_i \sin(\omega_i t + \alpha_i)$, where f_i is the amplitude, ω_i the forcing frequency, and α_i the phase angle of the generalized force in the ith coordinate direction. From (8.99) it is seen that the unforced system is described by

$$\{\dot{x}\} = [A]\{x\}$$

and the corresponding characteristic equation becomes

$$\det[\lambda I - [A]] = 0 \tag{8.130}$$

Since $[A]$ is a $2n \times 2n$ matrix where n is the number of degrees of freedom, it follows that there exist $2n$ eigenvalues. Some of the roots may be real and the remaining complex-conjugate pairs. A pair of real roots represents an over-damped quadratic and a pair of complex-conjugate roots an underdamped quadratic. Steady-state forced linear vibrations will not occur about an unstable equilibrium state. It is assumed here that all the roots of the characteristic equa-tion (8.130) have negative real part. It is further assumed here that matrix **A** has $2n$ distinct eigenvalues. In such a case, there exist $2n$ linearly independent eigenvectors as discussed in Chapter 6.

After determining these eigenvectors, we define a similarity transformation matrix $[P]$ as

$$[P] = [\{v_1\} \cdots \{v_{2n}\}] \tag{8.131}$$

The state variables $\{x\}$ are now transformed to normal state variables $\{y\}$ by the similarity transformation

$$\{x\} = [P]\{y\} \tag{8.132}$$

and (8.99) becomes transformed to

$$[P]\{\dot{y}\} = [A][P]\{y\} + [B]\{Q\}$$

or

$$\{\dot{y}\} = [P]^{-1}[A][P]\{y\} + [P]^{-1}[B]\{Q\} \tag{8.133}$$

where $[P]^{-1}[A][P] = \Lambda$, a diagonal matrix with the eigenvalues of **A** along its main diagonal. The equations in the normal state variables are now uncoupled and the diagonal state-transition matrix $\Phi(t)$ for (8.133) can be obtained readily. Thus, we obtain

$$\{y(t)\} = \Phi(t)\{y(0)\} + \int_0^t \Phi(t - t')[P]^{-1}[B]\{Q(t')\} \, dt' \tag{8.134}$$

Since all eigenvalues of matrix $[A]$ have a negative real part, the part of the response in (8.134) due to the initial conditions decays to zero with time and for steady-state forced vibrations, we get

$$\{y_{ss}(t)\} = \int_0^t \Phi(t - t')[P]^{-1}[B]\{Q(t')\} \, dt' \tag{8.135}$$

where the generalized force vector $\{Q(t)\}$ is harmonic with different forcing frequencies. For steady-state forced vibrations, the behavior of the state variables $\{x\}$ and generalized displacements $\{q\}$ is given by

$$\begin{aligned} \{x_{ss}(t)\} &= [P]\{y_{ss}(t)\} \\ \{q_{ss}(t)\} &= [E][P]\{y_{ss}(t)\} \end{aligned} \tag{8.136}$$

where matrix $[E]$ is defined by (8.100).

For large degrees of freedom systems, the use of a computer becomes a necessity for the determination of the eigenvalues, eigenvectors, the similarity transformation matrix, and for the solution of the convolution integral in (8.135). The same result can be obtained with much less computational effort by employing the frequency-response method, which is discussed in the next section.

Example 8.11

We consider the forced vibrations of the sprung mass of the automobile of Example 8.10, but now the damping matrix $[C]$ is not neglected and in addition we include sinusoidal force and moment on the right-hand side of (8.129) given by

$$\{Q\} = \begin{Bmatrix} a_1 \sin(\omega_1 t + \alpha_1) \\ a_2 \sin(\omega_2 t + \alpha_2) \end{Bmatrix}$$

Let the numerical values of $-[M]^{-1}[K]$ and $-[M]^{-1}[C]$ matrices be given such that $[A]$ matrix of (8.99) becomes

$$[A] = \begin{bmatrix} 0 & 0 & 1.0 & 0 \\ 0 & 0 & 0 & 1.0 \\ -0.25333 & -0.00178 & -0.03294 & -0.01098 \\ -0.00638 & -0.01184 & -0.03921 & -0.01307 \end{bmatrix} \tag{8.137}$$

From the definition of $[B]$ matrix in (8.99) for this example, we obtain

$$[B] = \begin{bmatrix} 0 & 0 \\ 0 & 0 \\ \dfrac{1}{m} & 0 \\ 0 & \dfrac{1}{I} \end{bmatrix} \tag{8.138}$$

The eigenvalues of matrix $[A]$ of (8.137) have been obtained by Hurty and Rubinstein [6] and are given by

$$\lambda_1, \lambda_2 = -0.016747 \pm j0.50265$$
$$\lambda_1, \lambda_2 = -0.0062573 \pm j0.108509$$

It is noted that there are two complex-conjugate pairs of eigenvalues. The corresponding complex eigenvectors are

$$\{v_1\}, \{v_2\} = \begin{Bmatrix} 1.67595 \\ 1.06079 \\ 6.31318 \\ 0.086550 \end{Bmatrix} \pm j \begin{Bmatrix} -12.61563 \\ -0.20743 \\ 1.05368 \\ 0.53809 \end{Bmatrix}$$

$$\{v_3\}, \{v_4\} = \begin{Bmatrix} -0.008194 \\ 0.293722 \\ -0.000788 \\ 0.134031 \end{Bmatrix} \pm j \begin{Bmatrix} 0.007513 \\ -1.252144 \\ -0.000938 \\ 0.039707 \end{Bmatrix}$$

The complex (4×4) similarity transformation matrix $[P]$ can be obtained as in (8.131) and employed to transform the equations to the normal state-variable equation

(8.133). The state transition matrix $\Phi(t)$ of (8.134) is a diagonal matrix given by

$$\Phi(t) = \begin{bmatrix} e^{\lambda_1 t} & 0 & 0 & 0 \\ 0 & e^{\lambda_2 t} & 0 & 0 \\ 0 & 0 & e^{\lambda_3 t} & 0 \\ 0 & 0 & 0 & e^{\lambda_4 t} \end{bmatrix}$$

This state transition matrix is employed in (8.135) to obtain the steady-state vibrations of the normal state variables. The steady-state vibrations of x and θ are then obtained from (8.136). It can be verified that x and θ are real quantities. The calculations are very lengthy and have been omitted here. It becomes obvious that the frequency-response method that is covered next is computationally much simpler and is preferable to the modal decomposition method for analysis of forced vibrations.

8.8.2 Forced Vibration Analysis by the Frequency-Response Method

We now describe the frequency-response method of analysis of steady-state forced vibrations. This method employs the harmonic-response function matrix and is the generalization of the frequency-response domain techniques discussed earlier for single-degree-of-freedom systems. For a system with n degrees of freedom and n forcing functions, the transfer matrix is described by (8.101b) and (8.102). Let Δ denote the characteristic determinant of the system. The characteristic equation then becomes

$$\Delta(s) = \det\left[s^2[M] + s[C] + [K]\right] = 0 \qquad (8.139)$$

Letting $G_{ij}(s)$ denote the numerator of an element of the transfer function matrix, we get

$$[G(s)] = \begin{bmatrix} \dfrac{G_{11}(s)}{\Delta(s)} & \dfrac{G_{12}(s)}{\Delta(s)} & \cdots & \dfrac{G_{1n}(s)}{\Delta(s)} \\ \cdot & & & \\ \cdot & & & \\ \cdot & & & \\ \dfrac{G_{n1}(s)}{\Delta(s)} & \dfrac{G_{n2}(s)}{\Delta(s)} & \cdots & \dfrac{G_{nn}(s)}{\Delta(s)} \end{bmatrix} \qquad (8.140)$$

Then, from (8.102), we obtain

$$\hat{q}_i(s) = \frac{G_{i1}(s)}{\Delta(s)}\hat{Q}_1(s) + \cdots + \frac{G_{in}(s)}{\Delta(s)}\hat{Q}_n(s) \qquad (8.141)$$

For forced vibrations, the exciting forces are harmonic. Let $Q_1 = a_1 \sin(\omega_1 t + \alpha_1)$ and in general $Q_k = a_k \sin(\omega_k t + \alpha_k)$, where a_k is the amplitude, ω_k the frequency, and α_k the phase angle. For simplicity, we have assumed here that each exciting force is simple harmonic. But general periodic forces can be easily accommodated by Fourier series expansion and superposition, as in the case of a single-degree-of-freedom systems. When all the eigenvalues of the characteristic equation (8.139) have a negative real part, it follows from the frequency-domain techniques discussed for single-degree-of-freedom systems that for steady-state vibrations, we get

$$q_{i,ss} = u_{i1} \sin(\omega_1 t + \alpha_1 + \psi_1) + \cdots + u_{in} \sin(\omega_n t + \alpha_n + \psi_n) \qquad (8.142)$$

where

$$u_{i1} = a_1 \left| \frac{G_{i1}(j\omega)}{\Delta(j\omega)} \right|, \qquad \psi_1 = \measuredangle \frac{G_{i1}(j\omega)}{\Delta(j\omega)}$$

$$u_{in} = a_n \left| \frac{G_{in}(j\omega)}{\Delta(j\omega)} \right|, \qquad \psi_n = \measuredangle \frac{G_{in}(j\omega)}{\Delta(j\omega)}$$

It is important to verify that all eigenvalues [i.e., roots of the characteristic equation (8.139)] have a negative real part, as otherwise the equilibrium may be unstable and the results of (8.142) invalid. The determination of the eigenvalues is not required and the Routh criterion discussed in Chapter 9 can be used for this purpose. Routh criterion provides this information readily without actually determining the roots of a high-order polynomial. Hence, it is seen that the frequency-response method is computationally much simpler than the time-domain method employing modal decomposition. The computations involve only the determination of the magnitudes and phase angles of complex quantities.

The matrix $[G(j\omega)]$ is called the harmonic-response function matrix. Bode diagrams, which have been extensively discussed for single degree-of-freedom systems, can be employed for each element $G_{ik}(j\omega)/\Delta(j\omega)$ of this harmonic-response function matrix. In this way, the amplitude magnification or attenuation, filter characteristics, and system identification from experimental frequency response can be investigated.

Example 8.12

We illustrate the techniques by considering the example of a vibration absorber. An engine of mass m_1 is mounted on a frame with stiffness k_1 and damping coefficient c_1. In order to absorb the forced vibrations, a mass m_2 is attached to m_1 through a spring of stiffness k_2 and structural damping c_2 as shown in Fig. 8.31. A sinusoidal force $Q_1 = a_1 \sin \omega t$ is acting on mass m_1.

A free-body diagram for this two-degree-of-freedom system is shown in Fig. 8.32. The equations of motion are as follows:

$$m_1 \ddot{q}_1 + k_1 q_1 + c_1 \dot{q}_1 + k_2(q_1 - q_2) + c_2(\dot{q}_1 - \dot{q}_2) = Q_1$$

$$m_2 \ddot{q}_2 - c_2(\dot{q}_1 - \dot{q}_2) - k_2(q_1 - q_2) = 0$$

These equations may be expressed as

$$\begin{bmatrix} m_1 & 0 \\ 0 & m_2 \end{bmatrix} \begin{Bmatrix} \ddot{q}_1 \\ \ddot{q}_2 \end{Bmatrix} + \begin{bmatrix} c_1 + c_2 & -c_2 \\ -c_2 & c_2 \end{bmatrix} \begin{Bmatrix} \dot{q}_1 \\ \dot{q}_2 \end{Bmatrix} + \begin{bmatrix} k_1 + k_2 & -k_2 \\ -k_2 & k_2 \end{bmatrix} \begin{Bmatrix} q_1 \\ q_2 \end{Bmatrix} = \begin{Bmatrix} Q_1 \\ 0 \end{Bmatrix} \qquad (8.143)$$

The mass, damping, and stiffness matrices are obvious from the foregoing equation. In the Laplace domain, (8.143) becomes

$$\begin{bmatrix} m_1 s^2 + (c_1 + c_2)s + k_1 + k_2 & -(c_2 s + k_2) \\ -(c_2 s + k_2) & m_2 s^2 + c_2 s + k_2 \end{bmatrix} \begin{Bmatrix} \hat{q}_1 \\ \hat{q}_2 \end{Bmatrix} = \begin{Bmatrix} \hat{Q}_1 \\ 0 \end{Bmatrix}$$

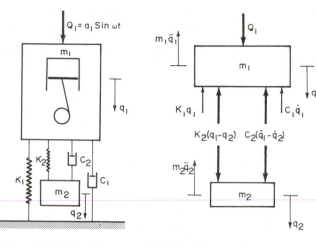

Figure 8.31 Vibration absorber.

Figure 8.32 Free-body diagram for system of Fig. 8.31.

Inverting the matrix in the foregoing equation, we obtain the 2×2 transfer function matrix as

$$\begin{Bmatrix} \hat{q}_1 \\ \hat{q}_2 \end{Bmatrix} = \begin{bmatrix} \dfrac{m_2 s^2 + c_2 s + k_2}{\Delta(s)} & \dfrac{c_2 s + k_2}{\Delta(s)} \\[2ex] \dfrac{c_2 s + k_2}{\Delta(s)} & \dfrac{m_1 s^2 + (c_1 + c_2)s + k_1 + k_2}{\Delta(s)} \end{bmatrix} \begin{Bmatrix} \hat{Q}_1 \\ 0 \end{Bmatrix} \qquad (8.144)$$

where the characteristic determinant Δ is given by

$$\Delta(s) = [m_1 s^2 + (c_1 + c_2)s + k_1 + k_2][m_2 s^2 + c_2 s + k_2] - (c_2 s + k_2)^2$$
$$= m_1 m_2 s^4 + (m_1 c_2 + m_2 c_1 + m_2 c_2)s^3 + (m_2 k_1 + m_2 k_2 + c_1 c_2 + m_1 k_2)s^2$$
$$+ (c_2 k_1 + c_1 k_2)s + k_1 k_2 \qquad (8.145)$$

The transfer function of (8.144) is shown in the block diagram of Fig. 8.33. It can be checked by the application of the Routh criterion, which is discussed in the next chapter, that all roots of the characteristic equation $\Delta(s) = 0$ have negative real parts. This implies that the equilibrium about which the vibrations occur is asymptotically stable. Hence, the harmonic response function matrix is obtained by substituting $j\omega$ for s in the transfer function matrix. Since, in this case, we have $Q_2 = 0$, it

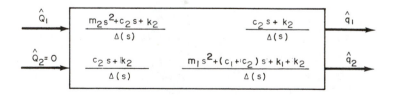

Figure 8.33 Block diagram.

follows from (8.144) that

$$\hat{q}_1 = \frac{m_2 s^2 + c_2 s + k_2}{\Delta(s)} \hat{Q}_1$$

$$\hat{q}_2 = \frac{c_2 s + k_2}{\Delta(s)} \hat{Q}_1$$

When $Q_1 = a_1 \sin \omega t$, for steady-state forced vibrations we obtain $q_1 = u_1 \sin(\omega t + \psi_1)$ and $q_2 = u_2 \sin(\omega t + \psi_2)$, where

$$u_1 = a_1 \left| \frac{-m_2\omega^2 + k_2 + jc_2\omega}{\Delta(j\omega)} \right|, \qquad \psi_1 = \measuredangle \frac{-m_2\omega^2 + k_2 + jc_2\omega}{\Delta(j\omega)}$$

$$u_2 = a_1 \left| \frac{k_2 + jc_2\omega}{\Delta(j\omega)} \right|, \qquad \psi_2 = \measuredangle \frac{k_2 + jc_2\omega}{\Delta(j\omega)} \qquad (8.146)$$

The vibration absorber is tuned such that $k_2/m_2 = \omega^2$. It then follows from (8.146) that in the ideal case where $c_2 = 0$, we get $u_1 = 0$; that is, the mass m_1 on which the exciting force is acting does not vibrate at all. It can be checked by the application of Routh's criterion that when $c_2 = 0$, all roots of the characteristic equation $\Delta(s) = 0$ of (8.145) still have a negative real part. Hence, we can set $c_2 = 0$ and still obtain valid results.

When $c_2 = 0$ and $k_2/m_2 = \omega^2$, it can be shown from (8.145) that $\Delta(j\omega) = -k_2^2$. Then from (8.146) it is seen that $u_2 = a_1/k_2$ and $\psi_2 = -180°$. Hence, for steady-state vibrations, we obtain $q_2 = a_1/k_2 \sin(\omega t - 180°) = -a_1/k_2 \sin \omega t$. The result that the mass m_1 which is acted upon by the exciting force does not vibrate at all can be easily explained. The spring force $k_2 q_2$ acting on mass m_1 is seen to be $-a_1 \sin \omega t$ (i.e., it is equal and opposite to the exciting force). Hence, the net force acting on mass m_1 is zero. Since the absorber is to be tuned to the forcing frequency, it is useful in cases where the exciting frequency is constant as in electrical motors and some machines. It is, in fact, employed in many applications, such as hair-cutting shears, in order to isolate the frame from vibrations. Fig. 8.34 shows a rotating machinery mounted on a beam. It is acted upon by a sinusoidal force due to unbalance. The vibration absorber consists of a double cantilever beam with a mass at each end. However, there are some applications such as internal combustion engines where the speed and hence the exciting frequency are variable and the vibration absorber would be out of tune.

The assumption that mass m_2 can be attached to mass m_1 with only a spring where the damping coefficient $c_2 = 0$ is only an idealization. In fact, the structural

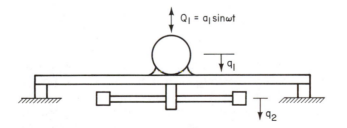

Figure 8.34 Tuned vibration absorber.

damping can be minimized but not completely eliminated. In this case, if the absorber is tuned such that $k_2/m_2 = \omega^2$, the amplitude u_1 of mass m_1 will be given by

$$u_1 = a_1 \frac{c_2 \omega}{|\Delta(j\omega)|}$$

Consider the element $(c_2 s + k_2)/\Delta(s)$ of the transfer function matrix which determines the forced vibrations of mass m_2. After dividing $\Delta(s)$ throughout by $k_1 k_2$ and denoting c_2/k_2 by the time constant τ, we obtain

$$G_{21}(s) = \frac{(1/k_1)(\tau s + 1)}{[(1/\omega_{n_1}^2)s^2 + (2\zeta_1/\omega_{n_1})s + 1][(1/\omega_{n_2}^2)s^2 + (2\zeta_2/\omega_{n_2})s + 1]} \qquad (8.147)$$

where ω_{n1} and ω_{n2} are the two natural frequencies and ζ_1 and ζ_2 are the two damping ratios. It may be desirable to plot the Bode diagram for (8.147) in order to determine attenuation and filter characteristics. Such a diagram is shown in Fig. 8.35 for the case $k_1 = 1$ and $1/\tau < \omega_{n1} < \omega_{n2}$.

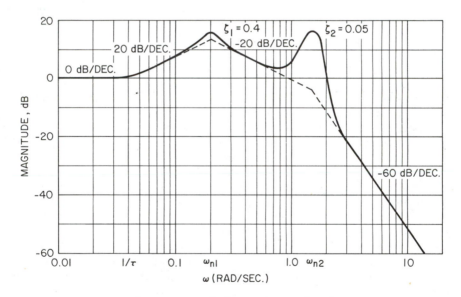

Figure 8.35   Bode diagram for Eq. (8.147).

8.9 SUMMARY

This chapter has dealt with vibrations of linear dynamic systems. When the nonlinearities are analytic functions of their arguments, it is possible to linearize the equations of motion for small displacements about an equilibrium. However, there are many nonlinear phenomena, such as frequency entrainment, jump phenomenon, synchronization, and limit cycle vibrations, that cannot be explained by the linear theory since they belong to nonlinear behavior.

In the first part of the chapter, we covered single-degree-of-freedom systems and the results were then generalized to multiple-degree-of-freedom systems. Even though free vibrations of conservative systems are not encountered in practice, it should be noted that in many lightly damped systems, a good approximation of the damped natural frequencies can be obtained by determining the undamped natural frequencies. Two methods have been employed for the analysis of forced vibrations of damped systems. The first method is a time-domain method and employs the state transition matrix and convolution integral. In order to use this method for multiple-degree-of-freedom systems, we have employed modal decomposition. The second method employs the frequency-response function or matrix and is a frequency-domain method.

Vibrations of flexible bodies or continuous systems that are described by partial differential equations have not been included here. The reader is referred to several references [1–4] dealing with this topic. A book that discusses many practical problems is reference [5]. Bode diagrams are used extensively in control engineering and several books in that area, such as reference [7], would be useful for further study of Bode diagrams.

It is clear from our discussion that to avoid vibrations or to attenuate them, the following measures may be undertaken:

1. If possible, eliminate the exciting force by balancing rotating components. If the exciting force is being transmitted from other equipment, vibration-isolation techniques can be used.

2. Attenuate the vibrations by proper choice of parameters such that the exciting force is filtered out.

3. Employ vibration absorbers. This method is sometimes called passive control of vibrations.

4. Use active control systems where the vibrations are sensed and a force is generated to oppose the exciting force.

This last method is not discussed here and the reader may consult any of several books in control engineering, such as reference [7], for this purpose.

PROBLEMS

8.1. Obtain frequency ω and period of oscillation T for the system shown in Fig. P8.1. The spring is linear and has a stiffness, $k = 5$ kN/cm. The pulley has a radius of 50 cm and its mass moment of inertia about O is 7000 N-cm·s². The mass m is 40 kg.

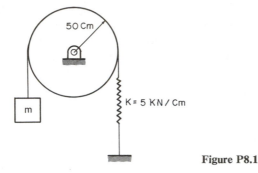

Figure P8.1

8.2. A vehicle traveling over a bridge is idealized by the system shown in Fig. P8.2. The bridge profile irregularities are represented by a sine function. Assuming that the vehicle travels with a uniform velocity $V =$ constant, calculate the response $q(t)$ and force transmitted to the vehicle. Also calculate the steady-state vertical motion of the vehicle using the following numerical data:

$$W = 20 \text{ kN}$$

$$k = 2500 \text{ N/cm}$$

$$y_0 = 3 \text{ cm}$$

$$L = 12 \text{ m}$$

$$V = 70 \text{ km/h}$$

$$\zeta = 40\% \text{ of critical damping}$$

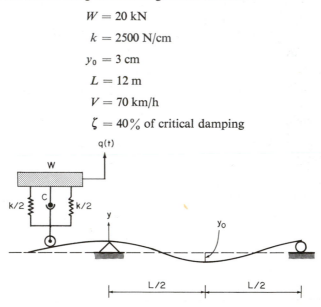

Figure P8.2

8.3. (a) Develop the equations of motion for the spring–mass system shown in Fig. P8.3.

(b) For $m_1 = m_3 = m$, $m_2 = 2m$, $k_1 = k_2 = 2k$, and $k_3 = k$:
  (1) Determine frequencies and mode shapes.
  (2) Obtain the generalized mass and stiffness matrices $[M^*]$ and $[K^*]$.
  (3) Establish the orthogonality conditions for the mode shapes.
  (4) Obtain the normal modal matrix $[\phi]$.
  (5) Develop uncoupled equations of motion.

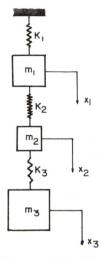

Figure P8.3

8.4. For the system shown in Fig. P8.4, find the position of the masses at time t when subjected to the forcing functions F_1 and F_3.

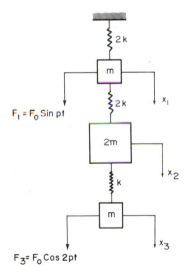

Figure P8.4 $F_3 = F_0 \cos 2pt$

8.5. For the vibrating system whose mass and stiffness matrices are

$$[M] = m \begin{bmatrix} 1 & 0 & 0 \\ 0 & 2 & 0 \\ 0 & 0 & 1 \end{bmatrix}$$

$$[K] = k \begin{bmatrix} 4 & -2 & 0 \\ -2 & 3 & -1 \\ 0 & -1 & 1 \end{bmatrix}$$

Compute the frequency and mode shape of the highest mode.

8.6. Obtain the equations of motion of the system shown in Fig. P8.6.

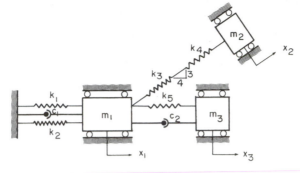

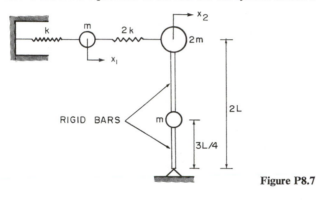

Figure P8.6

8.7. Obtain the equations of motion for the system shown in Fig. P8.7.

Figure P8.7

REFERENCES

1. Tse, F. S., Morse, I. E., and Hinkle, R. T., *Mechanical Vibrations: Theory and Applications*, 2nd ed., Allyn and Bacon, Inc., Boston, 1978.

2. Dimarogonas, A. D., *Vibration Engineering*, West Publishing Co., New York, 1976.

3. Meirovitch, L., *Analytical Methods in Vibrations*, Macmillan Publishing Co., Inc., New York, 1967.

4. Timoshenko, S., Young, D. H., and Weaver, W., *Vibration Problems in Engineering*, 4th ed., John Wiley & Sons, Inc., New York, 1974.

5. Den Hartog, J. P., *Mechanical Vibrations*, 4th ed., McGraw-Hill Book Company, New York, 1956.

6. Hurty, W. C., and Rubinstein, M. F., *Dynamics of Structures*, Prentice-Hall, Inc., Englewood Cliffs, N.J., 1965.

7. Raven, F. H., *Automatic Control Engineering*, 3rd ed., McGraw-Hill Book Company, New York, 1978.

8. Bauer, H. F., *Mechanical Analogy of Fluid Oscillations in Cylindrical Tanks with Circular and Annular Cross-sections*, Report MTP-AERO-61-4, Jan. 1961.

STABILITY OF MOTION

9.1 INTRODUCTION

When the equations of motion of a system are nonlinear, the principle of super-position is not applicable. Hence, if a motion is stable for a given set of initial conditions and input, it cannot be implied that the motion will remain stable for other sets of initial conditions and inputs. Hence, great care is required in the stability analysis of nonlinear equations of motion and in this connection it is useful to employ an appropriate stability theory.

There exist several concepts of stability such as stability in the sense of Lagrange, Poincaré, Lyapunov, boundedness of response, and input–output stability. For nonlinear systems, these different concepts of stability may not be identical. For a particular application, one concept may be unduly restrictive, whereas another may have no physical significance. The choice of one of these concepts of stability depends on its physical significance in a particular application. This chapter is concerned mainly with the stability analysis in the sense of Lyapunov.

To investigate the stability of a particular motion, it is first perturbed and the perturbation equations are analyzed further in order to examine whether the perturbations grow or decay with time. The definitions of stability are stated in the next section. When the perturbations are sufficiently small and the non-linearities are analytic functions of their arguments, it is possible to linearize the perturbation equations in the first approximation. These topics are discussed in the earlier part of the chapter. For those motions that are stable for small perturbations, it is of interest to determine the domain of stability or the size of perturbations for stable behavior. This topic is called stability in the large and is discussed in the latter part of the chapter. Here, the success depends on the selection of a suitable function called a Lyapunov function. The choice of a suitable Lyapunov function is not always obvious and there is no general procedure for its generation.

9.2 PERTURBATION EQUATIONS AND DEFINITIONS OF STABILITY

The equations of motions have been formulated in Chapters 3, 4, and 5. It has also been shown that by a suitable choice of state variables, the equations of motion can be expressed as a set of first-order coupled equations in the form

$$\{\dot{x}\} = \{f(x_1, \ldots, x_n, Q_i, \ldots, Q_m, t)\} \tag{9.1}$$

where $\{x\}$ is an n-dimensional column matrix of state variables and Q_i $(i = 1, \ldots, m)$ are input forces and moments. If a system has k degrees of freedom and the state variables include all the generalized coordinates and generalized velocities or momenta, the dimension n of the state variables is given by $n = 2k$. However, some of the coordinates may be ignorable and in that case, $n < 2k$. For example, let a rigid body have only three degrees of rotational freedom. Then $k = 3$ and the Euler's equations of motion expressed in the form of state variables are given by (4.52). When the applied moments M_1, M_2, and M_3 are not functions of the angular displacements, the angular displacements are ignorable coordinates and from (4.52) the dimension of $\{x\}$ is given by $n = k = 3$. In general, it can be stated that $n \leq 2k$.

For given inputs Q_i^* and initial conditions $\{x_0\}$ at time t_0, the solution of (9.1) yields the nominal motion $\{x^*\}$ assuming that (9.1) satisfies the existence and uniqueness conditions of Theorem 6.1. This nominal motion $\{x^*\}$ in the n-dimensional state space may be stable or unstable. When it is unstable, the motion is not realizable in practice. In order to study the stability of the nominal motion $\{x^*\}$, we consider the effect of perturbations $\{\Delta x_0\}$ on the initial state. The inputs Q_i^* are not perturbed and this restriction is necessary since the concept of stability in the sense of Lyapunov does not admit perturbation in the inputs. The nominal motion is perturbed only in the initial conditions, which may be caused by impulsive changes in the inputs or disturbances at the initial time. Some other concept of stability, such as input–output stability, may also require perturbations in the inputs.

Consider the effect of perturbations $\{\Delta x_0\}$ in the initial conditions and let $\{x\}$ be the resultant perturbed motion. The n-dimensional column matrix $\{\Delta x\}$ of perturbed state variables is defined by

$$\{\Delta x\} = \{x\} - \{x^*\}; \quad \text{that is,} \quad \{x\} = \{x^*\} + \{\Delta x\} \tag{9.2}$$

Now, substituting for $\{x\}$ from (9.2) in (9.1), we obtain

$$\{\dot{x}^*\} + \{\Delta\dot{x}\} = \{f(x_1^* + \Delta x_1, \ldots, x_n^* + \Delta x_n, Q_1^*, \ldots, Q_m^*, t)\} \qquad (9.3)$$

and since the nominal motion satisfies the equation

$$\{\dot{x}^*\} = \{f(x_1^*, \ldots, x_n^*, Q_1^*, \ldots, Q_m^*, t)\} \qquad (9.4)$$

it follows that the differential equations in the perturbations are described by

$$\{\Delta\dot{x}\} = \{f(x_1^* + \Delta x_1, \ldots, x_n^* + \Delta x_n, Q_1^*, \ldots, Q_m^*, t)\}$$
$$- \{f(x_1^*, \ldots, x_n^*, Q_1^*, \ldots, Q_m^*, t)\} \qquad (9.5)$$

In case the functions $\{f\}$ are continuously differentiable with respect to $\{x\}$, the right-hand side of (9.3) may be expanded in a Taylor series about the nominal motion $\{x^*\}$ as was done in Section 6.4. This procedure yields

$$\{\dot{x}^*\} + \{\Delta\dot{x}\} = \{f(x_1^*, \ldots, Q_1^*, \ldots, t)\} + \frac{\partial \mathbf{f}}{\partial \mathbf{x}}\bigg|_{\substack{\mathbf{x}=\mathbf{x}^* \\ \mathbf{Q}=\mathbf{Q}^*}} \Delta\mathbf{x} + \{h(\Delta x_1, \ldots, \Delta x_n, t)\} \qquad (9.6)$$

where the functions $\{h\}$ contain all the remaining terms of the Taylor series expansion. We let the Jacobian matrix in (9.6) be denoted by $\mathbf{A}(t)$ as defined by (6.29). Since the nominal motion satisfies (9.4) it follows from (9.6) that

$$\{\Delta\dot{x}\} = \mathbf{A}(t)\{\Delta x\} + \{h(\Delta x_1, \ldots, \Delta x_n, t)\} \qquad (9.7)$$

The original problem of determining the stability of the nominal motion $\{x^*\}$ is now equivalent to the problem of determining the stability of the null [i.e., trivial] solution of (9.7). If the perturbation described by (9.7) with initial conditions $\{\Delta x_0\}$ decay to zero with time, we say that the nominal motion $\{x^*\}$ is asymptotically stable. However, formal definitions of stability are given later.

We now consider a special case where the parameters in the equations of motion (9.1) are time invariant so that the functions f_i are not explicit functions of time. In addition, the input forces and moments are zero or constants, so that (9.1) reduces to

$$\{\dot{x}\} = \{f(x_1, \ldots, x_n)\} \qquad (9.8)$$

When $\{\dot{x}\} = \{0\}$, the nominal motion $\{x^*\}$ represents a stationary motion or an equilibrium $\{x_e\}$ which can be determined from the solution of the nonlinear algebraic equations

$$\{f(x_1, \ldots, x_n)\} = \{0\} \qquad (9.9)$$

After employing Taylor series expansion about the stationary motion or equilibrium $\{x_e\}$, the perturbation equation (9.7) becomes

$$\{\Delta\dot{x}\} = \mathbf{A}\{\Delta x\} + \{h(\Delta x_1, \ldots, \Delta x_n)\} \qquad (9.10)$$

where $\mathbf{A}$ is a $n \times n$ constant matrix and h_i are not explicit functions of time. As mentioned in Chapter 6, the perturbations equation (9.10) is called autonomous, whereas (9.7) is called nonautonomous. The stationary motion or equilibrium is denoted by the symbol $\{x_e\}$ to indicate that it is an equilibrium point in the state space rather than a time-varying trajectory $\{x^*(t)\}$. The transformation $\{\Delta x\} = \{x\} - \{x_e\}$ is now merely a transformation of coordinates such that the origin of the state space of perturbations $\{\Delta x\}$ is equivalent to the equilibrium point $\{x_e\}$ of the state space of $\{x\}$ as shown in Fig. 9.1. It should be noted that in the state space of perturbation variables $\{\Delta x\}$, the origin $\{\Delta x\} = \{0\}$ is an equilibrium.

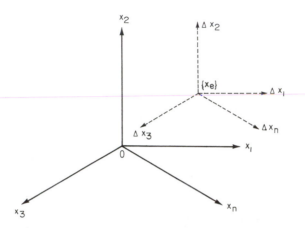

Figure 9.1   Stationary motion or equilibrium point in state space.

The definitions of stability are given next. For the simplicity of notation, we let $\{\Delta x\} = \{y\}$ and denote it by the symbol $\mathbf{y}$. The norm of $\mathbf{y}$ in Euclidean space is denoted by

$$\|\mathbf{y}\| = [y_1^2 + y_2^2 + \cdots + y_n^2]^{1/2} \tag{9.11}$$

Definition 9.1.   The nominal motion $\{x^*\}$ is stable in the sense of Lyapunov if for every $\epsilon > 0$, there exists a $\delta > 0$ where δ depends on ϵ and possibly on t_0 such that $\|\mathbf{y}(t_0)\| \leq \delta$ implies that $\|\mathbf{y}(t)\| < \epsilon$ for all $t > t_0$.

Definition 9.2.   The nominal motion $\{x^*\}$ is asymptotically stable if (a) it is stable, and (b) $\lim_{t \to \infty} \|\mathbf{y}(t)\| = 0$.

Definition 9.3.   The nominal motion $\{x^*\}$ is unstable if there is an ϵ such that no δ can be found to satisfy the condition of Definition 9.1.

These definitions of stability are in the sense of Lyapunov. In order to demonstrate that the nominal motion is stable, it is required that for *every* ϵ that is given, a δ must be found such that if the perturbation is initially in the δ neighborhood of the motion, the perturbation will never leave the ϵ neighborhood. Definitions 1 to 3 are illustrated in Figs. 9.2(a)–(c), respectively, for the autonomous case but only for a two-dimensional state space. For the n-dimensional state space, the circles become hyperspheres of radius δ and ϵ, respectively.

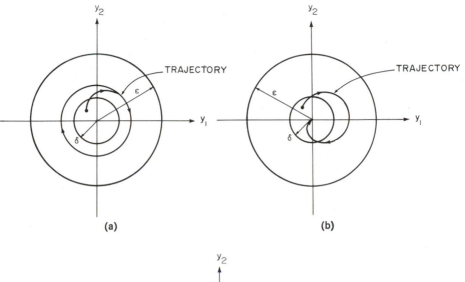

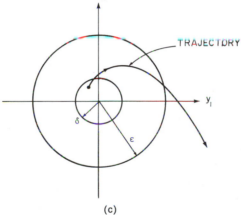

Figure 9.2   (a) Stable; (b) asymptotically stable; (c) unstable.

Example 9.1

We consider a mass, linear damping, and nonlinear spring of Example 6.7 described by

$$m\ddot{x} + c\dot{x} + k\left(x - \frac{x^3}{6}\right) = 0 \tag{9.12}$$

Choosing the state variables as $x_1 = x$ and $x_2 = \dot{x}$, the state equation representation becomes

$$\dot{x}_1 = x_2$$

$$\dot{x}_2 = -\frac{k}{m}\left(x_1 - \frac{x_1^3}{6}\right) - \frac{c}{m}x_2$$

In Example 6.7 it was shown that this system has three isolated equilibrium states given by

$$\{x_e\} = \begin{Bmatrix} 0 \\ 0 \end{Bmatrix}, \qquad \begin{Bmatrix} \sqrt{6} \\ 0 \end{Bmatrix}, \qquad \begin{Bmatrix} -\sqrt{6} \\ 0 \end{Bmatrix}$$

We first consider the equilibrium $\begin{Bmatrix} 0 \\ 0 \end{Bmatrix}$ and let $\Delta x_1 = y_1$ and $\Delta x_2 = y_2$ be the perturbations in the state variables about this equilibrium. The Jacobian matrix for this equilibrium was obtained in (6.61) and (6.62). The perturbation equations are

$$\begin{Bmatrix} \dot{y}_1 \\ \dot{y}_2 \end{Bmatrix} = \begin{bmatrix} 0 & 1 \\ -\dfrac{k}{m} & -\dfrac{c}{m} \end{bmatrix} \begin{Bmatrix} y_1 \\ y_2 \end{Bmatrix} + \dfrac{k}{m} \begin{Bmatrix} 0 \\ \dfrac{y_1^3}{6} \end{Bmatrix} \tag{9.13}$$

These equations are autonomous and the matrix $\mathbf{A}$ and function $\{h\}$ of (9.7) for this example are obvious. A typical trajectory in a sufficiently small neighborhood of the origin is shown in Fig. 9.3(a). For any given ϵ, a δ that depends on ϵ can be easily found to satisfy Definition 9.1. Furthermore, Definition 9.2 can also be satisfied and we conclude that the equilibrium $\{x_e\} = \begin{Bmatrix} 0 \\ 0 \end{Bmatrix}$ is asymptotically stable (see Example 9.15).

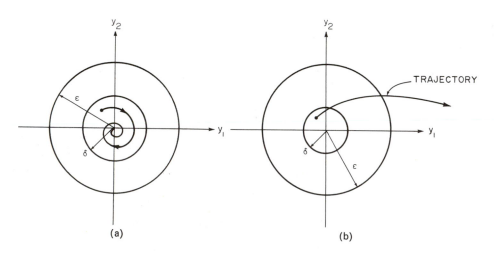

(a)   (b)

Figure 9.3   (a) Typical trajectory in the neighborhood of equilibrium $\begin{Bmatrix} 0 \\ 0 \end{Bmatrix}$; (b) unstable equilibrium $\begin{Bmatrix} \sqrt{6} \\ 0 \end{Bmatrix}$.

Next, we consider the equilibrium state $\{x_e\} = \begin{Bmatrix} \sqrt{6} \\ 0 \end{Bmatrix}$. The Jacobian matrix for this equilibrium was obtained in (6.63). The perturbation equations about this equilibrium are given by

$$\begin{Bmatrix} \dot{y}_1 \\ \dot{y}_2 \end{Bmatrix} = \begin{bmatrix} 0 & 1 \\ \dfrac{2k}{m} & -\dfrac{c}{m} \end{bmatrix} \begin{Bmatrix} y_1 \\ y_2 \end{Bmatrix} + \begin{Bmatrix} 0 \\ \dfrac{k}{m}\dfrac{3}{\sqrt{6}}\, y_1^2 + \dfrac{k}{m}\dfrac{1}{6}\, y_1^3 \end{Bmatrix} \tag{9.14}$$

A typical trajectory in a sufficiently small neighborhood of the origin is shown in Fig. 9.3(b). It is obvious that for any given ϵ, no δ can be found to satisfy the conditions of Definition 9.1. Hence, the equilibrium state $\begin{Bmatrix} \sqrt{6} \\ 0 \end{Bmatrix}$ is unstable. Similarly, it can be shown that the equilibrium state of $\begin{Bmatrix} -\sqrt{6} \\ 0 \end{Bmatrix}$ is also unstable (see Example 9.15).

Example 9.2

Consider the Van der Pol equation

$$\frac{d^2 x}{dt^2} + \mu(x^2 - 1)\frac{dx}{dt} + x = 0 \tag{9.15}$$

Choosing the state variables as $x_1 = x$ and $x_2 = \dot{x}$, the state equation is represented by

$$\begin{aligned} \dot{x}_1 &= x_2 \\ \dot{x}_2 &= -x_1 - \mu(x_1^2 - 1)x_2 \end{aligned} \tag{9.16}$$

It can be easily verified that (9.16) has only one equilibrium state ($x_1 = 0$, $x_2 = 0$). Letting $\Delta x_1 = y_1$ and $\Delta x_2 = y_2$, the perturbation equations about this equilibrium become

$$\begin{Bmatrix} \dot{y}_1 \\ \dot{y}_2 \end{Bmatrix} = \begin{bmatrix} 0 & 1 \\ -1 & \mu \end{bmatrix} \begin{Bmatrix} y_1 \\ y_2 \end{Bmatrix} - \begin{Bmatrix} 0 \\ \mu y_1^2 y_2 \end{Bmatrix} \tag{9.17}$$

Examination of (9.15) reveals that if $|x| < 1$, the damping is negative, whereas if $|x| > 1$, the damping becomes positive. This equation exhibits limit cycle (i.e., self-excited) oscillations which are represented by a closed trajectory in state space enclosing the origin as shown in Fig. 9.4(a). For a given ϵ_1 in Fig. 9.4(b), a δ can be found such that if the initial perturbation is inside the circle of radius δ, it does not leave the circle of radius ϵ_1. In this case, any $\delta < \epsilon_1$ will suffice. However, for given ϵ_2, no such

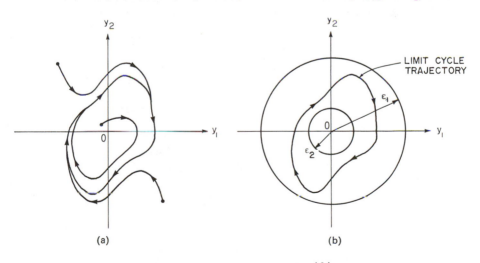

(a)   (b)

Figure 9.4   Unstable equilibrium $\begin{Bmatrix} 0 \\ 0 \end{Bmatrix}$.

δ can be found and we conclude that the equilibrium state $(0, 0)$ of (9.15) is unstable. It becomes clear that if there exists one value of ϵ for which no δ can be found, the equilibrium is unstable.

Example 9.3

We consider a mass and nonlinear hard spring without damping. This system is described by

$$m\ddot{x} + k_1 x + k_2 x^3 = 0$$

Again choosing the state variables as $x_1 = x$ and $x_2 = \dot{x}$, the state equations are described by

$$\dot{x}_1 = x_2$$
$$\dot{x}_2 = -\frac{k_1}{m}x_1 - \frac{k_2}{m}x_1^3 \tag{9.18}$$

There is only one equilibrium state given by $x_1 = 0$, $x_2 = 0$. Letting $\Delta x_1 = y_1$ and $\Delta x_2 = y_2$, the perturbation equations about this equilibrium become

$$\dot{y}_1 = y_2$$
$$\dot{y}_2 = -\frac{k_1}{m}y_1 - \frac{k_2}{m}y_1^3 \tag{9.19}$$

This system is conservative and (9.19) represents a nonlinear oscillation, which is a closed trajectory in the state space around the origin, and depends on the initial perturbation $\{y(t_0)\}$ since the initial energy is conserved. Hence, for any given ϵ a value of δ can be found depending on ϵ such that if the initial perturbation lies inside the circle of radius δ, the closed trajectory is enclosed inside the circle of radius ϵ. This is illustrated in Fig. 9.5.

From Example 9.2, it is seen that if there exists a self-excited oscillation around an equilibrium state, that equilibrium is unstable. In a conservative system, the equilibrium is stable in the sense of Lyapunov, but it is not asymptotically stable. The

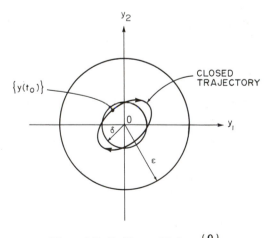

Figure 9.5   Stable equilibrium $\begin{Bmatrix} 0 \\ 0 \end{Bmatrix}$.

difference between self-excited and conservative oscillations is the following. A self-excited oscillation is independent of the initial perturbation. The initial energy is not conserved and the energy drawn from a nonoscillating source is just balanced by the energy dissipated per cycle. The frequency and amplitude of self-excited oscillations depend on this energy balance. On the other hand, in a conservative system the initial energy is conserved and the amplitude and frequency of the oscillation depend on the initial perturbation.

Definition 9.4.   The nominal motion $\{x^*\}$ is quasi-asymptotically stable if property (b) of Definition 9.2 is satisfied but not property (a).

We have observed from Definition 9.2 that a nominal motion must first be stable before it can qualify to be asymptotically stable. This requirement is to prevent a perturbed motion from straying far from the nominal motion before converging toward it. However, there are some pathological cases that satisfy Definition 9.4, as illustrated by the following example.

Example 9.4

Let the perturbation equations about an equilibrium or nominal motion be described by

$$\dot{y}_1 = 2y_1 y_2$$
$$\dot{y}_2 = y_2^2 - y_1^2 \tag{9.20}$$

The only equilibrium of (9.20) is the null or trivial solution ($y_1 = 0$, $y_2 = 0$). A nontrivial solution of (9.20) is a one-parameter family of circles described by

$$(y_1 - c)^2 + y_2^2 = c^2 \tag{9.21}$$

passing through the origin with radius c and center at $(c, 0)$, as shown in Fig. 9.6.

Starting at any initial condition $\{y(t_0)\}$, the circle through that point ultimately terminates at the origin and hence condition (b) of Definition 9.2 is satisfied. But condition (a) is not met because for a given ϵ, no δ can be found to satisfy the requirement of Definition 9.1.

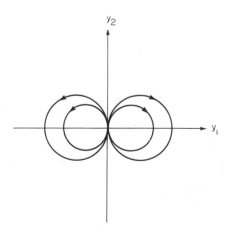

Figure 9.6 Quasi-asymptotically stable equilibrium.

In the examples we have considered so far, the nominal motion is either an equilibrium or stationary motion and the perturbation equations in $\{y\}$ are autonomous. In such cases, we find that δ of Definition 9.1 depends only on ϵ and not on the initial time t_0. For a time-varying nominal motion $\{x^*(t)\}$, the perturbation equations in $\{y\}$ are nonautonomous and δ may also depend on t_0. The following two definitions refine the notions of stability stated by Definitions 9.1 and 9.2 to nonautonomous cases.

Definition 9.5. The nominal motion $\{x^*\}$ is uniformly stable if it is stable in such a way that δ of Definition 9.1 does not depend on t_0.

Definition 9.6. The nominal motion $\{x^*\}$ is uniformly asymptotically stable if (a) it is uniformly stable and (b) perturbations with $\|\mathbf{y}(t_0)\| < \delta$ are such that $\lim_{t\to\infty} \|\mathbf{y}(t)\| \longrightarrow 0$ uniformly in both $\{y(t_0)\}$ and t_0.

When δ does not depend on t_0 but only on ϵ, the Lyapunov stability is independent of the initial time t_0 at which the perturbation occurs and we say that the nominal motion is uniformly stable. The expression $\lim_{t\to\infty} \|\mathbf{y}(t)\| \longrightarrow 0$ uniformly in $\{y(t_0)\}$ means that the convergence is only a function of the magnitude or norm of $\{y(t_0)\}$ and is not a function of the sense or direction of $\{y(t_0)\}$ from the origin. In nonlinear nonautonomous equations, the rate of convergence may depend on the direction of the perturbation from the origin, and in particular there may exist directions or hyperplanes along which the rate of convergence becomes very slow. Hence, the expression $\lim_{t\to\infty} \|\mathbf{y}(t)\| \longrightarrow 0$ uniformly in both $\{y(t_0)\}$ and t_0 means that the convergence is independent of both the direction of the initial perturbation and initial time at which it occurs. It should be noted that when an autonomous system is asymptotically stable, it implies that it is uniformly asymptotically stable.

Example 9.5

To illustrate the distinction between stability and uniform stability, we consider an example given by Hsu and Meyer [1] and also by Vidyasagar [2]. Let a scalar perturbation equation be described by

$$\dot{y} = (6t \sin t - 2t)y$$

The trivial solution $y = 0$ is an equilibrium of this equation and by separation of variables, the solution is obtained as

$$y(t) = y(t_0) \exp [6 \sin t - 6t \cos t - t^2 - 6 \sin t_0 + 6t_0 \cos t_0 + t_0^2]$$

We show that the trivial solution is stable but not uniformly stable. For $t_0 \geq 0$ and $t - t_0 > 6$, the ratio $|y(t)/y(t_0)|$ is bounded by $\exp [12 + T(6 - T)]$, where $T = t - t_0$. Defining

$$c(t_0) = \sup_{t \geq t_0} \exp [6 \sin t - 6t \cos t - t^2 - 6 \sin t_0 + 6t_0 \cos t_0 + t_0^2]$$

we know that $c(t_0)$ is a finite number for any fixed t_0. Thus, given any $\epsilon > 0$, we can choose $\delta(\epsilon, t_0) = \epsilon/c(t_0)$ to satisfy Definition 9.1, showing that the trivial solution $y = 0$ is stable for all $t_0 > 0$. On the other hand, if we choose $t_0 = 2n\pi$, the solution yields

$$y[(2n + 1)\pi] = y(2n\pi) \exp [(4n + 1)(6 - \pi)\pi]$$

This shows that

$$c(2n\pi) > \exp [(4n + 1)(6 - \pi)\pi]$$

Hence $c(t_0)$ is unbounded as a function of t_0. Thus, given $\epsilon > 0$, it is not possible to choose a single $\delta(\epsilon)$, independent of the initial time, to satisfy Definition 9.5. Therefore, the trivial solution $y = 0$ is not uniformly stable.

The foregoing definitions pertain to stability in the sense of Lyapunov. However, there are some applications where the concept of stability in the sense of Lyapunov is not appropriate. Specially, this concept is too stringent when applied to closed trajectories, as illustrated by the following two examples.

Example 9.6

Consider the Van der Pol equation (9.15) of Example 9.2. It was seen that the equilibrium state $(x_1 = 0, x_2 = 0)$ is unstable in the sense of Lyapunov and that there exists a self-excited oscillation around the origin. Let $\{x^*(t)\}$ denote this particular closed trajectory whose stability is to be investigated, as shown in Fig. 9.7(a). Let

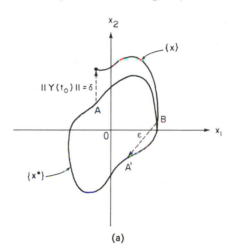

(a)

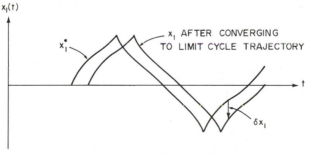

(b)

Figure 9.7   Lyapunov stable limit cycle.

$\{y(t_0)\}$ be the initial perturbation and $\{x(t)\}$ the perturbed trajectory. As time evolves, the perturbed trajectory $\{x\}$ converges to the same limit cycle trajectory $\{x^*\}$ at point B. However, the reference point A has now moved to A' and B and A' do not coincide at the same instant of time. For a given ϵ, one can find δ to satisfy Definition 9.1. But from Fig. 9.7(b) it is seen that after the perturbed trajectory converges to the limit cycle frequency, $y_1 = \delta x_1$ does not go to zero. Hence, any self-excited oscillation cannot be asymptotically stable in the sense of Lyapunov.

Example 9.7

The orbit of one body around another as governed by Newton's law of gravitation has been discussed in Section 3.8. From (3.71), the equation of motion in polar coordinates is given by

$$\ddot{r} - r\dot{\theta}^2 = -\frac{G(m_1 + m_2)}{r^2} \tag{9.22}$$

with $r^2\dot{\theta} = $ constant. We consider a circular orbit of radius r_0. For a circular orbit, $\ddot{r} = \dot{r} = 0, r = r_0$, a constant, and $\dot{\theta} = \omega$, a constant. Hence, from (9.22) we obtain

$$\omega = \left[\frac{G(m_1 + m_2)}{r^3}\right]^{1/2} \tag{9.23}$$

Now let the body's position and velocity be perturbed so that the resultant orbit is another circular orbit of radius $r_0 + \delta r$. We see that for the unperturbed satellite, the angular velocity ω is proportioned to $r_0^{-3/2}$, whereas for the perturbed body, it is proportioned to $(r_0 + \delta r)^{-3/2}$. The two orbits will therefore be traversed at different periods. As time evolves, the distance between the perturbed body and the reference body will increase to about $2r_0$, however small the value of δr may be. Hence, the body's orbit is unstable in the sense of Lyapunov, even though it is well behaved.

For a closed trajectory, a more appropriate definition of stability is orbital stability, which is also called stability in the sense of Poincaré. It is concerned with stability relative to the closed trajectory itself and is not concerned with any reference point traveling along the trajectory. Let $\rho(\mathbf{x}, C)$ be the minimum Euclidean distance from a point $\mathbf{x}$ to a closed curve C.

Definition 9.7. A closed trajectory C of a system $\{\dot{x}\} = \{f(x_1, \ldots, x_n, t)\}$ is orbitally stable if for every $\epsilon > 0$ there is a $\delta > 0$ where δ depended on ϵ and possibly on t_0 such that every solution of the system $\{x(t)\}$ with $\rho(\mathbf{x}(t_0), C) < \delta$ satisfies $\rho(\mathbf{x}(t), C) < \epsilon$ for all $t > t_0$.

Definition 9.8. A closed trajectory C of a system $\{\dot{x}\} = \{f(x_1, \ldots, x_n, t)\}$ is orbitally asymptotically stable if it is (a) orbitally stable, and (b) for all trajectories that are sufficiently close to C, $\rho(\mathbf{x}(t), C) \rightarrow 0$ as $t \rightarrow \infty$.

It is now seen that the limit cycle oscillation of Example 9.5 is orbitally asymptotically stable and the circular orbit of Example 9.6 is orbitally stable but not orbitally asymptotically stable.

9.3 STABILITY OF AUTONOMOUS MOTION FOR SMALL PERTURBATIONS

In this section it is assumed that the equations of motion are autonomous as given by (9.8) and the nominal motion $\{x^*\}$ whose stability is to be investigated is either an equilibrium or a stationary motion which we represent by $\{x_e\}$. The perturbation equations (9.5) about $\{x_e\}$ may be represented as in (9.10). Again for simplicity of notation, we let $\{\Delta x\} = \{y\}$ and represent (9.10) as

$$\{\dot{y}\} = \mathbf{A}\{y\} + \{h(y_1, \ldots, y_n)\} \tag{9.24}$$

It should be noted again that for the perturbation equations to be autonomous as in (9.24), it is necessary that the original equations of motion be autonomous and the nominal motion whose stability is to be investigated be a constant and not function of time. When the nonlinear functions $\{f\}$ in (9.8) are analytic functions of their arguments so that Taylor series expansion (9.10) is possible, we note that $\{h\}$ in (9.24) consists of terms which are of order higher than the first. Then in (9.24), the nonlinear terms $\{h\}$ are dropped by assuming that the perturbations are small and the stability of the linear approximation $\{\dot{y}\} = \mathbf{A}\{y\}$ is investigated.

This approach yields stability information in the small, that is, when the perturbations are sufficiently small but there is no indication of the magnitude of perturbations that could be considered as small. Hence, when asymptotic stability exists, the size of the region of asymptotic stability is not known. The determination of the size of this region will be studied in Section 9.5 by choosing a Lyapunov function. The theorem employed for stability investigation for small perturbations is stated in the following.

Theorem 9.1.   Consider the autonomous perturbations equation (9.24) about an equilibrium or stationary motion $\{x_e\}$. If $\lim_{\|y\|\to 0} \|\mathbf{h}(\mathbf{y})\|/\|\mathbf{y}\| = 0$, then:

1. If the linearized system $\{\dot{y}\} = \mathbf{A}\{y\}$ has only eigenvalues with negative real parts, $\{x_e\}$ is asymptotically stable in the small.
2. If the linearized system $\{\dot{y}\} = \mathbf{A}\{y\}$ has one or more eigenvalues with positive real parts, $\{x_e\}$ is unstable in the small.
3. If the linearized system $\{\dot{y}\} = \mathbf{A}\{y\}$ has one or more eigenvalues with zero real parts and the remaining eigenvalues have negative real parts, the stability of $\{x_e\}$ cannot be ascertained in the small by studying the linearized system alone.

This theorem is sometimes called the "principle of stability in the first approximation." It is also sometimes called "Lyapunov's first method." The proof of this theorem is based on Lyapunov's second or direct method and is given in Section 9.5. It is noted that when the functions $\{h\}$ in (9.24) consist of terms which are of an order higher than the first, the limit of $\|\mathbf{h}\|$ required by this theorem is always satisfied.

The determination of the eigenvalues of matrix $\mathbf{A}$ can be avoided by application of the Routh criterion. The application of this criterion is as follows. The characteristic equation of matrix $\mathbf{A}$ is first obtained as

$$|\lambda\mathbf{I} - \mathbf{A}| = a_n\lambda^n + a_{n-1}\lambda^{n-1} + \cdots + a_1\lambda + a_0 = 0 \qquad (9.25)$$

Next, the coefficients of the characteristic equation are arranged in the following Routh array:

a_n	a_{n-2}	a_{n-4}	a_{n-6}	$\cdots$	0	1st row
a_{n-1}	a_{n-3}	a_{n-5}	a_{n-7}	$\cdots$	0	2nd row
b_1	b_2	b_3	b_4	$\cdots$	0	
c_1	c_2	c_3	$\vdots$			
$\vdots$						
d_1	d_2	0				
e_1	e_2	0				
f_1	0					
g_1	0					$(n+1)$th row

After arranging the first two rows of this array from (9.25), the remaining rows are obtained from the previous two rows. The row of b terms is obtained as follows:

$$b_1 = \frac{a_{n-1}a_{n-2} - a_n a_{n-3}}{a_{n-1}}$$

$$b_2 = \frac{a_{n-1}a_{n-4} - a_n a_{n-5}}{a_{n-1}} \qquad (9.26)$$

$$b_3 = \frac{a_{n-1}a_{n-6} - a_n a_{n-7}}{a_{n-1}}$$

By dropping down a row, the same pattern is used to obtain the c terms as

$$c_1 = \frac{b_1 a_{n-3} - a_{n-1}b_2}{b_1}$$

$$\qquad (9.27)$$

$$c_2 = \frac{b_1 a_{n-5} - a_{n-1}b_3}{b_1}$$

This process is continued until it is terminated at the $(n+1)$ row, where n is the order of the system. Routh's criterion states that a necessary and sufficient condition for all eigenvalues of $\mathbf{A}$ to have negative real parts is that all coefficients in the first column of the array have the same sign. Furthermore, the number of changes of sign of the coefficient in the first column of the array is equal to the number of eigenvalues of **A** with positive real parts. The appearance of one or more zeros in the first column may signify that one or more eigenvalues of **A** have zero real parts and requires a further check. A proof of this criterion is given by Routh [3] and is omitted here.

Example 9.8

The tumbling motion of an orbiting rigid body satellite about its center of mass, where the tumbling rate far exceeds the orbiting rate, is described by the Euler equations (4.47) with $M_1 = M_2 = M_3 = 0$. These equations may be represented in state-variable form as

$$\dot{\omega}_1 = \frac{1}{I_1}(I_2 - I_3)\omega_2\omega_3$$

$$\dot{\omega}_2 = \frac{1}{I_2}(I_3 - I_1)\omega_3\omega_1 \tag{9.28}$$

$$\dot{\omega}_3 = \frac{1}{I_3}(I_1 - I_2)\omega_1\omega_2$$

In addition to the equilibrium state $(0, 0, 0)$, the three possible steady motions are the following:

$$(1) \quad \omega_1 = C_1, \quad \omega_2 = 0, \quad \omega_3 = 0$$

$$(2) \quad \omega_2 = C_2, \quad \omega_1 = 0, \quad \omega_3 = 0$$

$$(3) \quad \omega_3 = C_3, \quad \omega_1 = 0, \quad \omega_2 = 0$$

We wish to investigate the stability of these three stationary motions by application of Theorem 9.1. We first consider the stationary motion $(C_1, 0, 0)$ and introduce the perturbed motion

$$\omega_1 = C_1 + y_1, \qquad \omega_2 = 0 + y_2, \qquad \omega_3 = 0 + y_3 \tag{9.29}$$

where $\{y\}$ is the perturbation about the nominal motion. From (9.28) and (9.29), the differential equations for the perturbations are given by

$$\dot{y}_1 = \frac{1}{I_1}(I_2 - I_3)y_2y_3$$

$$\dot{y}_2 = \frac{1}{I_2}(I_3 - I_1)C_1y_3 + \frac{1}{I_2}(I_3 - I_1)y_1y_3 \tag{9.30}$$

$$\dot{y}_3 = \frac{1}{I_3}(I_1 - I_2)C_1y_2 + \frac{1}{I_3}(I_1 - I_2)y_1y_2$$

Considering small perturbations, the higher-order nonlinear terms in (9.30) are ignored to obtain the linearized equations

$$\dot{y}_1 = 0$$

$$\dot{y}_2 = \frac{1}{I_2}(I_3 - I_1)C_1y_3 \tag{9.31}$$

$$\dot{y}_3 = \frac{1}{I_2}(I_1 - I_2)C_1y_2$$

Here, it is seen that $\lim_{\|y\| \to \infty} \| h(y) \| / \| y \| = 0$ and the A matrix is obtained from (9.31) as

$$A = \begin{bmatrix} 0 & 0 & 0 \\ 0 & 0 & \dfrac{1}{I_2}(I_3 - I_1)C_1 \\ 0 & \dfrac{1}{I_3}(I_1 - I_2)C_1 & 0 \end{bmatrix}$$

The characteristic equation of this A matrix is given by

$$\lambda \left[\lambda^2 - C_1^2 \frac{(I_1 - I_2)(I_3 - I_1)}{I_2 I_3} \right] = 0 \tag{9.32}$$

If $I_2 > I_1 > I_3$ or $I_3 > I_1 > I_2$, the characteristic equation has a positive real root and the stationary motion $(C_1, 0, 0)$ is unstable. Otherwise, the roots have zero real parts and Theorem 9.1 fails to yield any stability information. We have, however, proved that steady rotation about the intermediate principal axis is unstable. The same information can be obtained by considering the other two steady motions $(0, C_2, 0)$ and $(0, 0, C_3)$. The stability of steady rotation about the largest and smallest axes will be investigated in Section 9.5 by Lyapunov's second method.

Here, Routh's criterion was not employed since the roots of (9.32) can be obtained by inspection. It is now employed for the purpose of illustration. Letting

$$\frac{(I_1 - I_2)(I_3 - I_1)}{I_2 I_3} = a$$

The characteristic equation may be written as

$$\lambda^3 - C_1^2 a \lambda = 0$$

The Routh array is

1	$-C_1^2 a$	1st row
$\epsilon \approx 0$	0	2nd row
$-C_1^2 a$	0	3rd row
0		4th row

The zero in the first column and second row has been replaced by ϵ, where $1 \gg \epsilon > 0$ in order to compute the subsequent rows. The zero in the fourth row and first column indicates that (9.32) has a root at the origin. If $a > 0$, then all the coefficients in the first column do not have the same sign and hence the origin of (9.31) is unstable. If $a < 0$, then the zero in the second row and first column indicates that (9.32) has a pair of purely imaginary roots.

Example 9.9

The equation of motion of a bead sliding on a circular hoop rotating at a constant angular velocity has been derived in Example 5.8. The equilibrium positions of the bead have also been determined in Example 5.6 by the application of the principle of virtual work. In this example, we investigated the stability of these equilibriums for small perturbations. The equation of motion as given by (5.76) is

$$\ddot{\theta} + \omega_0^2 \cos\theta \sin\theta + \frac{g}{c} \cos\theta = 0 \tag{9.33}$$

Choosing the state variables as $x_1 = \theta$ and $x_2 = \dot{\theta}$, the state equations become

$$\dot{x}_1 = x_2 \qquad (9.34)$$
$$\dot{x}_2 = -\omega_0^2 \cos x_1 \sin x_1 - \frac{g}{c} \cos x_1$$

The four distinct equilibria of (9.34) as determined in Example 5.6 are given by

(1) $x_{1e} = \dfrac{\pi}{2}$, $x_{2e} = 0$ (9.35)

(2) $x_{1e} = \dfrac{3\pi}{2}$, $x_{2e} = 0$ (9.36)

(3) $x_{1e} = -\sin^{-1}\left(\dfrac{g}{\omega_0^2 c}\right)$, $x_{2e} = 0$ (9.37)

(4) $x_{1e} = -\sin^{-1}\left(\dfrac{g}{\omega_0^2 c}\right) - \dfrac{\pi}{2}$, $x_{2e} = 0$ (9.38)

with the constraint that $\omega_0^2 c > g$. These four equilibria are shown in Figure 9.8.

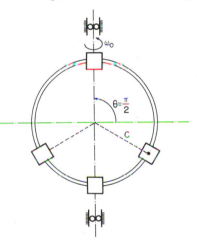

Figure 9.8 The four equilibrium positions of the bead.

Considering perturbations $\Delta x_1 = y_1$ and $\Delta x_2 = y_2$ about an equilibrium, the Jacobian matrix $\mathbf{A}$ of the Taylor series expansion about the equilibrium is obtained as

$$\mathbf{A} = \frac{\delta \mathbf{f}}{\delta \mathbf{x}}\bigg|_{\mathbf{x}=\mathbf{x}_e} = \begin{bmatrix} 0 & 1 \\ \omega_0^2 \sin^2 x_1 - \omega_0^2 \cos^2 x_1 + \dfrac{g}{c} \sin x_1 & 0 \end{bmatrix}_{x_1 = x_{1e}}$$

For the equilibrium (9.35), we get

$$\mathbf{A} = \begin{bmatrix} 0 & 1 \\ \omega_0^2 + \dfrac{g}{c} & 0 \end{bmatrix}$$

and the characteristic equation becomes

$$|\lambda \mathbf{I} - \mathbf{A}| = \lambda^2 - \left(\omega_0^2 + \frac{g}{c}\right) = 0$$

with roots

$$\lambda_{1,2} = \pm\left[\omega_0^2 + \frac{g}{c}\right]^{1/2}$$

Since one of the roots is positive, the equilibrium (9.35) is unstable according to Theorem 9.1.

For the equilibrium (9.36), we obtain

$$\mathbf{A} = \begin{bmatrix} 0 & 1 \\ \omega_0^2 - \dfrac{g}{c} & 0 \end{bmatrix}$$

and the characteristic equation becomes

$$|\lambda\mathbf{I} - \mathbf{A}| = \lambda^2 - \left(\omega_0^2 - \frac{g}{c}\right) = 0$$

with roots

$$\lambda_{1,2} = \pm\left[\omega_0^2 - \frac{g}{c}\right]^{1/2}$$

Since $\omega_0^2 c > g$, one of these roots is again positive and equilibrium (9.36) is also unstable.

For the equilibrium (9.37), we get

$$\mathbf{A} = \begin{bmatrix} 0 & 1 \\ -\omega_0^2\left(1 - \dfrac{g^2}{\omega_0^4 c^2}\right) & 0 \end{bmatrix}$$

and the characteristic equation becomes

$$\lambda^2 + \omega_0^2\left(1 - \frac{g^2}{\omega^4 c^2}\right) = 0$$

with roots

$$\lambda_1, \lambda_2 = \pm j\omega_0\left(1 - \frac{g^2}{\omega_0^4 c^2}\right)^{1/2}$$

Since $\omega_0^2 c > g$, both these roots are purely imaginary and according to Theorem 9.1, the stability of the equilibrium cannot be investigated from the linearized equations even for small perturbations. The same conclusion is arrived at for the equilibrium (9.38).

Example 9.10

We consider the two-body problem discussed in Chapter 3 but let the attractive central force be given by

$$f = -\frac{Gm_1 m_2}{r^n} \tag{9.39}$$

where n is an integer. We note that for Newton's law of gravitation, $n = 2$. From (3.71) and (3.72) the equation of motion can be obtained as

$$\ddot{r} - r\dot{\theta}^2 = \frac{k}{r^n} \tag{9.40}$$

$$r^2\dot{\theta} = h, \quad \text{constant} \tag{9.41}$$

where $k = G(m_1 + m_2)$. Substituting for $\dot{\theta}$ from (9.41) in (9.40), we obtain

$$\ddot{r} - \frac{h^2}{r^3} = -\frac{k}{r^n} \tag{9.42}$$

Choosing the state variables as $x_1 = r$ and $x_2 = \dot{r}$, the state equations are obtained as

$$\dot{x}_1 = x_2$$
$$\dot{x}_2 = \frac{h^2}{x_1^3} - \frac{k}{x_1^n} \tag{9.43}$$

We now consider a stationary nominal motion, namely, a circular orbit. This motion is obtained from the solution of the nonlinear algebraic equations when the left-hand sides of (9.43) are set equal to zero. Hence, this nominal motion is described by

$$x_{1e} = \left(\frac{k}{h^2}\right)^{1/(n-3)}, \qquad x_{2e} = 0$$

To study the stability of this stationary motion, we let $x_1 = x_{1e} + y_1$ and $x_2 = x_{2e} + y_2$, where y_1 and y_2 are the perturbations. The linearized equations in the perturbations become

$$\dot{y}_1 = y_2$$
$$\dot{y}_2 = \frac{h^2}{x_{1e}^4}(-3 + n)y_1 \tag{9.44}$$

The **A** matrix is given by

$$\mathbf{A} = \begin{bmatrix} 0 & 1 \\ \dfrac{h^2}{x_{1e}^4}(-3 + n) & 0 \end{bmatrix}$$

and its eigenvalues are obtained as

$$\lambda_{1,2} = \pm j\left[\frac{h^2}{x_{1e}^4}(3 - n)\right]^{1/2} \qquad \text{if } n < 3$$

$$\lambda_{1,2} = \pm\left[\frac{h}{x_{1e}^4}(n - 3)\right]^{1/2} \qquad \text{if } n > 3$$

If $n > 3$, from Theorem 9.1 we conclude that any circular orbit is unstable. If $n < 3$, **A** has purely imaginary eigenvalues and the stability of the circular orbit cannot be studied from the linearized equations.

9.4 STABILITY OF NONAUTONOMOUS MOTION FOR SMALL PERTURBATIONS

This section is concerned with the stability investigation of a general time-varying motion for small perturbations. A nominal motion $\{x^*\}$, obtained from the solution of (9.1) for given forces, is now perturbed by initial conditions at initial time $t_0 \geq 0$. The perturbation equations are expressed by (9.5). Letting $\{\Delta x\} = \{y\}$, the perturbation equations may be written as

$$\{\dot{y}\} = \mathbf{A}(t)\{y\} + \{h(y_1, \ldots, y_n, t)\} \tag{9.45}$$

and the linearized equations become

$$\{\dot{y}\} = \mathbf{A}(t)\{y\} \tag{9.46}$$

We now state a theorem for nonautonomous systems which is analogous to Theorem 9.1 for autonomous systems when the perturbations are sufficiently small.

Theorem 9.2.   Assuming that $\|\mathbf{h}(\mathbf{y}, t)\|/\|\mathbf{y}\| \to 0$ uniformly in t as $\|\mathbf{y}\| \to 0$, uniform asymptotic stability of the origin of the linearized system (9.46) implies that the nominal motion $\{x^*\}$ is also uniformly asymptotically stable for small perturbations.

The proof of this theorem is given by Hsu and Meyer [1, Chap. 11] and is omitted here. It should be noted that Theorem 9.2 states sufficient conditions for uniform asymptotic stability for small perturbations. However, unlike linear autonomous systems, the problem is now to establish uniform asymptotic stability for the time-varying linear system of (9.46). One approach is to employ Lyapunov's direct method to be discussed later. However, it will be realized that the selection of a suitable Lyapunov function is a very difficult task for nonautonomous systems.

Another approach is based on some conditions satisfied by the state transition matrix $\mathbf{\Phi}$ of system (9.46). In Chapter 6 it was shown that the solution of (9.46) can be written as

$$\{y(t)\} = \mathbf{\Phi}(t, t_0)\{y(t_0)\}$$

where the state transition matrix $\mathbf{\Phi}$ is obtained from the solution of (6.46) with conditions (6.47). But in general it is not possible to derive an analytic expression for $\mathbf{\Phi}$ in the case of time-varying parameter linear systems. Hence, this approach also has computational difficulties but is stated here for conceptual value. The induced norm of the state transition matrix is defined by

$$\|\mathbf{\Phi}\| = \sup_{\|\mathbf{x}\| \neq 0} \frac{\|\mathbf{\Phi x}\|}{\|\mathbf{x}\|} = \sup_{\|\mathbf{x}\| = 1} \|\mathbf{\Phi x}\| = \sup_{\|\mathbf{x}\| \leq 1} \|\mathbf{\Phi x}\|$$

The necessary and sufficient conditions for uniform asymptotic stability of the origin of (9.46) are now stated as follows. The origin of (9.46) is uniformly asymptotically stable for $t_0 \leq t < \infty$ and $t_0 \geq 0$ if and only if

$$\sup_{t_0 \geq 0} \sup_{t \geq t_0} \|\mathbf{\Phi}(t, t_0)\| < \infty$$

and

$$\|\mathbf{\Phi}(t, t_0)\| \to \infty \text{ as } t \to \infty \text{ uniformly in } t_0$$

An alternative necessary and sufficient condition is that there exist positive constants m and λ such that

$$\|\mathbf{\Phi}(t, t_0)\| \leq m e^{-\lambda(t-t_0)} \qquad \text{for all } t_0 \geq 0 \text{ and all } t \geq t_0$$

The proof of the conditions is given by Vidyasagar [2, p. 170]. Because an analytic expression for the state transition matrix is not available, there are computational difficulties in employing these conditions.

An approach that is sometimes employed is called the "freezing-time" method. At each instant of time $t = t_0, t_1, \ldots, t_k, \ldots$, the time-varying parameters are fixed at their current values and the matrix $A(t_i)$ is treated as a constant for the interval t_i to t_{i+1}. The condition for the constant matrix $A(t_i)$, with $t_i = t_0, t_t, \ldots, t_k, \ldots$, to have eigenvalues with negative real parts can then be investigated by employing Routh's criterion.

The following question then arises. If all eigenvalues of the matrix $A(t_i)$ have negative real parts for $t_i = t_1, t_2, \ldots, t_k, \ldots$, does it mean that the origin of the linear system of (9.46) is uniformly asymptotically stable? The answer to this question is not always in the affirmative, as demonstrated by the following two counterexamples.

Example 9.11

This example is quoted by Aggarwal and Infante [4] and attributed to Marcus and Yamabe. Consider the system $\{\dot{y}\} = A(t)\{y\}$, where $A(t)$ is given by

$$A(t) = \begin{bmatrix} -1 + a\cos^2 t & 1 - a\sin t \cos t \\ -1 - a\sin t \cos t & -1 + a\sin^2 t \end{bmatrix} \tag{9.47}$$

with $a > 0$. When the determinant $|\lambda I - A|$ is considered, it is found that the eigenvalues of $A(t)$ are given by

$$\lambda_{1,2} = \frac{a - 2 \pm \sqrt{a^2 - 4}}{2} \tag{9.48}$$

and are time invariant. They have negative real parts for $a < 2$. A closed-form solution of this equation is possible and is given by

$$\begin{Bmatrix} y_1(t) \\ y_2(t) \end{Bmatrix} = \begin{bmatrix} e^{(a-1)t}\cos t & e^{-t}\sin t \\ -e^{(a-1)t}\sin t & e^{-t}\cos t \end{bmatrix} \begin{Bmatrix} y_1(0) \\ y_2(0) \end{Bmatrix} \tag{9.49}$$

which shows that asymptotic stability requires that $a < 1$. Hence, if $a = 1.5$, then the freezing-time method indicates that the origin of the system of (9.47) is asymptotically stable, whereas it is actually unstable.

Example 9.12

This example is quoted by Hsu and Meyer [1] and is attributed to Vinogradov. Consider the system $\{\dot{y}\} = A(t)\{y\}$ with $A(t)$ given by

$$A(t) = \begin{bmatrix} -1 - 9\cos^2 6t + 12\sin 6t \cos 6t & 12\cos^2 6t + 9\sin 6t \cos 6t \\ -12\sin^2 6t + 9\sin 6t \cos 6t & -1 - 9\sin^2 6t - 12\sin 6t \cos 6t \end{bmatrix} \tag{9.50}$$

This example is also so contrived that the eigenvalues of $A(t)$ are

$$\lambda_{1,2} = -1, -10 \tag{9.51}$$

and are time invariant. A closed-form solution of this equation is available and is given by

$$\begin{Bmatrix} y_1(t) \\ y_2(t) \end{Bmatrix} = \begin{bmatrix} e^{2t}(\cos 6t + 2\sin 6t) & e^{-13t}(\sin 6t - 2\cos 6t) \\ e^{2t}(2\cos 6t - \sin 6t) & e^{-13t}(2\sin 6t + \cos 6t) \end{bmatrix} \begin{Bmatrix} y_1(0) \\ y_2(0) \end{Bmatrix} \qquad (9.52)$$

The presence of the term e^{2t} in (9.52) shows that the origin of (9.50) is actually unstable, whereas the freezing-time method indicates asymptotic stability.

These two examples show that the eigenvalues of $\mathbf{A}(t)$ do not carry a great deal of information regarding stability. Yet the freezing time method has sometimes been successfully employed in the aerospace industry for the design of autopilots for aircraft and missiles. It appears that if the elements of $\mathbf{A}(t)$ are periodic in time or if its eigenvalues are near the imaginary axis, as in the foregoing examples, then the stability results based on the eigenvalues of $\mathbf{A}(t)$ may be invalid.

9.5 STABILITY IN THE LARGE OF AUTONOMOUS SYSTEMS

This section deals with the second or direct method of Lyapunov for the investigation of stability of equilibrium states or stationary motions where the perturbation equation (9.24) is autonomous. The objective is also to determine the size of the region of stability around an equilibrium state or stationary motion (i.e., the size of the perturbations that can be tolerated). Hence, this analysis is also called the investigation of stability in the large.

The first step is to investigate the stability for small perturbations by application of Theorem 9.1. However, this theorem is sometimes not applicable because the condition that $\lim \|\mathbf{h}\|/\|\mathbf{y}\| = 0$ as $\|\mathbf{y}\| \to 0$ is not satisfied. This can happen when the nonlinearities are not analytic functions of their arguments, as, for example, in the case of Coulomb friction. Also, as shown in some of the previous examples, Theorem 9.1 fails to reveal any stability information when the matrix $\mathbf{A}$ has one or more eigenvalues with zero real parts and the remaining eigenvalues have negative real parts.

Even when Theorem 9.1 is applicable and reveals that a particular equilibrium or stationary motion is asymptotically stable for small perturbations, the size of the region of stability may be too small for practical considerations. For example, let a scalar perturbation equation be given by $\dot{y} = -0.01y + y^3$. From Theorem 9.1, the origin of this equation is asymptotically stable for small perturbation. However, when $|y(0)| > 0.1$, we can show by directly integrating the equation, after separating the variables, that the perturbation grows without bound. For the foregoing reasons, there is a strong motivation to employ the second method of Lyapunov. The success of the method depends on the selection of a suitable function called a Lyapunov function which has to satisfy certain sign definiteness properties. We consider a function $V(y_1, \ldots, y_n)$ of n

variables, where n is the order of the dynamic system (i.e., the number of state variables). Throughout this section, V is not an explicit function of time t.

Definition 9.9.   A scalar function $V(y_1, \ldots, y_n)$ is called positive definite in a region Ω containing the origin if $V(0, \ldots, 0) = 0$ and $V(y_1, \ldots, y_n) > 0$ for $\|\mathbf{y}\| \neq 0$ in Ω.

Definition 9.10.   A scalar function $V(y_1, \ldots, y_n)$ is called negative definite in a region Ω containing the origin if $V(0, \ldots, 0) = 0$ and $V(y_1, \ldots, y_n) < 0$ for $\|\mathbf{y}\| \neq 0$ in Ω.

Definition 9.11.   A scalar function $V(y_1, \ldots, y_n)$ is called positive semi-definite in a region Ω containing the origin if $V(0, \ldots, 0) = 0$ and $V(y_1, \ldots, y_n) \geq 0$ for $\|\mathbf{y}\| \neq 0$ in Ω.

Definition 9.12.   A scalar function $V(y_1, \ldots, y_n)$ is called negative semi-definite in a region Ω containing the origin if $V(0, \ldots, 0) = 0$ and $V(y_1, \ldots, y_n) \leq 0$ for $\|\mathbf{y}\| \neq 0$ in Ω.

Definition 9.13.   A scalar function $V(y_1, \ldots, y_n)$ that does not satisfy any one of Definitions 9.9 to 9.12 in a region Ω containing the origin is called sign indefinite.

Example 9.13

We consider a dynamic system described by three state variables so that the state space is three-dimensional. Let

$$V(y_1, y_2, y_3) = y_1^2 + y_2^2 + y_3^2$$

This function is positive definite and the region Ω is the entire state space. Now, let

$$V(y_1, y_2, y_3) = -y_1^2 - y_2^2 - y_3^2$$

This V function is negative definite throughout the state space. Hence, it is obvious that V is negative definite if $-V$ is positive definite. For the three-dimensional state space, let

$$V(y_1, y_2, y_3) = y_1^2 + y_2^2$$

This V function is positive semidefinite since it is zero not only at the origin but also along the y_3 axis.

We consider a dynamic system with two state variables and choose

$$V(y_1, y_2) = y_1^2 - y_2^2$$

As shown in Fig. 9.9, there is no region surrounding the origin in which this V function is sign definite. Hence, this V function is indefinite.

The foregoing V functions are simple enough that their sign definiteness can be determined by inspection. Unfortunately, there is no general method for determining whether any given function is sign definite or semidefinite except in

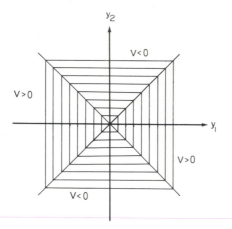

Figure 9.9 Sign indefinite function.

the case of quadratic functions. A quadratic function is in the form

$$V(y_1, \ldots, y_n) = \{y\}^T \mathbf{P}\{y\}$$

$$= \sum_{i=1}^{n} \sum_{j=1}^{n} p_{ij} y_i y_j \tag{9.53}$$

where without loss of generality it can be assumed that matrix $\mathbf{P}$ is symmetric. The following theorem may then be employed.

Theorem 9.3: Sylvester's Theorem.   A necessary and sufficient condition for a quadratic function $\{y\}^T \mathbf{P}\{y\}$ to be positive definite is that the following n determinants are all positive:

$$D_1 = p_{11} > 0, \qquad D_2 = \begin{vmatrix} p_{11} & p_{12} \\ p_{12} & p_{22} \end{vmatrix} > 0$$

$$D_n = \begin{vmatrix} p_{11} & p_{12} & \cdots & p_{1n} \\ & & \vdots & \\ & \vdots & & \\ p_{1n} & & \cdots & p_{nn} \end{vmatrix} > 0$$

A proof of this theorem is given by Bellman [5]. The converse of this theorem is not true; that is, a quadratic function need not be negative definite if all the n determinants are negative. There is another theorem to prove that a quadratic function is negative definite. However, we can prove that V is negative definite by showing that $-V$ is positive definite.

Example 9.14

Let a quadratic function be given by

$$V(y_1, y_2, y_3) = 2y_1^2 + 4y_1 y_3 + 3y_2^2 + 6y_2 y_3 + y_3^2$$

which can be written in the form

$$V(y_1, y_2, y_3) = \lfloor y_1 \quad y_2 \quad y_3 \rfloor \begin{bmatrix} 2 & 0 & 2 \\ 0 & 3 & 3 \\ 2 & 3 & 1 \end{bmatrix} \begin{Bmatrix} y_1 \\ y_2 \\ y_3 \end{Bmatrix} \tag{9.54}$$

The three determinants are obtained as

$$D_1 = 2 > 0, \qquad D_2 = \begin{vmatrix} 2 & 0 \\ 0 & 3 \end{vmatrix} = 6 > 0$$

$$D_3 = \begin{vmatrix} 2 & 0 & 2 \\ 0 & 3 & 3 \\ 2 & 3 & 1 \end{vmatrix} = -24 < 0$$

Since D_3 is negative, this quadratic function is not positive definite.

We now state a basic theorem of the Lyapunov second method for the stability investigation of autonomous systems. It should be noted again that in this section, the equations of motion are autonomous as expressed by (9.8):

$$\{\dot{x}\} = \{f(x_1, \ldots, x_n)\} \tag{9.8}$$

We are considering the stability of an equilibrium or stationary motion $\{x_e\}$ of (9.8) and denote the perturbation $\{\Delta x\}$ about $\{x_e\}$ by $\{y\}$. Let the perturbation equation (9.10) be expressed by

$$\{\dot{y}\} = \{g(y_1, \ldots, y_n)\} \tag{9.55}$$

where g_i are nonlinear functions of their arguments. The stability of $\{x_e\}$ is now equivalent to the stability of the null or trivial solution of (9.55) (i.e., its origin $\{y\} = \{0\}$).

Theorem 9.4.   Let $V(y_1, \ldots, y_n)$ be a scalar function with continuous first partial derivatives. Let Ω_k designate a bounded region about the origin of (9.55) in which $V(y_1, \ldots, y_n) < k$, where k is a constant. If in Ω_k:

1. $V(\mathbf{y})$ is positive definite and
2a. $\dot{V}(\mathbf{y})$ evaluated along the trajectory of (9.55) is negative semidefinite

then the origin of (9.55) is stable in Ω_k.

Or 2b. $\dot{V}(\mathbf{y})$ evaluated along the trajectory of (9.55) is negative definite,

then the origin of (9.55) is asymptotically stable in Ω_k.

Or 2c. $\dot{V}(\mathbf{y})$ evaluated along the trajectory of (9.55) is negative semidefinite and the trajectory of (9.55) cannot stay forever at the points $\dot{V} = 0$ within Ω_k other than at the origin, then the origin of (9.55) is asymptotically stable in Ω_k.

It is noted that Ω_k is the domain of stability or asymptotic stability. In the latter case, it is also called the domain of attraction. The V function is called the Lyapunov function.

A proof of this theorem can be given from geometrical considerations. We assume that a V function has been found that is positive definite and $V(y_1, \ldots, y_n) < k$ in a bounded region Ω_k. Then for such a function, $V = c$, where c is a constant, represents a closed hypersurface. For the two-dimensional case, these hypersurfaces become closed curves, as shown in Fig. 9.10. Now, $\dot{V}$ evaluated along the trajectory of (9.55) is obtained as

$$\dot{V} = \sum_{i=1}^{n} \frac{\partial V}{\partial y_i} \dot{y}_i = \{\text{grad } V\}^T \{\dot{y}\} = \{\text{grad } V\}^T \{g\} \qquad (9.56)$$

where the last equality is obtained by employing (9.55). Hence, the theorem requires that the V function selected should have continuous first partial derivatives.

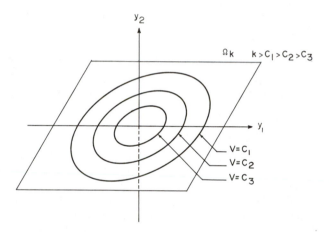

Figure 9.10   Contours of $V = c$ within k.

Case 2a.   Let $\dot{V}$ of (9.56) be negative semidefinite. Then we can show that the origin of (9.55) is stable since its trajectory which originates in Ω_k at time $t_0 \geq 0$ with initial condition $\{y(t_0)\}$ is always proceeding in a direction such that the associated V function never increases. More precisely, we must show that given any $\epsilon > 0$, there is a $\delta(\epsilon) > 0$ such that $\|\mathbf{y}(t_0)\| \leq \delta$ implies that $\|\mathbf{y}(t)\| < \epsilon$ for all $t > t_0$, where $\{y(t)\}$ belongs to Ω_k according to Definition 9.1.

For any such ϵ, let $V(y_1, \ldots, y_n) \geq c$ for $\|\mathbf{y}\| = \epsilon$, where $c > 0$ since V is positive definite. We choose a $\delta \leq \epsilon$ such that $V(y_1, \ldots, y_n) < c$ for $\|\mathbf{y}\| < \delta$. This is possible since V is continuous and is zero only at the origin. Since $\dot{V} \leq 0$, it follows that $V(\mathbf{y}(t)) \leq V(\mathbf{y}(t_0)) < c$ for $t > t_0$. Thus $\|\mathbf{y}(t)\| < \epsilon$ for all $t > t_0$.

Case 2b.   Let $\dot{V}$ of (9.56) be negative definite. It is obvious from the proof of the previous case that the origin of (9.55) is stable. Moreover, a trajectory of (9.55) which originates in Ω_k is always proceeding in a direction such that the associated V function is decreasing monotonically. Hence, the trajectories must end at the origin, which is the only absolute minimum point for V. It follows that $\lim_{t \to \infty} \| \mathbf{y}(t) \| \longrightarrow 0$.

Case 2c.   In this case $\dot{V}$ is negative semidefinite with the additional requirement for asymptotic stability that no trajectory of (9.55) can stay forever at the points at which $\dot{V} = 0$ other than at the origin. Since $\dot{V}$ is negative semidefinite, the origin of (9.55) is stable. As discussed in Examples 9.2 and 9.3, an oscillatory or periodic motion is represented by a closed hypersurface in state space. Also, as discussed in Example 9.3, the amplitude and frequency of oscillation in a conservative system depend on the initial conditions and the origin is stable. It has been observed that in a bounded region Ω_k, $V = c$, where c is a constant, represents a closed hypersurface in state space, as shown in Fig. 9.10. In case $V = c$ also represents a hypersurface of periodic motion, then $\dot{V} = 0$ along the system trajectory since c is a constant and the trajectory will remain forever on this hypersurface. The additional requirement that no trajectory of (9.55) can stay forever at the point, at which $\dot{V} = 0$ other than at the origin, rules out the case $V = c > 0$ representing a hypersurface of periodic motion and it follows that $\lim_{t \to \infty} \| \mathbf{y}(t) \| \longrightarrow 0$. Case 2c is also referred to as LaSalle's theorem, and its rigorous proof is given by LaSalle and Lefschetz [6].

The requirement of Theorem 9.4 that Ω_k be a bounded region around the origin is to assure that $V = c$, where c is a constant, is a closed hypersurface in the state space. If Ω_k is not a finite region, it is possible that far from the origin, $V = c$ can represent an open hypersurface. Then it is possible for the trajectory of (9.55) to escape toward infinity, even when $V(y_1, \ldots, y_n)$ is positive definite and $\dot{V}(y_1, \ldots, y_n)$ is negative definite, as illustrated in Fig. 9.11 for a second-order system.

In case the region Ω_k is not bounded but extends to infinity, it is required

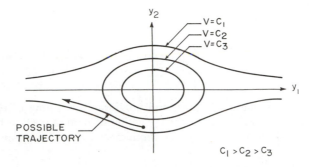

Figure 9.11   Open contour of $V = c_1$ and possible escape trajectory.

in Theorem 9.4 that $\lim V(y_1, \ldots, y_n) \to \infty$ as $\| \mathbf{y} \| \to \infty$. This requirement is to assure that $V(y_1, \ldots, y_n) = c$ represent closed hypersurfaces in state space. For example, in the case of two state variables, a V function is chosen as

$$V(y_1, y_2) = \frac{y_1^2}{1 + y_1^2} + \frac{y_2^2}{1 + y_2^2}$$

This V function is positive definite but $\lim V$ does not tend to infinity as $\| \mathbf{y} \| \to \infty$. In case the region Ω_k of Theorem 9.4 is not bounded but $\lim V \to \infty$, the origin of (9.55) is said to be globally stable or globally asymptotically stable, as the case may be. This is also referred to as the theorem of Barbashin and Krasovskii. It should be noted that the equations of motion may not be valid throughout the state space because of some assumptions and approximations. In that case, global asymptotic stability may have no practical significance.

A point to be noted is that Theorem 9.4 provides only sufficient conditions and in case a suitable Lyapunov function cannot be found, it cannot be implied that the origin of (9.55) is unstable. A Lyapunov function may be considered as a generalized energy function and Theorem 9.4 as a generalization of the energy method of stability investigation. For stability (or asymptotic stability) it is not necessary that the generalized energy of the perturbation evaluated along the trajectory be nonincreasing (or monotonically decreasing) at every instant of time. It is possible for the generalized energy of the perturbation along the trajectory to increase momentarily with time and yet to decay to zero with time as shown in Fig. 9.12.

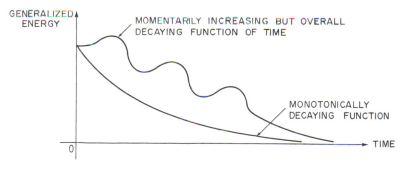

Figure 9.12   Generalized energy versus time.

Example 9.15

We consider the mass, linear damping, and nonlinear soft spring of Example 6.7, which is also studied in Example 9.1. The equation of motion is given by (9.12). In Examples 6.7 and 9.1, the three isolated equilibrium states have been found as

$$x_e = \begin{Bmatrix} 0 \\ 0 \end{Bmatrix}, \quad \begin{Bmatrix} \sqrt{6} \\ 0 \end{Bmatrix}, \quad \begin{Bmatrix} -\sqrt{6} \\ 0 \end{Bmatrix}$$

Stability of equilibrium state (0, 0).   The perturbation equations about this equilibrium are given by (9.13), from which the **A** matrix is obtained as

$$\mathbf{A} = \begin{bmatrix} 0 & 1 \\ -\dfrac{k}{m} & -\dfrac{c}{m} \end{bmatrix}$$

The characteristic equation becomes

$$|\lambda \mathbf{I} - \mathbf{A}| = \lambda^2 + \frac{c}{m}\lambda + \frac{k}{m} = 0 \tag{9.57}$$

The Routh array is given by

$$1 \qquad \frac{k}{m}$$

$$\frac{c}{m} \qquad 0$$

$$\frac{k}{m}$$

Since all three elements of the first column have the same sign, we conclude that both eigenvalues of (9.57) have negative real parts. From (9.13) we note that $\lim \|\mathbf{h(y)}\|/\|\mathbf{y}\| = 0$ as $\|\mathbf{y}\| \to 0$. Hence, from Theorem 9.1 we conclude that equilibrium (0, 0) is asymptotically stable for sufficiently small perturbations. We now determine the size of the region of asymptotic stability around this equilibrium state by application of Theorem 9.4. A candidate for the Lyapunov function is the total mechanical energy of the perturbations. The kinetic and potential energies are obtained as

$$T = \frac{1}{2}m\dot{y}^2$$

$$U = \int_0^y k\left(y - \frac{y^3}{6}\right)dy$$

$$= k\left(\frac{y^2}{2} - \frac{y^4}{24}\right)$$

The V function in terms of the state variable ($y_1 = y$, $y_2 = \dot{y}$) becomes

$$V(y_1, y_2) = T + U$$

$$= \frac{1}{2}my_2^2 + \frac{1}{2}ky_1^2\left(1 - \frac{y_1^2}{12}\right) \tag{9.58}$$

This V function is positive definite for $|y_1| < \sqrt{12}$. However, in order to determine the region Ω_k of stability, it is required to determine the region where

$$V = \frac{1}{2}my_2^2 + \frac{1}{2}ky_1^2\left(1 - \frac{y_1^2}{12}\right) - c$$

represents a closed curve. Now, $\frac{1}{2}my_2^2 = c$ has a solution for all values of y_2 and hence $V = c$ does not open along the y_2 axis. But $\frac{1}{2}ky_1^2[1 - (y_1^2/12)] = c$ has a real solution up to a maximum value of y_1. Differentiating this expression with respect to y_1 and solving for y_1, this maximum value of y_1 is $\pm\sqrt{6}$. Hence, $V = c$ represents closed curves for

$$c = \frac{1}{2}k6\left(1 - \frac{6}{12}\right) = \frac{3}{2}k$$

The bounded region Ω_k in which $V = c$ represents closed curves is given by

$$\Omega_k = \frac{1}{2}my_2^2 + \frac{1}{2}ky_1^2\left(1 - \frac{y_1^2}{12}\right) \leq \frac{3}{2}k \qquad (9.59)$$

Differentiating (9.58) with respect to time, we obtain

$$\dot{V} = my_2\dot{y}_2 + k\left(y - \frac{y^3}{6}\right)\dot{y}_1 \qquad (9.60)$$

In order to evaluate $\dot{V}$ along the system trajectory, we substitute for $\dot{y}_1$ and $\dot{y}_2$ in (9.60) from the right-hand side of (9.13) and get

$$\dot{V} = my_2\left(-\frac{k}{m}y_1 + \frac{k}{6m}y_1^3 - \frac{c}{m}y_2\right) + k\left(y_1 - \frac{y_1^3}{6}\right)y_2 = -cy_2^2 \qquad (9.60a)$$

This $\dot{V}$ is negative semidefinite and $\dot{V} = 0$ all along the y_1 axis. Now, $y_2 = 0$ and $y_1 = f(t)$ does not satisfy (9.13) except for $y_1 = 0$; that is, a trajectory of (9.13) cannot remain forever on the y_1 axis except at the origin. We therefore conclude from Theorem 9.4 that the origin of (9.13) [i.e., the equilibrium state $(x_1 = 0, x_2 = 0)$] is asymptotically stable and the region of asymptotic stability is given by (9.59), which is shown in Fig. 9.13.

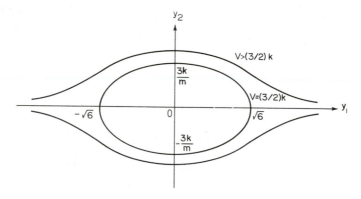

Figure 9.13   Region of asymptotic stability.

Stability of equilibrium states $(\sqrt{6}, 0)$ and $(-\sqrt{6}, 0)$.   The perturbation equations about these equilibrium states are given by (9.14), from which the **A** matrix is obtained as

$$\mathbf{A} = \begin{bmatrix} 0 & 1 \\ \dfrac{2k}{m} & -\dfrac{c}{m} \end{bmatrix}$$

The characteristic equation is obtained as

$$|\lambda\mathbf{I} - \mathbf{A}| = \lambda^2 + \frac{c}{m}\lambda - \frac{2k}{m} = 0 \qquad (9.61)$$

The Routh array becomes

$$1 \qquad -2\frac{k}{m}$$

$$\frac{c}{m} \qquad 0$$

$$-2\frac{k}{m}$$

There is one change in sign in the first column of this array, and we conclude that (9.61) has one root with positive real part. Hence, from Theorem 9.1, the origin of (9.14) [i.e., equilibrium states $(\sqrt{6}, 0)$ and $(-\sqrt{6}, 0)$] are unstable. The application of Theorem 9.4 for these equilibrium states is therefore meaningless.

Example 9.16

The tumbling motion of a rigid-body satellite was discussed in Example 9.8 and the equations of motion given by (9.28). In order to study the stability of a steady rotation about one of the axes, the motion was perturbed and the perturbation equations given by (9.30). By application of Theorem 9.1 it was shown in Example 9.8 that a steady rotation about the intermediate principal axis is unstable. However, Theorem 9.1 does not yield any information on the stability of steady rotation about the largest or smallest principal axis, and this is studied in this example by application of Theorem 9.4. A function $h(y_1, y_2, y_3)$ which becomes a constant when y_1, y_2, and y_3 represent solutions of (9.30) is called a first integral of the differential equations (9.30). Thus, we have

$$h(y_1, y_2, y_3) = \text{const.} \tag{9.62}$$

$$\dot{h} = \frac{\partial h}{\partial y_1}\dot{y}_1 + \frac{\partial h}{\partial y_2}\dot{y}_2 + \frac{\partial h}{\partial y_3}\dot{y}_3 = 0 \tag{9.63}$$

It is known [7] that (9.30) has two first integrals given by

$$h = \frac{I_2 - I_1}{I_3}y_2^2 + \frac{I_3 - I_1}{I_2}y_3^2 \pm (I_1 y_1^2 + I_2 y_2^2 + I_3 y_3^2 + 2c_1 I_1 y_1)^2 \tag{9.64}$$

We choose one of these as a candidate for the Lyapunov function and obtain

$$V(y_1, y_2, y_3) = \frac{I_2 - I_1}{I_3}y_2^2 + \frac{I_3 - I_1}{I_2}y_3^2 + (I_1 y_1^2 + 1_2 y_2^2 + I_3 y_3^2 + 2c_1 I_1 y_1)^2$$

This V function is positive definite throughout the entire state space if $I_1 < I_2 \leq I_3$. Furthermore, $\lim V \to \infty$ as $\|\mathbf{y}\| \to \infty$. Since this V function is the first integral, it follows from (9.63) that $\dot{V}$ evaluated along the trajectory of (9.30) yields $\dot{V} = 0$. Since $\dot{V} = 0$ throughout the entire state space, asymptotic stability cannot be proved and we conclude that a steady rotation about the smallest principal axis is globally stable. In order to investigate the stability of steady rotation about the largest principal axis, we choose a V function from (9.61) as

$$V(y_1, y_2, y_3) = -\left[\frac{I_2 - I_1}{I_3}y_2^2 + \frac{I_3 - I_1}{I_2}y_3^2 - (I_1 y_1^2 + I_2 y_2^2 + I_3 y_3^2 + 2c_1 I_1 y_1)^2\right]$$

This V function is positive definite throughout the state space if $I_1 > I_2 \geq I_3$, and $\dot{V}$ evaluated along the trajectory of (9.31) yields $\dot{V} = 0$ throughout the entire state space. Hence, steady rotation about the largest principal axis is globally stable.

Example 9.17

The equations of motion for a spring pendulum have been derived in Example 5.11. Now, let the spring be replaced by a rigid massless rod of length a. Then, from (5.103) the equation of motion for a damped simple pendulum becomes

$$ma^2\ddot{\theta} + c\dot{\theta} + mga\sin\theta = 0 \tag{9.65}$$

Letting $x_1 = \theta$ and $x_2 = \dot{\theta}$, we express (9.65) as

$$\dot{x}_1 = x_2$$
$$\dot{x}_2 = -\frac{g}{a}\sin x_1 - \frac{c}{ma^2}x_2 \tag{9.66}$$

The equilibrium states are obtained by setting the left-hand sides of the foregoing equations to zero (i.e., $x_{2e} = 0$ and $x_{1e} = \pm k\pi$, where $k = 0, 1, 2, \ldots$). These equilibrium states are shown in Fig. 9.14.

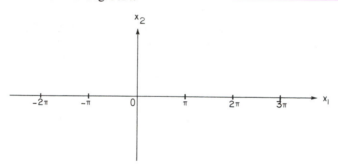

Figure 9.14   Equilibrium states of a simple pendulum.

Let us consider the stability of the equilibrium state $(0, 0)$. Denoting the perturbations about this equilibrium by y_1 and y_2, the perturbation equations become

$$\dot{y}_1 = y_2$$
$$\dot{y}_2 = -\frac{g}{a}\sin y_1 - \frac{c}{ma^2}y_2 \tag{9.67}$$

The Jacobian matrix $\mathbf{A}$ about this equilibrium is given by

$$\mathbf{A} = \begin{bmatrix} 0 & 1 \\ -\dfrac{g}{a} & -\dfrac{c}{ma^2} \end{bmatrix}$$

and the characteristic equation by

$$|\lambda\mathbf{I} - \mathbf{A}| = \lambda^2 + \frac{c}{ma^2}\lambda + \frac{g}{a} = 0$$

The Routh array is obtained as

$$
\begin{array}{cc}
1 & \dfrac{g}{a} \\[2ex]
\dfrac{c}{ma^2} & 0 \\[2ex]
\dfrac{g}{a} &
\end{array}
$$

Hence, both eigenvalues of **A** have negative real parts and according to Theorem 9.1, this equilibrium is asymptotically stable for sufficiently small perturbations. We now obtain the region of asymptotic stability around this equilibrium by employing Theorem 9.4. The total mechanical energy of the perturbations is chosen as a candidate for the Lyapunov function. Hence, we get

$$V(y_1, y_2) = T + U$$
$$= \tfrac{1}{2}ma^2 y_2^2 + mga(1 - \cos y_1) \tag{9.68}$$

This V function is positive definite for $|y_1| < 2\pi$. We have to determine the region in which $V = c$ represents a closed curve.

Now, $\tfrac{1}{2}ma^2 y_2^2 = c$ has a solution for all values of y_2 and hence $V = c$ does not open along the y_2 axis. But $mga(1 - \cos y_1) = c$ has a real solution for y_1 up to a maximum value of y_1 which is $\pm\pi$. The maximum value of c for which $V = c$ represents a closed curve is $2mga$. Therefore, the bounded region Ω_k in which $V = c$ represents a closed curve is given by

$$\Omega_k = \tfrac{1}{2}ma^2 y_2^2 + mga(1 - \cos y_1) \leq 2mga \tag{9.69}$$

Differentiating (9.68) with respect to time, we obtain

$$\dot{V} = ma^2 y_2 \dot{y}_2 + mga(\sin y_1)\dot{y}_1$$

In order to evaluate $\dot{V}$ along the system trajectory, we substitute for $\dot{y}_1$ and $\dot{y}_2$ in the foregoing equation from (9.67) and get

$$\dot{V} = ma^2 y_2\left(-\frac{g}{a}\sin y_1 - \frac{c}{ma^2}y_2\right) + mga(\sin y_1)y_2 = -cy_2^2$$

This $\dot{V}$ is negative semidefinite and $\dot{V} = 0$ all along the y_1 axis. Now, $y_2 = 0$ and $y_1 = f(t)$ does not satisfy (9.67) except for $y_1 = 0$ [i.e., a trajectory of (9.67) cannot remain forever on the y_1 axis except at the origin]. We therefore conclude from Theorem 9.4 that the origin of (9.67) [i.e., the equilibrium state $(x_1 = 0, x_2 = 0)$] is asymptotically stable and the region of asymptotic stability is given by (9.69), which is shown in Fig. 9.15. It should be noted that the region of asymptotic stability that is obtained is strongly dependent on the choice of the Lyapunov function.

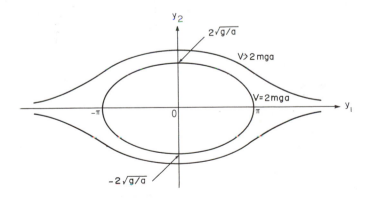

Figure 9.15   Region of asymptotic stability.

Let us consider the case of an undamped simple pendulum where $c = 0$ in (9.65) and (9.66). The equilibrium states as shown in Fig. 9.14 are unchanged. Choosing a V function as in (9.68), we find that this V function is positive definite for $|y| < 2\pi$ and $V = c$ represents a closed curve in the bounded region given by (9.69). However, $\dot{V}$ evaluated along the system trajectory becomes $\dot{V} = 0$ (i.e., negative semidefinite throughout the state space). Hence, we conclude that the equilibrium ($x_1 = 0, x_2 = 0$) is stable and (9.69) represents the region of stability. As is well known, the undamped simple pendulum is a conservative system and undergoes oscillations in the region Ω_k whose amplitudes and period depend on the initial conditions.

We now consider the equilibrium state ($x_1 = \pi, x_2 = 0$). The perturbation equations about this equilibrium state are given by

$$\dot{y}_1 = y_2$$

$$\dot{y}_2 = -\frac{g}{a} \sin(\pi + y_1) - \frac{c}{ma^2} y_2$$

The Jacobian matrix $\mathbf{A}$ about this equilibrium is given by

$$\mathbf{A} = \begin{bmatrix} 0 & 1 \\ \dfrac{g}{a} & -\dfrac{c}{ma^2} \end{bmatrix}$$

and the characteristic equation by

$$|\lambda \mathbf{I} - \mathbf{A}| = \lambda^2 + \frac{c}{ma^2}\lambda - \frac{g}{a} = 0$$

The Routh array is obtained as

$$
\begin{array}{cc}
1 & -\dfrac{g}{a} \\[2ex]
\dfrac{c}{ma^2} & \\[2ex]
-\dfrac{g}{a} &
\end{array}
$$

Hence, matrix $\mathbf{A}$ has one eigenvalue with positive real part and according to Theorem 9.1, the equilibrium ($x_1 = \pi, x_2 = 0$) is unstable. This equilibrium is also unstable for the undamped pendulum. Similarly, for the damped simple pendulum, we can show that the equilibrium states ($x_1 = \pm k\pi, x_2 = 0$) are asymptotically stable for $k = 0, 2, 4, \ldots$ and unstable for $k = 1, 3, 5, \ldots$

9.5.1 Generation of Lyapunov Function for Linearized Autonomous Systems

The perturbation equations about an equilibrium or stationary motion for the autonomous case are represented by (9.24). Considering small perturbations and letting $\{h\} \rightarrow \{0\}$, the linearized equations are obtained as

$$\{\dot{y}\} = \mathbf{A}\{y\} \tag{9.70}$$

For this linear autonomous system (9.70), a Lyapunov function that is a quadratic form in $y_1, \ldots, y_n$ is both necessary and sufficient for asymptotic stability of the origin of (9.70). If matrix $\mathbf{A}$ is nonsingular, then (9.70) has only one equilibrium state which is the origin and the region of asymptotic stability is global. But the global nature of asymptotic stability should not be taken literally because the perturbations are about a particular equilibrium, and linearization implies that the perturbations are small. The existence of a quadratic Lyapunov function is of little consequence for the stability investigation of the origin of (9.70) since the application of the Routh criterion is much simpler. However, use will be made later of this quadratic function for the stability investigation of the origin of the nonlinear system (9.24).

We chose a quadratic form as a candidate for the Lyapunov function and let

$$V(y_1, \ldots, y_n) = \{y\}^T \mathbf{P}\{y\} \tag{9.71}$$

where without loss of generality $\mathbf{P}$ is a symmetric matrix. Then,

$$\dot{V} = \{\dot{y}\}^T \mathbf{P}\{y\} + \{y\}^T \mathbf{P}\{\dot{y}\}$$

and after substituting for $\{\dot{y}\}$ from the right-hand side of (9.70), we obtain

$$\dot{V} = \{y\}^T \mathbf{A}^T \mathbf{P}\{y\} + \{y\}^T \mathbf{P}\mathbf{A}\{y\}$$
$$= \{y\}^T (\mathbf{A}^T \mathbf{P} + \mathbf{P}\mathbf{A})\{y\}$$
$$= -\{y\}^T \mathbf{Q}\{y\} \tag{9.72}$$

where matrix $\mathbf{Q}$ is defined by

$$\mathbf{A}^T \mathbf{P} + \mathbf{P}\mathbf{A} = -\mathbf{Q} \tag{9.73}$$

Equation (9.73) is known as the matrix Lyapunov equation. Because $\mathbf{P}$ is a symmetric matrix, it follows that

$$\mathbf{Q}^T = -(\mathbf{A}^T \mathbf{P} + \mathbf{P}\mathbf{A})^T = -(\mathbf{P}^T \mathbf{A} + \mathbf{A}^T \mathbf{P}^T)$$
$$= -(\mathbf{P}\mathbf{A} + \mathbf{A}^T \mathbf{P}) = \mathbf{Q}$$

(i.e., matrix $\mathbf{Q}$ is also symmetric). It can be proved [6, 8], that:

1. Equation (9.73) has a unique solution for $\mathbf{P}$ corresponding to every $\mathbf{Q}$ if and only if the sum of any two eigenvalues of $\mathbf{A}$ is not zero. (Note that if all eigenvalues of $\mathbf{A}$ have negative real parts, this condition is satisfied.)
2. If $\mathbf{Q}$ is positive definite and all eigenvalues of $\mathbf{A}$ have negative real parts, $\mathbf{P}$ is also positive definite.

It is noted that if $\mathbf{Q}$ is positive definite, the condition that $\mathbf{P}$ is positive definite is a sufficient condition for the stability of (9.70) since (9.71) becomes a Lyapunov function. In addition, the foregoing proposition 2 states that if $\mathbf{Q}$ is positive definite, a necessary condition for the stability of the origin of (9.70) is that $\mathbf{P}$ be positive definite. We select $\mathbf{Q}$ to be positive definite (in particular, $\mathbf{Q}$ may be selected as the identity matrix) and solve for $\mathbf{P}$ from (9.73). If this $\mathbf{P}$ is not positive definite, the origin (9.70) is unstable and if it is positive definite, the origin of (9.70) is globally asymptotically stable. The reverse is not true; that is, if $\mathbf{P}$ is chosen as positive definite, it does not follow that $\mathbf{Q}$ obtained from (9.73) will be positive definite when all eigenvalues of $\mathbf{A}$ have negative real parts. If the reverse were true, it would be far easier to solve for $\mathbf{Q}$ for a given $\mathbf{P}$ from (9.73) since it involves only matrix multiplication and addition. The determination of $\mathbf{P}$ from (9.73) for a given $\mathbf{Q}$ involves the solution of $\frac{1}{2}n(n+1)$ equations since matrices $\mathbf{P}$ and $\mathbf{Q}$ are symmetric.

Proof of Theorem 9.1.   In Section 9.3 it was mentioned that proof of Theorem 9.1, which is also call Lyapunov's first method, can be provided by Lyapunov's second method. The perturbation equations about an equilibrium or stationary motion for the autonomous case are represented by (9.24):

$$\{\dot{y}\} = \mathbf{A}\{y\} + \{h(y_1, \ldots, y_n)\} \tag{9.24}$$

where by assumption $\lim \|\mathbf{h}\|/\|\mathbf{y}\| = 0$ as $\|\mathbf{y}\| \to 0$ or $\|\mathbf{h}\|$ goes to zero faster than $\|\mathbf{y}\|$.

We choose a quadratic form as a candidate for the Lyapunov function and obtain

$$V(y_1, \ldots, y_n) = \{y\}^T \mathbf{P}\{y\} \tag{9.74}$$

where $\mathbf{P}$ is symmetric. Evaluating $\dot{V}$ along the trajectory of (9.24), we obtain

$$
\begin{aligned}
\dot{V} &= \{\dot{y}\}^T \mathbf{P}\{y\} + \{y\}^T \mathbf{P}\{\dot{y}\} \\
&= (\mathbf{A}\{y\} + \{h\})^T \mathbf{P}\{y\} + \{y\}^T \mathbf{P}(\mathbf{A}\{y\} + \{h\}) \\
&= \{y\}^T \mathbf{A}^T \mathbf{P}\{y\} + \{h\}^T \mathbf{P}\{y\} + \{y\}^T \mathbf{P}\mathbf{A}\{y\} + \{y\}^T \mathbf{P}\{h\} \\
&= \{y\}^T (\mathbf{A}^T \mathbf{P} + \mathbf{P}\mathbf{A})\{y\} + 2\{y\}^T \mathbf{P}\{h\} \\
&= -\{y\}^T \mathbf{Q}\{y\} + 2\{y\}^T \mathbf{P}\{h\} \tag{9.75}
\end{aligned}
$$

We choose $\mathbf{Q}$ as a positive-definite symmetric matrix. By assumption $\|\mathbf{h}\|$ goes to zero faster than $\|\mathbf{y}\|$. Thus in a sufficiently small region about the origin, the first term on the right-hand side of (9.75) will be dominating and $\dot{V}$ will be negative definite. If all eigenvalues of $\mathbf{A}$ have negative real parts, then after solving for $\mathbf{P}$ from (9.73) we will find that (9.74) is positive definite and hence a Lyapunov function. Then at least in a sufficiently small region, the origin of (9.24) will be asymptotically stable according to Theorem 9.4.

If $\mathbf{A}$ has at least one eigenvalue with positive real part, we give the proof here only for the case where the sum of any two eigenvalues of $\mathbf{A}$ is not zero. The proof for the general case can be obtained from the one given here by using continuity argument. In this case the equation $\mathbf{A}^T \mathbf{P} + \mathbf{P}\mathbf{A} = \mathbf{I}$ has a unique solution for $\mathbf{P}$ and this matrix $\mathbf{P}$ has at least one positive eigenvalue. Then by choosing $V_1 = \{y\}^T \mathbf{P}\{y\}$, there exist points arbitrarily close to the origin such that $V_1 > 0$ and $\dot{V}_1$ is positive definite. Hence, by Lyapunov's first instability theorem given later, the origin of (9.24) is unstable.

In case $\mathbf{A}$ has one or more eigenvalues with zero real parts and the remaining eigenvalues have negative real parts, the stability of the origin of (9.24) cannot be ascertained from the linearized equations.

Theorem 9.5: Lyapunov's First Instability Theorem. For the system (9.24), the origin is unstable if there exists a function $V_1(y_1, \ldots, y_n)$ with continuous first partial derivatives such that (a) $\dot{V}_1$ evaluated along the system trajectory is positive definite and (b) $V_1(0, \ldots, 0) = 0$ and there are points $\{y\}$ arbitrarily close to the origin such that $V_1(y_1, \ldots, y_n) > 0$.

The proof of this theorem is straightforward and is given by Hsu and Meyer [1] and by Vidyasagar [2].

9.5.2 Generation of Lyapunov Function for Nonlinear Autonomous Systems

The success of the Lyapunov second method depends on the selection of a suitable Lyapunov function. It is also clear that the estimate of the domain of stability or asymptotic stability is directly dependent on the choice of the Lyapunov function. An interesting question is whether the total mechanical energy of a dynamic system is useful as a Lyapunov function. The total mechanical energy is given by $E = T + U$, where T is kinetic energy and U is the potential energy. As seen in Chapter 5, a general expression for the kinetic energy is of the form $T = T_2 + T_1 + T_0$, where T_2 is a quadratic function of the generalized velocities, T_1 is a linear function of generalized velocities, and T_0 is not a function of the generalized velocities but only of the generalized coordinates. If the total mechanical energy is selected as the Lyapunov function, this V function will not in general be a quadratic form and there will be serious difficulties in determining its sign definiteness. We note that Sylvester's theorem (Theorem 9.3) is applicable to the determination of sign definiteness of quadratic forms only and that for general nonquadratic functions, such techniques are not available.

In case $T_1 = T_0 = 0$ and $T = T_2$, the difficulties are considerably reduced but not completely eliminated because the potential energy U need not be a quadratic function of the generalized coordinates. In Chapter 5 it is shown that a general Hamiltonian function is defined by $H = T_2 - T_0 + U$ and is not the total mechanical energy. In case $T_1 = T_0 = 0$, then $H = T + U = E$. When the generalized coordinates and generalized momenta are selected as state variables and the equations of motion are expressed by Hamilton's equations, let the perturbation equations about an equilibrium or stationary motion be described by

$$\dot{q}_i = \frac{\partial H}{\partial p_i}$$

$$\dot{p}_i = -\frac{\partial H}{\partial q_i} + Q_i, \qquad i = 1, \ldots, m$$

(9.76)

It is assumed that the generalized forces are dissipative forces given by $Q_i = -c_i \dot{q}_i$, where $c_i \geq 0$ but there is at least one $c_i > 0$. It is further assumed that $H = T_2 + U$. In case we can show that H is positive definite in a bounded region Ω_k around the origin in which $H < k$ where $k > 0$, we choose H as the Lyapunov function. Now $\dot{V}$ evaluated along a trajectory of (9.76) is obtained as

$$\dot{V} = \dot{H} = \sum_i \frac{\partial H}{\partial q_i} \dot{q}_i + \sum_i \frac{\partial H}{\partial p_i} \dot{p}_i + \frac{\partial H}{\partial t} \tag{9.77}$$

But in this case, $\partial H / \partial t = 0$ and substituting for $\dot{q}_i$ and $\dot{p}_i$ in (9.77) from (9.76), we get

$$\dot{V} = \sum_i \left[\frac{\partial H}{\partial q_i} \frac{\partial H}{\partial p_i} - \frac{\partial H}{\partial p_i} \frac{\partial H}{\partial q_i} + Q_i \dot{q}_i \right] = -\sum_i^m c_i \dot{q}_i^2 \tag{9.78}$$

In case all $c_i > 0$, then this $\dot{V}$ is negative definite and the origin of (9.76) is asymptotically stable and Ω_k is an estimate of the region of attraction. In case some but not all c_i are zero, $\dot{V}$ becomes negative semidefinite and if we can show that a trajectory of (9.76) cannot remain forever at the points where $\dot{V} = 0$ except at the origin, we can still conclude that the origin of (9.76) is asymptotically stable in Ω_k.

Alternatively, when generalized coordinates and generalized velocities are chosen as the state variables, let the perturbation equations be described by

$$\frac{d}{dt}\left(\frac{\partial L}{\partial \dot{q}_i} \right) - \frac{\partial L}{\partial q_i} = Q_i, \qquad i = 1, \ldots, m$$

where Q_i is restricted as in the foregoing. Choosing the Lyapunov function as

$$V = \sum_{i=1}^m \frac{\partial L}{\partial \dot{q}_i} \dot{q}_i - L(q_1, \ldots, q_m, \dot{q}_i, \ldots, \dot{q}_m) \tag{9.79}$$

we obtain

$$\dot{V} = \sum_{i=1}^m \left[\frac{d}{dt}\left(\frac{\partial L}{\partial \dot{q}_i} \right) \dot{q}_i + \frac{\partial L}{\partial \dot{q}_i} \ddot{q}_i - \frac{\partial L}{\partial q_i} \dot{q}_i - \frac{\partial L}{\partial \dot{q}_i} \ddot{q}_i \right]$$

$$= \sum_{i=1}^m \left[\frac{d}{dt}\left(\frac{\partial L}{\partial \dot{q}_i} \right) - \frac{\partial L}{\partial q_i} \right] \dot{q}_i$$

$$= \sum_{i=1}^m Q_i \dot{q}_i \tag{9.80}$$

This procedure may be considered as the generalization of the procedure employed in Examples 9.15 and 9.17, where total mechanical energy was employed as a Lyapunov function. For conservative systems, $Q_i = 0$ and hence it follows that $\dot{V} = 0$ (i.e., $\dot{V}$ is negative semidefinite throughout the state space). Assuming that $T = T_2$, we choose a Lyapunov function as $V = T + U$. If the potential energy U has a local minimum at the origin, there is a region around the origin in which $U > 0$. Hence, there exists a region Ω_k around the origin in which V is positive definite whereas $\dot{V}$ is negative semidefinite and the origin is stable. This special case is known as Lagrange's theorem, which states that in conservative systems when $T = T_2$, an equilibrium state is stable if the potential energy at that point is a local minimum.

The technique of generating Lyapunov functions has received considerable attention in the past and several methods have been proposed in the literature

[9–11]. Most of the methods are not very general and are restricted to very special system configuration, such as a single-degree-of-freedom system with nonlinearities of a special kind. A discussion of these methods is not given here. In engineering systems with multiple degrees of freedom, closed-form expressions defining the stability boundary can rarely be obtained. An effective computational procedure is desireable for estimating the domain of attraction.

In the following, a computational procedure is described for quadratic estimation of the domain of attraction of the origin of (9.24) when the associated linearized system (9.70) is found to have asymptotically stable origin by application of Theorem 9.1. A quadratic estimate yields a hyperellipsoid for the domain of attraction which can be visualized much more readily than the hypervolumes of the higher-order estimates. A quadratic function is chosen as a Lyapunov function:

$$V = \{y\}^T \mathbf{P}\{y\} \tag{9.81}$$

Then $\dot{V}$ evaluated along a trajectory of (9.24) yields

$$\dot{V} = \{y\}^T(\mathbf{A}^T\mathbf{P} + \mathbf{P}\mathbf{A})\{y\} + 2\{y\}^T\mathbf{P}\{h\} \tag{9.82}$$

Let

$$\mathbf{A}^T\mathbf{P} + \mathbf{P}\mathbf{A} = -\mathbf{Q} \tag{9.83}$$

We select $\mathbf{Q}$ as positive definite. Then $\mathbf{P}$ obtained from the solution of (9.83) is also positive definite since the origin is assumed to be asymptotically stable for sufficiently small perturbations. Also, $\dot{V}$ will be negative definite at least in a sufficiently small region around the origin since $\{h\}$ consists of higher-order terms in $y_1, \ldots, y_n$. Since $\mathbf{A}$ is a stable matrix, every positive definite $\mathbf{Q}$ will result in a positive definite $\mathbf{P}$. Theorem 9.4 requires that $\dot{V}$ be negative within a region Ω_k, exclusive of the origin. The selection of the largest such Ω_k is equivalent to finding the minimum of V on the surface $\dot{V} = 0$ (i.e., minimum $V = c$ subject to the constraint $\dot{V} = 0$, $\{y\} \neq 0$).

At the point of tangency of the V and $\dot{V} = 0$ surfaces, the gradients of V and $\dot{V}$ are in the same direction as shown in Fig. 9.16, Hence, we obtain

$$\vec{\nabla} V = k \, \vec{\nabla}\dot{V} \tag{9.84}$$

where k is an unknown constant. In addition, the equation $\dot{V} = 0$ has to be satisfied. Hence, we obtain

$$\frac{\partial V}{\partial y_1} = k \, \frac{\partial \dot{V}}{\partial y_1}$$

$$\vdots$$

$$\tag{9.85}$$

$$\frac{\partial V}{\partial y_n} = k \, \frac{\partial \dot{V}}{\partial y_n}$$

$$\dot{V} = 0$$

These are $(n + 1)$ nonlinear algebraic equations in the $(n + 1)$ unknowns $y_1, \ldots, y_n, k$. A solution of this set of nonlinear algebraic equations may be

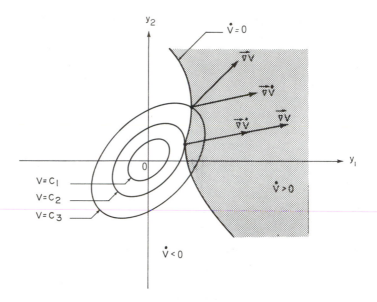

Figure 9.16 Solution for the tangent point.

obtained by employing an appropriate digital computer program such as the Newton–Raphson method. Let the minimum value of V evaluated at the solution point be c. Then, $V = \{y\}^T \mathbf{P}\{y\} = c$ is the hyperellipsoid which is an estimate of the domain of attraction Ω_k. This domain is a function of $\mathbf{P}$ or equivalently of $\mathbf{Q}$ and it is desirable to find the optimal value of this domain. The optional estimate is given by that choice of $\mathbf{Q}$ which maximizes the volume of Ω. This volume is proportional to

$$J(\mathbf{Q}) = \left[\frac{c^n}{\prod_{i=1}^{n} \lambda_i(\mathbf{P})} \right]^{1/2} = \left[\frac{c^n}{\det \mathbf{P}} \right]^{1/2} \tag{9.86}$$

where $\lambda_i(\mathbf{P})$ are the eigenvalues of $\mathbf{P}$, n is the dimension of $\{y\}$, and the expression $J(\mathbf{Q})$ is the product of the principal axes of the hyperellipsoid Ω_k. The optional quadratic estimate of the domain of attraction is given by that choice of $\mathbf{Q}$ that maximizes the volume of Ω_k.

Summarizing, the procedure involves the following steps:

1. Generate arbitrary elements of the $\mathbf{Q}$ matrix.
2. Solve the Lyapunov matrix equation (9.83) for $\mathbf{P}$.
3. Find c which is the minimum value of V subject to the constraint that $\dot{V} = 0$ by solving the nonlinear algebraic equations (9.85).
4. Optimize $\mathbf{Q}$ by calculating $J(\mathbf{Q})$ from (9.86) and repeating with a new positive $\mathbf{Q}$ until no further improvement is made in $J(\mathbf{Q})$.

9.6 STABILITY IN THE LARGE OF NONAUTONOMOUS SYSTEMS

It has been observed earlier that the perturbation equations become autonomous when the equations of motion are autonomous and the stability under consideration is that of an equilibrium or stationary motion. When the stability of a time-varying motion is to be considered, the perturbation equations as given by (9.45) are nonautonomous. Let the perturbation equation be described by

$$\{\dot{y}\} = \{g(y_1, \ldots, y_n, t)\} \tag{9.87}$$

where g_i are nonlinear functions of their arguments and are explicit functions of time. The stability of the time-varying motion $\{x^*(t)\}$ is new equivalent to the stability of the origin of (9.87). Even when a V function is chosen that is not an explicit function of time, the time derivative $\dot{V}$ evaluated along a trajectory of (9.87) will be an explicit function of time. Because $\dot{V}$ and possibly V are explicit functions of time, some modifications are required to the definitions of sign definiteness and to Theorem 9.4. The functions V and $\dot{V}$ are suitably bound by scalar functions that are not explicit functions of time.

Definition 9.14.   A time-varying scalar function $V(y_1, \ldots, y_n, t)$ is positive definite in a region Ω containing the origin if $V(0, \ldots, 0, t) = 0$ and if a continuous and nondecreasing function ϕ exists such that $\phi(0) = 0$ and $V(y_1, \ldots, y_n, t) > \phi(\|\mathbf{y}\|)$ in Ω, where $\phi(x) > 0$ for $x > 0$. A strictly increasing function $\phi(x)$ obeys the property that for $x_2 > x_1$, $\phi(x_2) > \phi(x_1)$. A function $V(y_1, \ldots, y_n, t)$ is called negative definite if $-V(y_1, \ldots, y_n, t)$ is positive definite.

Definition 9.15.   A time-varying scalar function $V(y_1, \ldots, y_n, t)$ is called decrescent in a region Ω containing the origin if a continuous and nondecreasing function ψ exists such that $\psi(0) = 0$ and

$$V(y_1, \ldots, y_n, t) \leq \psi(\|\mathbf{y}\|) \text{ in } \Omega, \qquad \text{where } \psi(x) > 0 \text{ for } x > 0$$

We note that a positive-definite decrescent function $V(y_1, \ldots, y_n, t)$ must dominate the function ϕ and be dominated by the function ψ as shown in Fig. 9.17. A function that is positive definite but not decrescent can become arbitrarily large for $\|\mathbf{y}\|$ arbitrarily small. Some authors use the more stringent requirement that ϕ and ψ in Definitions 9.14 and 9.15, respectively, are strictly increasing functions.

Example 9.18

Consider the following function for a system with two state variables:

$$V(y_1, y_2, t) = y_1^2(1 + \sin^2 t) + y_2^2(1 + \cos^2 t) \tag{9.88}$$

We may choose the ϕ and ψ functions as

$$\phi(\|\mathbf{y}\|) = y_1^2 + y_2^2 = \|\mathbf{y}\|^2$$
$$\psi(\|\mathbf{y}\|) = 2y_1^2 + 2y_2^2 = 2\|\mathbf{y}\|^2$$

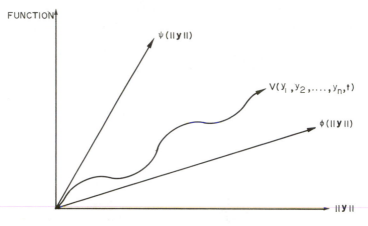

Figure 9.17 Positive-definite decrescent function, $V(y_1, \ldots, y_n, t)$.

Hence, the V function (9.88) is positive definite and decrescent. Now let

$$V(y_1, y_2, t) = y_1^2 + (1 + t)y_2^2, \qquad t > 0 \tag{9.89}$$

We choose the same ϕ function but a ψ function cannot be found since this V function can become arbitrarily large for arbitrarily small $\|\mathbf{y}\| \neq 0$. Hence, the V function of (9.89) is positive definite but not decrescent. When $V(y_1, \ldots, y_n)$ is not an explicit function of time, positive definiteness requires only that $V = 0$ for $\|\mathbf{y}\| = 0$ and $V > 0$ for $\|\mathbf{y}\| \neq 0$. For V, which is an explicit function of time, $\dot{V}$ evaluated along a trajectory of (9.87) becomes

$$\dot{V} = \sum_{i=1}^{n} \frac{\partial V}{\partial y_i} g_i + \frac{\partial V}{\partial t}$$

$$= \{\mathbf{\nabla} V\}^T \{g\} + \frac{\partial V}{\partial t} \tag{9.90}$$

We now state the following theorems concerning the stability of the origin of (9.87).

Theorem 9.6. If there exists a continuously differentiable positive-definite scalar function $V(y_1, \ldots, y_n, t)$ in a region Ω containing the origin such that $\dot{V}(y_1, \ldots, y_n, t) \leq 0$ in Ω, then the origin of (9.87) is stable at time $t_0 \geq 0$ and Ω is the domain of stability for $t \geq t_0$.

Theorem 9.7. If there exists a continuously differentiable positive-definite and decrescent function $V(y_1, \ldots, y_n, t)$ in a region Ω containing the origin such that $\dot{V}(y_1, \ldots, y_n, t) \leq 0$ in Ω, then the origin of (9.87) is uniformly stable at time $t_0 \geq 0$ and Ω is the domain of uniform stability for $t \geq t_0$.

Theorem 9.8. If there exists a continuously differentiable positive-definite and decrescent function $V(y_1, \ldots, y_n, t)$ in a region Ω containing the origin such that $\dot{V}(y_1, \ldots, y_n, t)$ is negative definite in Ω, then the origin of (9.87) is uniformly asymptotically stable at time $t_0 \geq 0$ and Ω is the domain of uniform asymptotic stability for $t \geq t_0$.

Theorem 9.9.   If in Theorem 9.8, the region Ω is the entire state space, and in addition the function $\phi(\|\mathbf{y}\|)$ dominated by $V(y_1, \ldots, y_n, t)$ is such that $\phi(\|\mathbf{y}\|) \to \infty$ as $\|\mathbf{y}\| \to \infty$ and the function $\psi(\|\mathbf{y}\|)$ dominating $V(y_1, \ldots, y_n, t)$ is such that $\psi(\|\mathbf{y}\|) \to \infty$ as $\|\mathbf{y}\| \to \infty$, then the origin of (9.87) is globally uniformly asymptotically stable.

It should be noted that these theorems provide only sufficient conditions. In Theorem 9.6, merely requiring that $V > 0$ for all $\{y\} \neq 0$ and all $t \geq t_0$ is not sufficient to guarantee stability even when $\dot{V} \leq 0$. When V is an explicit function of time, as long as $\partial V / \partial t$ in (9.90) is negative and less than $\{\vec{\nabla} V\}^T \{g\}$, $\dot{V}$ will be negative. It is possible to have $\dot{V} < 0$ while the trajectory moves outside the bounding region Ω. With $V(y_1, \ldots, y_n, t) > \phi\ (\|\mathbf{y}\|)$, this behavior can no longer occur. In order to prove Theorem 9.6, we have to show that given any $\epsilon > 0$, there can be found a δ which may be a function of ϵ and t_0 such that Definition 9.1 is satisfied.

To prove Theorem 9.7, we have to show that δ does not depend on t_0 and hence require that V be also decrescent. To prove Theorem 9.8, we need to show that in addition every trajectory of (9.87) converges to the origin as $t \to \infty$ uniformly in t_0 and $\|\mathbf{y}(t_0)\|$ in Ω, and hence require that $\dot{V}$ be negative definite [i.e., there exists a nondecreasing function θ such that $\theta(0) = 0$, $\theta(x) > 0$ for $x > 0$ and $-\dot{V}(y_1, \ldots, y_n, t) \geq \theta(\|\mathbf{y}\|)$ in Ω]. The proofs of these theorems are given by Vidyasagar [2] and Hahn [9].

It is rather difficult to determine whether a given function is positive definite from Definitions 9.14 because of the need to exhibit the function ϕ. A continuous function $V(y_1, \ldots, y_n, t)$ which is an explicit function of time is positive definite if we can find a positive definite function $W(y_1, \ldots, y_n)$ which is not an explicit function of time such that $V(y_1, \ldots, y_n, t) \geq W(y_1, \ldots, y_n)$ in Ω for all $t \geq 0$. For W to be positive definite, it is only required that $W(0, \ldots, 0) = 0$ and $W(y_1, \ldots, y_n) > 0$ for $\|\mathbf{y}\| \neq 0$.

A final remark about Theorems 9.6 and 9.7 is that in case a trajectory of (9.87) cannot remain forever at points where $\dot{V} = 0$ except at the origin, then it is not true in general that the origin is asymptotically stable. We have seen from Case 2c of Theorem 9.4 (LaSalle's theorem) that it is true for autonomous systems. In the nonautonomous case, it is true when all the time-varying coefficients are periodic in time.

Example 9.19[2]

Consider a linear time-varying parameter system known as a damped Mathieu equation described by

$$\ddot{x} + \dot{x} + (4 + \sin t)x = 0, \qquad t \geq 0 \qquad (9.91)$$

In state-variable form, we have

$$\dot{x}_1 = x_2$$
$$\dot{x}_2 = -(4 + \sin t)x_1 - x_2 \tag{9.92}$$

The only equilibrium state is given by $x_{1e} = 0$, $x_{2e} = 0$ for all $t \geq 0$. Letting y_1 and y_2 be the perturbations about x_{1e} and x_{2e}, respectively, the perturbation equations are described by

$$\dot{y}_1 = y_2$$
$$\dot{y}_2 = -(4 + \sin t)y_1 - y_2 \tag{9.93}$$

We choose a candidate for the Lyapunov function as

$$V(y_1, y_2, t) = y_1^2 + \frac{y_2^2}{4 + \sin t}$$

This V function is continuously differentiable. Also, V dominates the positive-definite function

$$W_1(y_1, y_2) = y_1^2 + \frac{y_2^2}{5}$$

and is dominated by the positive-definite function

$$W_2(y_1, y_2) = y_1^2 + y_2^2$$

Hence, this V function is positive definite and decrescent. Then,

$$\dot{V}(y_1, y_2, t) = 2y_1\dot{y}_1 + \frac{2y_2\dot{y}_2}{4 + \sin t} - \frac{\cos t}{(4 + \sin t)^2}y_2^2$$

Substituting for $\dot{y}_1$ and $\dot{y}_2$ in the foregoing equation from (9.93), it follows that

$$\dot{V}(y_1, y_2, t) = 2yy_2 + \frac{2y_2}{4 + \sin t}[-(4 + \sin t)y_1 - y_2] - \frac{\cos t}{(4 + \sin t)^2}y_2^2$$
$$= -y_2^2 \frac{2(4 + \sin t) + \cos t}{(4 + \sin t)^2}$$
$$= -y_2^2 \frac{8 + 2\sin t + \cos t}{(4 + \sin t)^2}$$
$$\leq 0$$

Hence, according to Theorem 9.7 the origin of (9.93) [i.e., the equilibrium ($x_1 = 0$, $x_2 = 0$) of (9.92)] is uniformly stable. Now, $\dot{V} = 0$ all along the y_1 axis where $y_2 = 0$, which from (9.93) implies that y_1 is a constant and hence $(4 + \sin t)y_1 = 0$. Then $y_1 = 0$ and a trajectory of (9.93) cannot remain forever at points where $\dot{V} = 0$ except at the origin. Since the time-varying coefficient in (9.93) is periodic, we conclude that the origin of (9.93) is uniformly asymptotically stable. Furthermore, both $W_1 \to \infty$ and $W_2 \to \infty$ as $\|y\| \to \infty$. Hence, the origin of (9.93) is globally uniformly asymptotically stable.

Mathieu–Hill equation.   The example just considered belongs to a more general class of equations described by

$$\ddot{x} + a[1 + 2bp(t)]x = 0$$

where $p(t)$ is a periodic function of time and a and b are parameters reflecting the system properties. This equation is known as the undamped Hill equation or Mathieu–Hill equation. The corresponding linearly damped equation is described by

$$\ddot{x} + c\dot{x} + a[1 + 2bp(t)]x = 0$$

Mathieu–Hill type of equations are encountered in many areas of engineering, celestial mechanics, and theory of parametrically excited oscillations. For such second-order equations with periodic coefficients, Floquet's theory provides a form of solution which is useful to investigate the stability boundary in the parameter space. Interested readers may consult references [12] and [13] for this purpose.

Generation of Lyapunov functions for nonautonomous systems.

It has been observed earlier that there is a lack of general procedures for the generation of Lyapunov functions for nonlinear autonomous systems. This difficulty is compounded in the case of nonautonomous systems. It was shown that for linearized autonomous systems, a Lyapunov function is of the quadratic form. Even for the linearized nonautonomous equation $\{\dot{y}\} = \mathbf{A}(t)\{y\}$, a systematic procedure for the selection of a Lyapunov function is lacking. Some methods have been proposed in the literature but they yield very conservative results, if any. It is a small help to the designer of an automobile to know that a lane-change maneuver of an automobile is stable if the speed is 5 km/h. The designer would have to know the limit velocity at which the lane-change maneuver becomes unstable so that the design can be improved dynamically and the limit velocity can be raised.

Hence, direct integration of the equations of motion by computer simulation offers an attractive alternative for the stability investigation of a general time-varying motion. But computer simulation requires considerable caution, as seen in Chapter 7. The instability exhibited by the computer simulation may turn out to be numerical instability, especially in nonlinear systems, and not physical instability. Some integration techniques may introduce numerical damping and stabilize a physically unstable motion. In nonlinear systems, superposition becomes invalid and if a motion is stable for one set of initial conditions, it does not imply that the motion will remain stable when the initial conditions are changed. Hence, the determination of the domain of stability by computer simulation becomes a time-consuming task.

9.7 SUMMARY

In this chapter we have studied the stability in the sense of Lyapunov of equilibrium states, stationary motions, and time-varying motions. The three basic concepts of Lyapunov theory are stability, asymptotic stability, and instability. In the autonomous case, stability and asymptotic stability are always uniform, but in the nonautonomous case, a distinction must be made between stability and uniform stability. There are some applications where the concept of stability, in the sense of Lyapunov, is not an appropriate one. The concept of orbital stability has been introduced and it is appropriate for the stability of periodic motion and limit cycle oscillations. However, orbital stability has not been further employed in this chapter.

In the earlier part of this chapter, we studied a method whereby the stability of a nonlinear system is ascertained by assuming that the perturbations are sufficiently small and linearizing the perturbation equations. This method is also known as Lyapunov's first or indirect method. It is commonly employed in many fields of engineering. It is a recommended first step for autonomous systems when it is applicable.

The basic feature of Lyapunov's second or direct method is that stability is investigated without solving the system equations by selecting a suitable Lyapunov function. This Lyapunov function may be regarded as a generalized energy and in many dynamic systems, the total mechanical energy is a candidate for the Lyapunov function. A serious disadvantage of the method is that there is no systematic procedure for the generation of Lyapunov functions and this difficulty is compounded for nonautonomous systems. The theorems present only sufficient conditions for various forms of stability and if a suitable Lyapunov function cannot be found, no conclusions can be reached regarding stability. This is a very serious drawback of Lyapunov's second method.

PROBLEMS

9.1. The motion of a particle of mass m is resisted by a force that is proportional to the exponential function of its velocity v so that the equation of motion is described by

$$m\dot{v} + ce^v = 0$$

Investigate the equilibrium state(s). State, giving reasons, whether Lyapunov stability theory can be employed to investigate the stability of the equilibrium state(s).

9.2. Consider the system of Example 3.4 except that the Coulomb friction is replaced by viscous damping and $P = 0$. Hence, (3.26) is modified but (3.27) remains unchanged. Thus, the equations of motion are described by

$$m_1\ddot{x} + m_2\ddot{x}\sin^2\theta - m_2g\sin\theta\cos\theta - m_2b\dot{\theta}^2\sin\theta + c\dot{x} = 0$$

$$m_2\ddot{x}\cos\theta + m_2b\ddot{\theta} + m_2g\sin\theta = 0$$

(a) Determine the equilibrium states.

(b) By employing Theorem 9.1, investigate the stability of the equilibrium state ($x = $ constant, $\dot{x} = 0$, $\theta = \pi$, $\dot{\theta} = 0$) for small perturbations.

9.3. A boom that may be considered as a slender rod of length b and mass m is being transported on a crawler which is moving in a straight line at a constant velocity v (Fig. P9.3). The boom is pivoted at A, where there is viscous friction, and attached to the crawler frame at B by a linear spring of stiffness k. When the

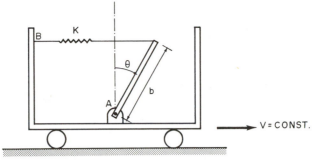

Figure P9.3

boom is displaced through an angle θ, the kinetic and potential energies are given by

$$T = \frac{1}{6} mb^2 \dot{\theta}^2$$

$$U = \frac{1}{2} kb^2 \sin^2 \theta - \frac{mgb}{2}(1 - \cos \theta)$$

(a) Obtain the equation of motion of the boom.
(b) Determine the equilibrium states.
(c) By employing Theorem 9.1, determine the condition relating k, m, b, and g for asymptotic stability of the equilibrium state ($\theta = 0$, $\dot{\theta} = 0$) for small perturbations.

9.4. The Lagrange equations of motion for the spring pendulum of Example 5.11 are given by (5.102) and (5.103).
(a) Obtain all the equilibrium states.
(b) By employing Theorem 9.1, investigate the stability of the equilibrium states for small perturbations.

9.5. Determine the sign definiteness of the following functions. In each case, n is the number of state variables.
(a) $V(x_1, x_2, x_3) = x_1^2 + 4x_1 x_2 + x_2^2 + x_2 x_3 + 3x_3^2$ $n = 3$
(b) $V(x_1, x_2, x_3) = x_1^4 + x_1^2 x_2^2$ $n = 3$
(c) $V(x_1, x_2) = x_1^4 + \frac{1}{2}x_1^2 + x_1 x_2 + x_2^2$ $n = 2$

9.6. A mass m is suspended between two linear springs, each of stiffness k (Fig. P9.6). The friction force is due to Coulomb friction. The equation of motion is given by

$$m\ddot{x} + c \operatorname{sgn} \dot{x} + 2kx = 0$$

By considering the total mechanical energy as a Lyapunov function, investigate the stability of the equilibrium zone and the size of the region of stability.

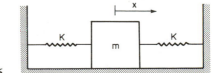

Figure P9.6

9.7. Consider the mass–spring system shown in Fig. P9.7(a). The spring characteristic shown in Fig. P9.7(b) can be modeled such that the spring force is given by $k\,|x|\,x$. The equation of motion becomes

$$m\ddot{x} + kx\,|x| = 0$$

Choosing the total mechanical energy as a Lyapunov function, show that the equilibrium state ($x = 0$, $\dot{x} = 0$) is stable and obtain the region of stability.

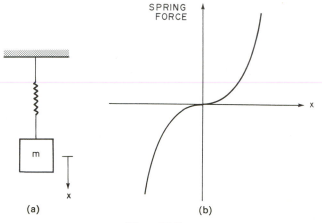

Figure P9.7

9.8. The Van der Pol equation is described by (9.16). By choosing a V_1 function as $V_1(x_1, x_2) = x_1^2 + x_2^2$ and employing Theorem 9.5 show that the origin ($x_1 = 0$, $x_2 = 0$) is an unstable equilibrium point.

9.9. The tumbling motion of an orbiting rigid-body satellite about its center of mass, where the tumbling rate far exceeds the orbiting rate, is described by the Euler equation (4.47). It is desired to stop the tumbling by applying control torques proportional to the angular velocities (i.e., $M_1 = -k_1\omega_1$, $M_2 = -k_2\omega_2$, $M_3 = -k_3\omega_3$, where $k_i > 0$). The equations of motion of the controlled satellite become

$$\dot{\omega}_1 = \frac{1}{I_1}(I_2 - I_3)\omega_2\omega_3 - \frac{k_1}{I_1}\omega_1$$

$$\dot{\omega}_2 = \frac{1}{I_2}(I_3 - I_1)\omega_1\omega_3 - \frac{k_2}{I_2}\omega_2$$

$$\dot{\omega}_3 = \frac{1}{I_3}(I_1 - I_2)\omega_1\omega_2 - \frac{k_3}{I_3}\omega_3$$

By choosing the square of the norm of the total angular momentum as a candidate Lyapunov function (i.e., $V = I_1^2\omega_1^2 + I_2^2\omega_2^2 + I_3^2\omega_3^2$), show that the origin is globally asymptotically stable.

9.10. Investigate whether the following time-varying V functions are positive definite and decrescent. In each case, the number of state variables is two.
(a) $V(x_1, x_2, t) = x_1^2 + (2 + e^{-3t})x_2^2$
(b) $V(x_1, x_2, t) = e^{-3t}(x_1^2 + x_2^2)$

(c) $V(x_1, x_2, t) = x_1^2 + tx_2^2$

(d) $V(x_1, x_2, t) = x_1^2 + (1 + \sin^2 t)x_2^2$

9.11. Let the perturbation equations be described by

$$\dot{x}_1 = x_2$$

$$\dot{x}_2 = -cx_2 - g(x_1, t)x_1$$

By choosing a V function as

$$V = \frac{1}{2}cax_1^2 + ax_1x_2 + \frac{1}{2}x_2^2 + \int_0^{x_1} g(x_1, t)x_1 \, dx_1$$

where $c > a > 0$, obtain the conditions that the function $g(x_1, t)$ must satisfy so that the origin is uniformly, asymptotically stable.

REFERENCES

1. Hsu, J. C., and Meyer, A. U., *Modern Control Principles and Applications*, Mc-Graw-Hill Book Company, New York, 1968.

2. Vidyasagar, M., *Nonlinear Systems Analysis*, Prentice-Hall, Inc., Englewood Cliffs, N.J., 1978.

3. Routh, E. J., *Dynamics of a System of Rigid Bodies*, Dover Publications, Inc., New York, 1955.

4. Aggarwal, J. K., and Infante, E. F., "Some Remarks on the Stability of Time-Varying Systems," *IEEE Transactions on Automatic Control*, Vol. AC-13, Dec. 1968, pp. 722–723.

5. Bellman, R. E., *Introduction to Matrix Analysis*, 2nd ed., McGraw-Hill Book Company, New York, 1970.

6. LaSalle, J., and Lefshetz, S., *Stability by Lyapunov's Direct Method*, Academic Press, Inc., New York, 1961.

7. Meirovitch, L., *Methods of Analytical Dynamics*, McGraw-Hill Book Company, New York, 1970.

8. Chen, C. T., *Introduction to Linear System Theory*, Holt, Rinehart and Winston, New York, 1971.

9. Hahn, W., *Theory and Application of Lyapunov's Direct Method*, Prentice-Hall, Inc., Englewood Cliffs, N.J., 1963.

10. Zubov, V. J., *Methods of A. M. Lyapunov and Their Applications*, P. Noordhoff, Groningen, The Netherlands, 1964.

11. Lefshetz, S., *Stability of Nonlinear Control Systems*, Academic Press, New York, 1965.

12. McLachlan, N. W., *Theory and Applications of Mathieu Functions*, Oxford University Press, New York, 1947.

13. Bolotin, V. V., *The Dynamic Stability of Elastic Systems*, Holden-Day, Inc., San Francisco, 1964.

Appendix A

MATRIX ALGEBRA

A matrix is defined as the assemblage of a set of numbers in a rectangular array of rows and columns. The size of a matrix is $m \times n$ if it has m rows and n columns, as shown in (A.1).

$$\mathbf{A} = [a_{ij}] = \begin{bmatrix} a_{11} & a_{12} & \cdots & a_{1n} \\ a_{21} & a_{22} & \cdots & a_{2n} \\ \cdot & \cdot & & \cdot \\ \cdot & \cdot & & \cdot \\ \cdot & \cdot & & \cdot \\ a_{m1} & a_{m2} & & a_{mn} \end{bmatrix} \tag{A.1}$$

The subscripts of a general term a_{ij} specify the position of the term. The first subscript is the row position and the second the column position. The number of columns need not be equal to the number of rows.

A matrix having a single column is said to be a column matrix written as

$$\mathbf{B} = \begin{Bmatrix} b_{11} \\ b_{21} \\ \cdot \\ \cdot \\ \cdot \\ b_{m1} \end{Bmatrix} \tag{A.2}$$

A matrix having a single row is called a row matrix, shown as

$$\mathbf{C} = \lfloor c_{11} \quad c_{12} \quad \cdots \quad c_{1n} \rfloor \tag{A.3}$$

A square matrix has the same number of rows and columns.

A diagonal matrix is a square matrix having all its elements zero except those on the leading diagonal. A unit matrix is a diagonal matrix whose diagonal elements are each equal to unity.

The transpose of a matrix is a matrix with the rows and columns interchanged from the original matrix:

$$\mathbf{A}^T = \begin{bmatrix} a_{11} & a_{21} & \cdots & a_{m1} \\ a_{12} & a_{22} & \cdots & a_{m2} \\ \cdot & \cdot & & \cdot \\ \cdot & \cdot & & \cdot \\ \cdot & \cdot & & \cdot \\ a_{1n} & a_{2n} & \cdots & a_{mn} \end{bmatrix} \tag{A.4}$$

A symmetrical matrix is a square matrix whose elements are symmetrical about its leading diagonal. A symmetrical matrix is equal to its transpose:

$$a_{ij} = a_{ji} \tag{A.5}$$

An antisymmetrical matrix is a square matrix whose elements are symmetrical but with opposite sign about its leading diagonal:

$$a_{ij} = -a_{ji} \tag{A.6}$$

A triangular matrix is a square matrix which has zero elements either below or above the leading diagonal.

Two matrices can be added together or subtracted from each other only when they have equal numbers of rows and columns. Each element of the resultant matrix is equal to the addition or subtraction of the corresponding elements of the two matrices. It follows that

$$\begin{aligned} \text{if } \mathbf{C} = \mathbf{A} + \mathbf{B}, \quad & c_{ij} = a_{ij} + b_{ij} \\ \text{if } \mathbf{D} = \mathbf{A} - \mathbf{B}, \quad & d_{ij} = a_{ij} - b_{ij} \end{aligned} \tag{A.7}$$

The following rules hold true for matrix addition:

1. Commutative law:
$$\mathbf{A} + \mathbf{B} = \mathbf{B} + \mathbf{A}.$$

2. Associative law:
$$\mathbf{A} + (\mathbf{B} + \mathbf{C}) = (\mathbf{A} + \mathbf{B}) + \mathbf{C}.$$

3. The sum of the transposes of two matrices is equal to the transpose of their sum.

4. Any square matrix can be broken into two parts, one symmetrical and one antisymmetrical.

Two matrices $\mathbf{A}$ and $\mathbf{B}$ can be multiplied only when $\mathbf{A}$ has the same number of columns as $\mathbf{B}$ has rows. The resultant matrix $\mathbf{C}$ has the same number of rows as $\mathbf{A}$ and the same number of columns as $\mathbf{B}$. We get

$$\mathbf{C} = [C_{pq}] = \mathbf{A} \times \mathbf{B} = [a_{pr}][b_{rq}] \tag{A.8}$$

The following rules hold true for matrix multiplication:

1. Premultiplication of **A** by **B** does not equal postmultiplication of **A** by **B**:

$$\mathbf{A} \times \mathbf{B} \neq \mathbf{B} \times \mathbf{A} \qquad (A.9)$$

2. Distributive law:

$$\mathbf{A(B + C)} = \mathbf{AB} + \mathbf{AC} \qquad (A.10)$$

3. Associative law:

$$\mathbf{A(BC)} = \mathbf{AB(C)} \qquad (A.11)$$

4. The product of two transposed matrices is equal to the transpose of the product of the original matrices in reverse order:

$$\mathbf{A}^T\mathbf{B}^T = \mathbf{(BA)}^T = \mathbf{C} \qquad (A.12)$$

5. Any matrix **A** multiplied by a unit matrix **I** gives a product identical with **A**:

$$\mathbf{AI} = \mathbf{A} \qquad (A.13)$$

There is no direct division of matrices. The operation of division is performed by inversion; if

$$\mathbf{PQ} = \mathbf{R} \qquad (A.14)$$

then

$$\mathbf{Q} = \mathbf{P}^{-1}\mathbf{R} \qquad (A.15)$$

when $\mathbf{P}^{-1}$ is called the inverse of matrix **P**. The requirements for obtaining a unique inverse of a matrix are:

1. The matrix is a square matrix.
2. The determinant of the matrix is not zero (the matrix is nonsingular).

The inverse of a matrix is also defined by the relationship that

$$\mathbf{P}^{-1}\mathbf{P} = \mathbf{I} \qquad (A.16)$$

The following are the properties of an inverted matrix:

1. The inverse of a matrix is unique.
2. The inverse of the product of two matrices is equal to the product of the inverse of the two matrices in reverse order:

$$\mathbf{(AB)}^{-1} = \mathbf{B}^{-1}\mathbf{A}^{-1} \qquad (A.17)$$

3. The inverse of a triangular matrix is itself a triangular matrix of the same type.
4. The inverse of a symmetrical matrix is itself a symmetrical matrix.
5. The negative powers of a nonsingular matrix are obtained by raising the inverse of the matrix to positive powers.

6. The inverse of the transpose of $\mathbf{P}$ is equal to the transpose of the inverse of $\mathbf{P}$:

$$(\mathbf{P}^T)^{-1} = (\mathbf{P}^{-1})^T \tag{A.18}$$

Characteristic Equation and Eigenvalues

We consider a set of linear simultaneous equations in the form

$$\mathbf{AX} = \lambda \mathbf{X} \tag{A.19}$$

where $\mathbf{A}$ is a square matrix, $\mathbf{X}$ is a column matrix, and λ is a number.

We can rewrite (A.19) as

$$\begin{bmatrix} a_{11} - \lambda & a_{12} & \cdots & a_{1n} \\ a_{21} & a_{22} - \lambda & \cdots & a_{2n} \\ \vdots & \vdots & & \vdots \\ a_{n1} & \cdots & \cdots & a_{nn} - \lambda \end{bmatrix} \begin{Bmatrix} x_1 \\ x_2 \\ \vdots \\ x_n \end{Bmatrix} = 0$$

or

$$(\mathbf{A} - \lambda \mathbf{I})\mathbf{X} = 0 \tag{A.20}$$

A nontrivial solution of (A.20) can exist only when the determinant of $(\mathbf{A} - \lambda \mathbf{I})$ vanishes, or

$$|\mathbf{A} - \lambda \mathbf{I}| = 0 \tag{A.21}$$

Equation (A.21) is called the *characteristic equation*. The roots $\lambda_1, \lambda_2, \ldots, \lambda_n$ of the characteristic equation are called the eigenvalues of matrix $\mathbf{A}$. When each root is substituted back into (A.19), we obtain a set of linear equations which are not all independent. By assuming value of one x, say x_1, and discarding one equation we can solve for the values of other x's. The column matrix obtained by this procedure is called a *characteristic vector* or eigenvector. Thus, there is one characteristic vector for each eigenvalue. Hence, only the direction of the eigenvectors is obtained and their length is arbitrary and may be normalized to unity. The following theorems are valid:

Theorem A.1. If a real matrix $\mathbf{A}$ has eigenvalues λ_i and characteristic vectors $\mathbf{X}_i$, then $\mathbf{A}^T$ has the same eigenvalues λ_i but with characteristic vectors $\mathbf{Y}_i$ orthogonal to $\mathbf{X}_i$:

$$\mathbf{Y}_j^T\mathbf{X}_i = \begin{cases} 1 & \text{when } i = j \\ 0 & \text{when } i \neq j \end{cases} \tag{A.22}$$

where the vectors $\mathbf{X}$ and $\mathbf{Y}$ are normalized.

Theorem A.2. If a matrix $\mathbf{A}$ is symmetric and all its elements are real numbers, then all its eigenvalues and characteristic vectors are real. Moreover, the characteristic vectors are orthogonal to each other:

$$\mathbf{X}_i^T \mathbf{X}_J = \begin{cases} 1 & \text{when } i = j \\ 0 & \text{when } i \neq j \end{cases} \tag{A.23}$$

Theorem A.3.   The determinant of a matrix is equal to the product of all its eigenvalues:

$$|\mathbf{A}| = \lambda_1 \lambda_2 \cdots \lambda_n \tag{A.24}$$

Theorem A.4.   The trace of a matrix which is the sum of the elements on the leading diagonal of a matrix is equal to the sum of its eigenvalues:

$$a_{11} + a_{22} + \cdots + a_{nn} = \lambda_1 + \lambda_2 + \cdots + \lambda_n \tag{A.25}$$

Theorem A.5.   If the eigenvalues and characteristic vectors of matrix $\mathbf{A}$ are λ_i and $\mathbf{X}_i$, then a matrix $\mathbf{B} = \mathbf{T}\mathbf{A}\mathbf{T}^{-1}$ has the same eigenvalues λ_i but characteristic vectors equal to $\mathbf{T}\mathbf{X}_i$, $\mathbf{T}$ being any nonsingular square matrix.

REFERENCES

1. Wang, P. C., *Numerical and Matrix Methods in Structural Mechanics with Applications to Computers*, John Wiley & Sons, Inc., New York, 1966.

2. Laursen, H. I., *Matrix Analysis of Structures*, McGraw-Hill Book Company, New York, 1966.

3. Kaplan, W., *Advanced Calculus*, Addision-Wesley Publishing Company, Inc., Reading, Mass., 1952.

Appendix B

VECTOR ALGEBRA AND ANALYSIS

It is very advantageous to employ vector notation in dynamics. First, it permits us to express many laws and formulas in a form independent of the coordinate system, and second, it facilitates a simple and compact description of many relations and their manipulation. The forces, moments, displacements, velocities, accelerations, and other quantities of dynamics such as linear and angular momenta are generally expressed by vectors in three-dimensional space. The state variables, on the other hand, form an n-dimensional vector space. Hence, it is understood when dealing with state variables as in Chapter 9 that the vector space is n-dimensional.

Definition B.1. An n-dimensional complex vector $\vec{a}$ is an ordered n-tuple of complex numbers $(a_1, a_2, \ldots, a_n)$ which form an n-dimensional vector space. If the ordered n-tuple $(a_1, a_2, \ldots, a_n)$ admits only real numbers, then we define an n-dimensional real vector $\vec{a}$.

Definition B.2. The vectors $\vec{a} = (a_1, a_2, \ldots, a_n)$ and $\vec{b} = (b_1, b_2, \ldots, b_n)$ are said to be equal if and only if $a_1 = b_1, a_2 = b_2, \ldots, a_n = b_n$. The sum of vectors $\vec{a} = (a_1, a_2, \ldots, a_n)$ and $\vec{b} = (b_1, b_2, \ldots, b_n)$ is the vector $\vec{a} + \vec{b} = (a_1 + b_1, \ldots, a_n + b_n)$. The product of a vector $\vec{a} = (a_1, a_2, \ldots, a_n)$ and a scalar number c is the vector $c\vec{a} = (ca_1, ca_2, \ldots, ca_n)$.

Laws of Vector Operation

Vector operations satisfy the following rules:

1. Commutative law: $\vec{a} + \vec{b} = \vec{b} + \vec{a}$.
2. Associative law: $(\vec{a} + \vec{b}) + \vec{c} = \vec{a} + (\vec{b} + \vec{c})$.
3. Distributive law: $c(\vec{a} + \vec{b}) = c\vec{a} + c\vec{b}$ and $(c + d)\vec{a} = c\vec{a} + d\vec{a}$, where c and d are numbers.
4. $c(d\vec{a}) = (cd)\vec{a}$; $0\vec{a} = \vec{0}$, where $\vec{0}$ is called the zero vector or null vector; $c\vec{0} = \vec{0}$.
5. The equality $c\vec{a} = \vec{0}$ holds if and only if $c = 0$ or $\vec{a} = \vec{0}$.
6. $-(c\vec{a}) = (-c)\vec{a} = c(-\vec{a})$.

Definition B.3. The vectors $\vec{a}_1, \vec{a}_2, \ldots, \vec{a}_k$ are said to be linearly dependent if there exist numbers $c_1, \ldots, c_k$, which are not all zero, such that

$$c_1\vec{a}_1 + c_2\vec{a}_2 + \cdots + c_k\vec{a}_k = \vec{0} \qquad (B.1)$$

If the vectors $\vec{a}_1, \ldots, \vec{a}_k$ are not linearly dependent, we say that they are linearly independent.

Example B.1

The vectors $\vec{a}_1 = (1, -1, 0)$, $\vec{a}_2 = (0, -2, 1)$, $\vec{a}_3 = (2, 4, -3)$, are linearly dependent, since

$$2\vec{a}_1 + (-3)\vec{a}_2 + (-1)\vec{a}_3 = \vec{0}$$

The vectors $\vec{i} = (1, 0, 0)$, $\vec{j} = (0, 1, 0)$, $\vec{k} = (0, 0, 1)$ are linearly independent.

Definition B.4. A vector $\vec{a}$ is said to be a linear combination of vectors $\vec{a}_1, \ldots, \vec{a}_k$ if numbers $c_1, \ldots, c_k$ exist such that

$$\vec{a} = c_1\vec{a}_1 + \cdots + c_k\vec{a}_k \qquad (B.2)$$

Vectors $\vec{a}_1, \ldots, \vec{a}_m$ are linearly dependent if and only if at least one of them can be expressed as a linear combination of the others.

Example B.2

The vectors $\vec{a}_1 = (3, 1, 2)$, $\vec{a}_2 = (-1, 0, 2)$, $\vec{a}_3 = (7, 2, 2)$ are linearly dependent, since $2\vec{a}_1 - \vec{a}_2 - \vec{a}_3 = \vec{0}$. From this equation it follows that

$$\vec{a}_1 = \tfrac{1}{2}\vec{a}_2 + \tfrac{1}{2}\vec{a}_3, \qquad \vec{a}_3 = 2\vec{a}_1 - \vec{a}_2, \qquad \vec{a}_2 = 2\vec{a}_1 - \vec{a}_3$$

so that each of them is a linear combination of the other two.

Definition B.5. The length of a vector $\vec{a} = (a_1, a_2, \ldots, a_n)$ is the non-negative number $[a_1^2 + a_2^2 + \cdots + a_n^2]^{1/2}$ which is denoted by $|\vec{a}|$ or a. A vector whose length equals unity is called a unit vector. The terms modulus, magnitude, norm, or absolute value of a vector are also used for the length of a vector.

Three-dimensional Vectors

In the rest of this appendix we shall restrict ourselves to three-component vectors, vectors in three-dimensional space. In rectangular Cartesian coordinate system, we employ x, y, and z to represent the three coordinates. The linearly independent unit vectors $\vec{i} = (1, 0, 0)$, $\vec{j} = (0, 1, 0)$, and $\vec{k} = (0, 0, 1)$ in the directions x, y, and z, respectively, as shown in Fig. B.1 are called coordinate vectors. In three-dimensional space, any four vectors are linearly dependent.

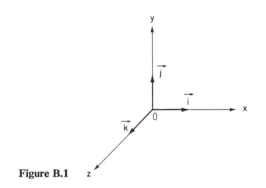

Figure B.1 z

Thus, every vector $\vec{a} = (a_1, a_2, a_3)$ can be expressed as a linear combination of the coordinate vectors $\vec{i}$, $\vec{j}$, and $\vec{k}$ as

$$\vec{a} = a_1\vec{i} + a_2\vec{j} + a_3\vec{k}$$

Two linearly dependent vectors are called collinear (parallel). Two vectors $\vec{a}$ and $\vec{b}$ are linearly dependent (parallel) if and only if one of them is a multiple of the other; that is, there is a number c such that $\vec{a} = c\vec{b}$. Three linearly dependent vectors are called coplanar.

Definition B.6.   The angle between two nonzero vectors $\vec{a}$ and $\vec{b}$ is the angle $\theta(0 \leq \theta \leq \pi)$ between the directed segments representing both vectors.

Scalar Product of Two Vectors

Definition B.7.   The scalar product or inner product or dot product of two vectors $\vec{a} = (a_1, a_2, a_3)$ and $\vec{b} = (b_1, b_2, b_3)$, written as $\vec{a} \cdot \vec{b}$, is defined by

$$\vec{a} \cdot \vec{b} = a_1b_1 + a_2b_2 + a_3b_3 \tag{B.3}$$

It can also be shown that

$$\vec{a} \cdot \vec{b} = |\vec{a}||\vec{b}| \cos \theta \tag{B.4}$$

where $|\vec{a}|$ and $|\vec{b}|$ are the magnitudes, respectively, and θ is the angle between the vectors.

Example B.3

Using the definition of the scalar product, we obtain

$$\vec{i} \cdot \vec{i} = 1, \qquad \vec{j} \cdot \vec{j} = 1, \qquad \vec{k} \cdot \vec{k} = 1$$
$$\vec{i} \cdot \vec{j} = 0, \qquad \vec{j} \cdot \vec{k} = 0, \qquad \vec{k} \cdot \vec{i} = 0$$

(B.5)

Example B.4

We compute the angle between two vectors which are given by their components $\vec{a} = (2, 1, 2)$ and $\vec{b} = (1, -1, 4)$. We have

$$\cos \theta = \frac{\vec{a} \cdot \vec{b}}{|\vec{a}||\vec{b}|} = \frac{2(1) + 1(-1) + 2(4)}{[2^2 + 1^2 + 2^2]^{1/2}[1^2 + (-1)^2 + 4^2]^{1/2}}$$

$$\cos \theta = \frac{1}{\sqrt{2}}$$

or

$$\theta = \frac{\pi}{4}$$

(B.6)

Perpendicular vectors. Two nonzero vectors $\vec{a}$ and $\vec{b}$ are perpendicular if and only if $\vec{a} \cdot \vec{b} = 0$.

Properties of scalar product of vectors. The scalar product of vectors satisfies the relations:

1. $\vec{a} \cdot \vec{b} = \vec{b} \cdot \vec{a}$ (i.e., it is commutative).
2. $(\vec{a} + \vec{b}) \cdot \vec{c} = \vec{a} \cdot \vec{c} + \vec{b} \cdot \vec{c}$ (i.e., it is distributive).
3. $\vec{a} \cdot \vec{a} = |\vec{a}|^2$.

Definition B.8. The angles that a nonzero vector makes with the coordinate vectors and thus with the coordinate axes are called the direction angles and their cosines the direction cosines of the given vector.

Let α, β, and γ be the direction angles of a nonzero vector $\vec{a} = (a_1, a_2, a_3)$. The direction cosines obey the rules

1. $\cos \alpha = a_1/|\vec{a}|$, $\cos \beta = a_2/|\vec{a}|$, $\cos \gamma = a_3/|\vec{a}|$ (B.7)
2. $\cos^2 \alpha + \cos^2 \beta + \cos^2 \gamma = 1$

Vector Product

Definition B.9. The vector product or cross product or outer product of vectors $\vec{a} = (a_1, a_2, a_3)$ and $\vec{b} = (b_1, b_2, b_3)$, denoted by $\vec{a} \times \vec{b}$, is a vector $\vec{c}$ defined by

$$\vec{a} \times \vec{b} = \vec{c}$$

(B.8)

where $\vec{c}$ is a vector perpendicular to both $\vec{a}$ and $\vec{b}$, is in the sense of a right-hand screw, and has magnitude $|\vec{a}||\vec{b}| \sin \theta$, as shown in Fig. B.2. Here, θ is the angle between $\vec{a}$ and $\vec{b}$. The vector product $\vec{a} \times \vec{b}$ may be written as

$$\vec{a} \times \vec{b} = (a_2 b_3 - a_3 b_2)\vec{i} + (a_3 b_1 - a_1 b_3)\vec{j} + (a_1 b_2 - a_2 b_1)\vec{k}$$

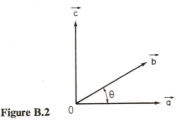

Figure B.2   0

This result is also obtained by the expansion of the determinant

$$\vec{a} \times \vec{b} = \begin{vmatrix} \vec{i} & \vec{j} & \vec{k} \\ a_1 & a_2 & a_3 \\ b_1 & b_2 & b_3 \end{vmatrix} \tag{B.9}$$

Properties of vector product.   The vector product satisfies the following relations:

1. It is not commutative (i.e., $\vec{a} \times \vec{b} \neq \vec{b} \times \vec{a}$ but $\vec{a} \times \vec{b} = -\vec{b} \times \vec{a}$.
2. It is distributive [i.e., $\vec{a} \times (\vec{b} + \vec{c}) = \vec{a} \times \vec{b} + \vec{a} \times \vec{c}$].
3. It is not associative [i.e., $\vec{a} \times (\vec{b} \times \vec{c}) \neq (\vec{a} \times \vec{b}) \times \vec{c}$].
4. $k\vec{a} \times \vec{b} = k(\vec{a} \times \vec{b})$, where k is a scalar number.

Example B.5
The vector product of two linearly dependent vectors is the zero vector. Hence, it follows that

$$\vec{i} \times \vec{i} = \vec{0}, \qquad \vec{j} \times \vec{j} = \vec{0}, \qquad \vec{k} \times \vec{k} = \vec{0} \tag{B.10}$$

Also, it can be seen that

$$\vec{i} \times \vec{j} = \vec{k}, \qquad \vec{j} \times \vec{k} = \vec{i}, \qquad \vec{k} \times \vec{i} = \vec{j} \tag{B.11}$$

Mixed Product

Definition B.10.   The mixed product or triple scalar product of three vectors $\vec{a}, \vec{b}, \vec{c}$ is a scalar denoted by $\vec{a} \cdot (\vec{b} \times \vec{c})$. The mixed product of three vectors is also sometimes called a trivector.

Properties of mixed product.   The mixed product satisfies the following relations:

$$\vec{a} \cdot (\vec{b} \times \vec{c}) = \vec{b} \cdot (\vec{c} \times \vec{a}) = \vec{c} \cdot (\vec{a} \times \vec{b})$$
$$= -\vec{a} \cdot (\vec{c} \times \vec{b}) = -\vec{c} \cdot (\vec{b} \times \vec{a}) = -\vec{b} \cdot (\vec{a} \times \vec{c})$$
$$= \begin{vmatrix} a_1 & a_2 & a_3 \\ b_1 & b_2 & b_3 \\ c_1 & c_2 & c_3 \end{vmatrix} \tag{B.12}$$

Derivative of a Vector

Definition B.11. In dynamics we deal with vectors whose components are functions of a scalar variable t. Thus, for every value of t in the domain under consideration we get, in general, a different vector. We deal with a vector field and denote the vector by $\vec{a}(t)$. The derivative of $\vec{a}(t)$ with respect to t is defined by

$$\frac{d}{dt}\vec{a}(t) = \lim_{\Delta t \to 0} \frac{\vec{a}(t + \Delta t) - \vec{a}(t)}{\Delta t}$$

$$= \dot{\vec{a}}(t) \tag{B.13}$$

Similarly, we can define higher derivatives. For example,

$$\frac{d^2}{dt^2}\vec{a}(t) = \lim_{\Delta t \to 0} \frac{\dot{\vec{a}}(t + \Delta t) - \dot{\vec{a}}(t)}{\Delta t} = \ddot{\vec{a}}(t) \tag{B.14}$$

The components of a vector can in general be functions of several variables. The corresponding partial derivatives can be defined in a similar manner.

Rules of vector differentiation.   It can be shown that

1. $\frac{d}{dt}(c\vec{a}) = c\dot{\vec{a}} + \dot{c}\vec{a}$, where c is a scalar function of t.

2. $\frac{d}{dt}(\vec{a} + \vec{b}) = \dot{\vec{a}} + \dot{\vec{b}}$.

3. $\frac{d}{dt}(\vec{a} \cdot \vec{b}) = \dot{\vec{a}} \cdot \vec{b} + \vec{a} \cdot \dot{\vec{b}}$.

4. $\frac{d}{dt}(\vec{a} \times \vec{b}) = \dot{\vec{a}} \times \vec{b} + \vec{a} \times \dot{\vec{b}}$.

Gradient of a Scalar Function

Definition B.12. Let a scalar field u be defined in a certain domain of space. The gradient of the given scalar field u is defined by

$$\text{grad } u = \vec{\nabla} u \tag{B.15}$$

where $\vec{\nabla}$ is called the nabla operator or "del" and is defined in the following with respect to a chosen coordinate system. The gradient of a scalar function is a vector whose magnitude and direction give the maximum space rate of increase of the function.

Figure B.3 illustrates the three commonly employed coordinate systems: the Cartesian coordinate system x, y, z; the cylindrical coordinate system r, θ, z; and the spherical coordinate system ρ, ϕ, θ. In the Cartesian coordinate system, we have

$$\vec{\nabla} = \vec{i}\,\frac{\partial}{\partial x} + \vec{j}\,\frac{\partial}{\partial y} + \vec{k}\,\frac{\partial}{\partial z} \tag{B.16}$$

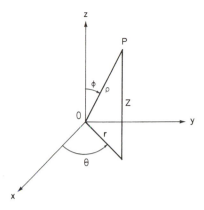

Figure B.3

and in that system it follows that

$$\text{grad } u = \vec{\nabla} u = \frac{\partial u}{\partial x}\vec{i} + \frac{\partial u}{\partial y}\vec{j} + \frac{\partial u}{\partial z}\vec{k} \tag{B.17}$$

In cylindrical coordinates, $x = r \cos \theta$, $y = r \sin \theta$, and $z = z$, which yields

$$\text{grad } u = \vec{\nabla} u = \frac{\partial u}{\partial r}\vec{i}_r + \frac{1}{r}\frac{\partial u}{\partial \theta}\vec{i}_\theta + \frac{\partial u}{\partial z}\vec{k} \tag{B.18}$$

where $\vec{i}_r$, $\vec{i}_\theta$, and $\vec{k}$ are the unit coordinate vectors. In spherical coordinates, $x = \rho \sin \phi \cos \theta$, $y = \rho \sin \phi \sin \theta$, and $z = \rho \cos \phi$, and we obtain

$$\text{grad } u = \vec{\nabla} u = \frac{\partial u}{\partial \rho}\vec{i}_\rho + \frac{1}{\rho}\frac{\partial u}{\partial \phi}\vec{i}_\phi + \frac{1}{\rho \sin \phi}\frac{\partial u}{\partial \theta}\vec{i}_\theta \tag{B.19}$$

where $\vec{i}_\rho$, $\vec{i}_\phi$, and $\vec{i}_\theta$ are the unit coordinate vectors.

Properties of gradient.   The gradient obeys the following rules:

1. $\text{grad } (u_1 + u_2 + \cdots + u_n) = \text{grad } u_1 + \text{grad } u_2 + \cdots + \text{grad } u_n$.
2. $\text{grad } (uv) = u \text{ grad } v + v \text{ grad } u$.
3. $\text{grad } f(u) = f'(u) \text{ grad } u$.

For a vector field $\vec{F}(x, y, z)$ given in a region, if there exists a function $u(x, y, z)$ in this region such that

$$\vec{F}(x, y, z) = \text{grad } u(x, y, z) \tag{B.20}$$

then this vector field is called irrotational or conservative and u is called the scalar potential. In an irrotational field, the work done by the force $\vec{F} = \text{grad } u$ along a curve C connecting two points A and B of this field does not depend on the path of this curve. In particular the work done along a closed curve is zero.

Divergence of a Vector

Definition B.13.   Let $\vec{u}$ be a vector field defined in a certain domain of space. The divergence of this vector field, denoted by $\text{div } \vec{u}$, is defined by

$$\text{div } \vec{u} = \vec{\nabla} \cdot \vec{u} \tag{B.21}$$

Hence, the divergence of a vector field is a scalar. In the Cartesian coordinate system, we get

$$\text{div } \vec{u} = \vec{\nabla} \cdot \vec{u} = \frac{\partial u_1}{\partial x} + \frac{\partial u_2}{\partial y} + \frac{\partial u_3}{\partial z} \tag{B.22}$$

where u_1, u_2, and u_3 are the components of $\vec{u}$ in the directions x, y, and z, respectively. In cylindical coordinates, we obtain

$$\text{div } \vec{u} = \vec{\nabla} \cdot \vec{u} = \frac{1}{r} \frac{\partial}{\partial r}(r u_r) + \frac{1}{r} \frac{\partial u_\theta}{\partial \theta} + \frac{\partial u_z}{\partial z} \tag{B.23}$$

In spherical coordinates, we obtain

$$\text{div } \vec{u} = \vec{\nabla} \cdot \vec{u} = \frac{1}{\rho^2} \frac{\partial}{\partial \rho}(\rho^2 u_\rho) + \frac{1}{\rho \sin \phi} \frac{\partial}{\partial \phi}(u_\phi \sin \phi) + \frac{1}{\rho \sin \phi} \frac{\partial u_\theta}{\partial \theta} \tag{B.24}$$

Properties of divergence. The divergence of a vector obeys the following rules:

1. $\text{div } (\vec{a} + \vec{b}) = \text{div } \vec{a} + \text{div } \vec{b}$.
2. $\text{div } (u\vec{a}) = u \text{ div } \vec{a} + \vec{a} \text{ grad } u$.

Curl of a Vector

Definition B.14. Let $\vec{u}$ be a vector field defined in a certain domain of space. The curl of this vector field, denoted by curl $\vec{u}$, is defined by

$$\text{curl } \vec{u} = \vec{\nabla} \times \vec{u} \tag{B.25}$$

Sometimes, the symbol rot $\vec{u}$ is also used for curl $\vec{u}$. In Cartesian coordinates, we have

$$\text{curl } \vec{u} = \vec{\nabla} \times \vec{u}$$

$$= \left(\frac{\partial u_3}{\partial y} - \frac{\partial u_2}{\partial z}\right)\vec{i} + \left(\frac{\partial u_1}{\partial z} - \frac{\partial u_3}{\partial x}\right)\vec{j} + \left(\frac{\partial u_2}{\partial x} - \frac{\partial u_1}{\partial y}\right)\vec{k}$$

$$= \begin{vmatrix} \vec{i} & \vec{j} & \vec{k} \\ \frac{\partial}{\partial x} & \frac{\partial}{\partial y} & \frac{\partial}{\partial z} \\ u_1 & u_2 & u_3 \end{vmatrix} \tag{B.26}$$

In cylindical coordinates, the components of curl $\vec{u}$ in the direction r, θ, and z, respectively, are given by

$$\frac{1}{r} \frac{\partial u_z}{\partial \theta} - \frac{\partial u_\theta}{\partial z}, \qquad \frac{\partial u_r}{\partial z} - \frac{\partial u_z}{\partial r}, \qquad \frac{1}{r} \frac{\partial}{\partial r}(r u_\theta) - \frac{1}{r} \frac{\partial u_r}{\partial \theta}$$

In spherical coordinates, the components of curl $\vec{u}$ in the direction ρ, ϕ, and θ, respectively, become

$$-\frac{1}{\rho \sin \phi}\left[\frac{\partial u_\theta}{\partial \theta} - \frac{\partial}{\partial \phi}(u_\theta \sin \phi)\right]$$

$$-\frac{1}{\rho}\left[\frac{\partial}{\partial \rho}(\rho u_\theta) - \frac{1}{\sin \phi}\frac{\partial u_\rho}{\partial \theta}\right]$$

$$-\left[\frac{1}{\rho}\frac{\partial u_\rho}{\partial \phi} - \frac{1}{\rho}\frac{\partial}{\partial \rho}(\rho u_\phi)\right]$$

Properties of curl.   The curl of a vector obeys the following rules:

1. curl $(\vec{a} + \vec{b}) = $ curl $\vec{a} + $ curl $\vec{b}$.
2. curl $(u\vec{a}) = u$ curl $\vec{a} - \vec{a} \times $ grad u.

Laplacian Operator

Definition B.15.   The scalar product of the operator $\vec{\nabla}$ with itself yields a scalar operator call the Laplacian and denoted by ∇^2 (i.e., $\nabla^2 = \vec{\nabla} \cdot \vec{\nabla}$). The Laplacian of a scalar function u can then be obtained in Cartesian coordinates x, y, and z as

$$\nabla^2 u = \frac{\partial^2 u}{\partial x^2} + \frac{\partial^2 u}{\partial y^2} + \frac{\partial^2 u}{\partial z^2} \tag{B.27}$$

in cylindical coordinates r, θ, and z as

$$\nabla^2 u = \frac{1}{r}\frac{\partial}{\partial r}\left(r\frac{\partial u}{\partial r}\right) + \frac{1}{r^2}\frac{\partial^2 u}{\partial \theta^2} + \frac{\partial^2 u}{\partial z^2} \tag{B.28}$$

and in spherical coordinates ρ, ϕ, and θ as

$$\nabla^2 u = \frac{1}{\rho^2}\frac{\partial}{\partial \rho}\left(\rho^2\frac{\partial u}{\partial \rho}\right) + \frac{1}{\rho^2 \sin \phi}\frac{\partial}{\partial \phi}\left(\sin \phi\frac{\partial u}{\partial \phi}\right) + \frac{1}{\rho^2}\frac{1}{\sin^2 \phi}\frac{\partial^2 u}{\partial \theta^2} \tag{B.29}$$

For further study of vector analysis and theorems of Gauss, Green, and Stokes, the reader should consult books on mathematics such as those given in references [1–3].

REFERENCES

1. Lass, H., *Vector and Tensor Analysis*, McGraw-Hill Book Company, New York, 1950.
2. Kaplan, W., *Advanced Calculus*, Addison-Wesley Publishing Company, Inc., Reading, Mass., 1952.
3. Rektorys, K., *Survey of Applicable Mathematics*, The MIT Press, Cambridge, Mass., 1969.

Appendix C

ANSWERS TO SELECTED PROBLEMS

CHAPTER 2

2.1. 12.1 km/h at 54.37° east of north   **2.2. (a)** $a_t = (\ddot{x}\dot{x} + \ddot{y}\dot{y} + \ddot{z}\dot{z})/(\dot{x}^2 + \dot{y}^2 + \dot{z}^2)^{1/2}$

2.3. (a) $\vec{\omega}_1 = -\dfrac{\omega_0 \cos\theta}{\left(\dfrac{L^2}{r^2} - \sin^2\theta\right)^{1/2}}\vec{k}$; **(b)** $\vec{v}_p = -r\omega_0 \sin\theta\left[1 + \dfrac{\cos\theta}{\left(\dfrac{L^2}{r^2} - \sin^2\theta\right)^{1/2}}\right]\vec{i}$

2.4. $\vec{v}_p = (u\cos\theta + \omega_0 L \sin\theta)\vec{i} + (u\sin\theta - \omega_0(R + L\cos\theta))\vec{j}$
$\vec{a}_p = (2\omega_0 u\sin\theta - \omega_0^2(R + L\cos\theta))\vec{i} - (2\omega_0 u\cos\theta + \omega_0^2 L \sin\theta)\vec{j}$

2.8. $\vec{a}_p = -\omega_0^2 b_2\vec{i} + (2\omega_0 v_0 - \omega_0^2 b_1)\vec{j}$

2.9. $\vec{a} = (-2.08\vec{i} + 16.58\vec{j} - 19.76\vec{k}) \times 10^{-3} \text{ m/s}^2$

CHAPTER 3

3.1. $m(\ddot{x} - \omega_0^2 x) + \dfrac{2cmx}{1 + 4c^2x^2}(2c\dot{x}^2 + g + 2c\omega_0^2 x^2) + \dfrac{\mu}{(1 + 4c^2x^2)^{1/2}}N \, \text{sgn} \, \dot{x} = 0$

where

$$N = \left[\frac{(2cm\dot{x}^2 + mg + 2cm\omega_0^2 x^2)^2}{1 + 4c^2x^2} + 4m^2\omega_0^2\dot{x}^2\right]^{1/2}$$

3.2. $\max P = (m_1 + m_2)g(\mu_1 + \mu_2)$

3.3. Velocity $= -\dfrac{m_2}{m_1 + m_2}v_0$; Distance $= -\dfrac{m_2}{m_1 + m_2}d$

3.4. (a) $\omega_f = \dfrac{\frac{1}{12}ML^2 + 2ma^2}{\frac{1}{12}ML^2 + 2m\frac{L^2}{4}}\omega_0$; **(b)** $\left(\dfrac{M}{24}L^2 + ma^2\right)\omega_0^2 - \dfrac{\frac{1}{24}ML^2 + ma^2}{\frac{1}{12}ML^2 + \frac{mL^2}{2}}\omega_0^2$

3.5. $\dfrac{x^2}{r_0^2} + \dfrac{y^2}{\frac{m}{c}v_0^2} = 1$ **3.9.** $e = 0.549$

CHAPTER 4

4.1. $\vec{v}_p = \omega(R \cos \theta - R_0) \, \vec{i} + \omega R \sin \theta \vec{j}$
$\vec{a}_p = (\dot{\omega} R \cos \theta - \omega^2 R \sin \theta - \dot{\omega} R_0) \vec{i} + (\omega^2 R \cos \theta + \dot{\omega} R \sin \theta) \vec{j}$
where inertial coordinates are used with x horizontal and y vertical.
4.2. 116,992.7; 150,000; 253,007.4   **4.3.** $-[(\frac{1}{3}mL^2 - \frac{1}{4}ma^2)\omega_0^2 \cos \beta \sin \beta - \frac{1}{2}mgL \sin \beta]\vec{k}$
4.4. $\vec{F} = -m\omega_1^2 L\vec{i} + mg\vec{j} - m\dot{\omega}_1 L\vec{k}$ $\vec{M} = \frac{1}{12} m(3R^2 + h^2)\dot{\omega}_1\vec{j} - \frac{1}{2}mR^2\omega_1\omega_2 k$
where L is the distance from 0 to disk center and h is the disk thickness.
4.5. $g \sin \theta = \omega_0^2 L(\frac{1}{2} \cos \theta + \frac{7}{12} \sin \theta \cos \theta)$ **4.8.** $\theta = 55.15°$

CHAPTER 5

5.1. (a) $x = 0$ or $\omega_0^2 = 2cg$ for any x **5.3.** See answer to problem 3.1.
5.4. $(m_1 + m_2)\ddot{x} + m_2 b\ddot{\theta} \cos \theta - m_2 b\dot{\theta}^2 \sin \theta + \mu N \operatorname{sgn} \dot{x} = P$
$m_2 b\ddot{x} \cos \theta + m_2 b^2\ddot{\theta} + m_2 gb \sin \theta = 0$
where $N = m_1 g + m_2(b\dot{\theta}^2 + g \cos \theta - \ddot{x} \sin \theta) \cos \theta$
Note that the first equation is equivalent to (3.26).
5.6. (a) $\sin \theta \, dx - \cos \theta \, dy + \dfrac{b}{2} d\theta = 0$; **(b)** $m\ddot{x} = \lambda \sin \theta$; $m\ddot{y} = -\lambda \cos \theta$; $I\ddot{\theta} = \lambda\dfrac{b}{2}$
5.7. $\ddot{x}_1 \cos \alpha + \frac{3}{2}\ddot{x}_2 = g \sin \alpha$; $(m_1 + m_2)\ddot{x}_1 + m_2\ddot{x}_2 \cos \alpha + \mu N_1 \operatorname{sgn} \dot{x}_1 = F(t)$
where $N_1 = (m_1 + m_2)g - m_2\ddot{x}_2 \sin \alpha$
5.10. $\dot{x} = \dfrac{m_2 b^2}{\Delta}p_1 - \dfrac{m_2 b \cos \theta}{\Delta}p_2$; $\dot{\theta} = -\dfrac{m_2 b \cos \theta}{\Delta}p_1 + \dfrac{m_1 + m_2}{\Delta}p_2$

$\dot{p}_1 = -\mu N \operatorname{sgn}\left(\dfrac{m_2 b^2}{\Delta} p_1 - \dfrac{m_2 b \cos \theta}{\Delta} p_2\right) + P$

$\dot{p}_2 = -m_2 b \sin \theta \left[g + \dfrac{1}{\Delta^2}(m_2 b^2 p_1 - m_2 b \cos \theta \, p_2)(-m_2 b \cos \theta \, p_1 + (m_1 + m_2)p_2)\right]$

where $\Delta = m_2 b^2(m_1 + m_2 \sin^2 \theta)$

CHAPTER 6

6.1. (a) Global existence and uniqueness

(b) Local existence and uniqueness in any finite region not containing the x_1 axis which is a singular region.

6.2. $t_1 = t_0 + \dfrac{1}{x(t_0)}$ **6.5.** $\Phi(t) = \begin{bmatrix} 1 & t & 0 & 0 \\ 0 & 1 & 0 & 0 \\ 0 & 0 & 1 & t \\ 0 & 0 & 0 & 1 \end{bmatrix}$

where x, $\dot{x}$, y and $\dot{y}$ are chosen as state variables.
Range $= x_0 + \dfrac{2v_0^2 \sin \alpha \cos \alpha}{g}$

6.6. $\Phi(t) = \begin{bmatrix} 1 & 0 & \dfrac{1}{c_3}(1 - e^{-c_3 t}) & 0 \\ 0 & 1 & \dfrac{c_4}{c_3^2}(1 - c_3 t - e^{-c_3 t}) & t \\ 0 & 0 & e^{-c_3 t} & 0 \\ 0 & 0 & -\dfrac{c_4}{c_3}(1 - e^{-c_3 t}) & 1 \end{bmatrix}$

where $c_3 = \dfrac{-3c}{\Delta}$, $c_4 = \dfrac{2c \cos \alpha}{\Delta}$
$\Delta = m_2(2 \cos^2 \alpha - 3) - 3m_1$

6.7. $\Phi(t) = \begin{bmatrix} 1 & 0 & 0 \\ 1 - e^{-t} & e^{-t} & 0 \\ -1 + 2e^{-t} - e^{-2t} & -2e^{-t} + 2e^{-2t} & e^{-2t} \end{bmatrix}$

CHAPTER 8

8.1. $\omega = 27.2$ rad/s $T = 0.231$ s

8.2. Steady state vertical motion is given by
$q = 3[1 - 0.9\sin(10.18\,t - 41°)]$ cm

8.3. (b) (1) $\omega_1 = 0.523\sqrt{k/m}$; $\omega_2 = 1.251\sqrt{k/m}$; $\omega_3 = 2.159\sqrt{k/m}$

$[\Phi] = \begin{bmatrix} 1 & 1 & 1 \\ 1.863 & 1.2175 & -.3305 \\ 2.566 & -2.155 & .09 \end{bmatrix}$

(2) $[M^*] = \begin{bmatrix} 14.53 & .0 & 0 \\ 0 & 8.61 & 0 \\ 0 & 0 & 1.227 \end{bmatrix}$; $[K^*] = \begin{bmatrix} 3.98 & 0 & 0 \\ 0 & 13.47 & 0 \\ 0 & 0 & 5.72 \end{bmatrix}$

(4) $[\bar{\phi}] = \dfrac{1}{\sqrt{m}} \begin{bmatrix} 0.262 & 0.341 & 0.903 \\ 0.489 & 0.415 & -0.298 \\ 0.673 & -0.734 & 0.0813 \end{bmatrix}$

8.5. $\omega = 2.159\sqrt{k/m}$; $\{\phi\} = \left\{ \begin{matrix} 1 \\ -0.3305 \\ .09 \end{matrix} \right\}$

8.7. $m\ddot{x}_1 + kx_1 + 2k(x_1 - x_2) = 0$; $\frac{73}{64}m\ddot{x}_2 + 2k(x_2 - x_1) = 0$

$[M] = \begin{bmatrix} m & 0 \\ 0 & \frac{73}{64}m \end{bmatrix}$; $[K] = \begin{bmatrix} 3k & -2k \\ -2k & 2k \end{bmatrix}$

CHAPTER 9

9.1. Lyapunov stability theory cannot be employed.

9.2. (a) $x_1 =$ constant, $x_2 = \pm n\pi(n = 0, 1, 2, \ldots)$, $x_3 = 0$, $x_4 = 0$; (b) unstable

9.3. (c) $k > \dfrac{1}{2}\dfrac{mg}{b}$ **9.4.** (a) $r = a \pm \dfrac{mg}{k}$, $\theta = \pm n\pi(n = 0, 1, 2, \ldots)$, $p_1 = 0, p_2 = 0$

9.5. (a) Sign indefinite; (b) Positive semidefinite

9.6. Equilibrium zone is globally, asymptotically stable.

9.7. Equilibrium zone is globally stable.

9.10. (c) Not positive definite; (d) Positive definite, and decrescent

INDEX